Opportunistic Fungi, Nematode and Plant Interactions

Mohd. Sayeed Akhtar
Editor

Opportunistic Fungi, Nematode and Plant Interactions

Interplay and Mechanisms

Springer

Editor
Mohd. Sayeed Akhtar
Department of Botany
Gandhi Faiz-e-Aam College
Shahjahanpur, Uttar Pradesh, India

ISBN 978-981-97-2044-6 ISBN 978-981-97-2045-3 (eBook)
https://doi.org/10.1007/978-981-97-2045-3

This Springer imprint is published by the registered company Springer Nature Singapore Pte Ltd.
The registered company address is: 152 Beach Road, #21-01/04 Gateway East, Singapore 189721, Singapore

This book is dedicated to

Abdul Ghaffar Khan (1890–1988)
(Frontier Gandhi)

Freedom fighter who fought for injustice with sadbhavana,
and founder of Khudai Khitmatgar

Foreword

Opportunistic fungi play a significant role in plant-nematode interactions. They can both positively and negatively impact the interactions between fungi and nematodes, as well as influence plant diseases. Fungi can attract or repel nematodes; nematodes can feed on fungi, acting as biological controls for plant diseases while simultaneously inducing suppressiveness in agricultural soils, affecting cyst and root-knot nematode population dynamics. Nematode feeding sites, such as syncytia or giant cells, provide favorable conditions for fungi to enter and infect host plants. The nematode-instigated physiological changes can at the same time promote various phases of fungal life cycles, such as germination, development, and reproduction. Opportunistic fungi are commonly found in the soil or in decomposing organic material and in their interaction with plant nematodes, these can engage in various modes, including direct parasitism or the infiltration of the nematode body for the purpose of tissue consumption. Occasionally, they produce toxins or enzymes that are capable of eliminating or debilitating the nematode, thereby rendering it more susceptible to other predatory organisms. Moreover, opportunistic fungi can contribute to safeguarding plants from nematode infection by out-competing the nematodes for resources and living space. The interaction between opportunistic fungi and nematodes is characterized by complexity and is contingent upon the particular species involved as well as the prevailing environmental conditions. For instance, *Fusarium oxysporum*, a pathogen responsible for causing wilt disease, as well as *Trichoderma harzianum*, a biocontrol fungus, possess the ability to parasitize root-knot nematodes. In contrast, *Paecilomyces lilacinus* parasitizes root-knot nematodes, cyst nematodes, and lesion nematodes. Overall, the interactions between opportunistic fungi and plant nematodes are complex and can have significant implications for plant health and disease management in agricultural ecosystems.

This book volume titled *Opportunistic Fungi, Nematode and Plant Interactions: Interplay and Mechanisms* includes 11 chapters contributed by academicians, scientists, and researchers from different parts of the world. Chapter 1 focuses on different opportunistic fungi, nematode interactions on host plants and their mechanisms of actions, whereas Chaps. 2 and 3 discuss the interaction between the opportunistic fungi, plant, and nematode in agricultural and vegetable crops. In Chap. 4 contributors highlight the biochemical changes in host-plant invaded by opportunistic fungi and nematodes, whereas Chap. 5 discusses the response of ROS in host-plant assaulted by opportunistic fungi and nematodes. The functions of flavonoids in plants, pathogen, and opportunistic fungal interactions have been evaluated in Chap. 6, whereas the occupancy of flavonoids in plants, pathogen, and opportunistic fungi interactions is outlined in Chap. 7. The mode of interactions between opportunistic fungi, nematodes, and other soil microbiota has been discussed at length in Chap. 8, whereas the molecular interactions and recent developments between the opportunistic fungi, nematodes, and plants are given in Chap. 9. Chapter 10 highlights the involvement of parasitism proteins in plant nematodes and opportunistic fungi connections. The role of lifestyle pattern and soil microbiome interaction of arthrobotrys species for the biostimulant production are summarized in Chap. 11.

This book covers the basic and applied aspects of opportunistic fungi and nematode interactions, physiological and molecular changes in host and pathogens, as well as mechanistic aspects. It has been written with the great aim of proving highly beneficial to academicians, researchers, and students.

Faculty of Science
Ege University,
İzmir, Turkey

Münir Öztürk

Preface

Plant parasitic nematodes (PPNs) are documented as the most destructive pests for numerous economically important crops worldwide. According to an estimate, the average crop lost caused by this notorious pest is about 12.3%. Over the past few decades, chemical nematicides have been frequently used for the management of nematode diseases; however, general concerns about human health and environmental safety have led to restrictions on chemical nematicides applications for the control of PPNs. Cultural practices are also used for nematode management, but extensive annual losses in crop yields and quality demonstrate a crucial need for an ecofriendly and environmentally safe approach that provides better management of PPNs.

Soil serves as an excellent niche for a diverse range of microorganisms. The microorganisms lying in the rhizosphere have the capability to act as frontline warriors against the pathogen attacks on root. Among the different types of beneficial rhizospheric microorganisms, opportunistic fungi (OP) such *as Arthrobotrys, Cylindrocarpon, Fusarium, Paecilomyces, Plectosphaerella, Pochonia, Rhizoctonia,* and *Trichoderma* are of prime importance and might play a very significant and crucial role in the management of plant diseases caused by PPNs. OP have the ability to directly parasitize PPNs by secreting metabolic compounds that affect the viability of one or more stages of the nematode life span, with obese female or cysts becoming highly susceptible to these fungi. Since opportunistic fungi and PPNs both share the common ecological niche and influence plant growth, it is a bit tough to generalize the series of events taking place during this underground assembly in one plane, but the recent development of sophisticated molecular tools has provided an impression of the interaction taking place between OP fungi, PPNs, plants health, and sustainable environmental conditions.

The present book *Opportunistic Fungi, Nematode and Plant Interactions: Interplay and Mechanisms* is a timely effort in the direction of describing the opportunistic fungi, nematode and plant interactions, fungal behavior and etiology, their management strategies, and the mechanisms involved against the unseen enemies. This book volume consists of 11 chapters from renowned scientists and scholars throughout the globe who contribute basic and applied aspects of opportunistic fungi and nematode interactions, physiological and molecular changes in hosts and pathogens, and mechanistic aspects. Besides this, the book also provides valuable information for students, teachers, and professionals involved in research related to plant pathology, ecology, soil science, host-plant interactions, and microbial ecology.

I am highly grateful to all our contributors for readily accepting our invitation to not only share their knowledge and research but also venerably integrate their expertise in dispersed information from diverse field in composing the chapters, and enduring editorial suggestions to finally produce this venture. I greatly appreciate their commitment. I am also thankful to Professor Munir Ozturk for his suggestion and writing the foreword for this volume. We also thank the team at Springer International, especially Rhea Dadra, Priyanga Kabali, and Sanchi Bhimrajka, for their generous cooperation at every stage of the publication.

Shahjahanpur, Uttar Pradesh, India Mohd. Sayeed Akhtar

Contents

Editor and Contributors

About the Editor

Mohd. Sayeed Akhtar is Senior Assistant Professor in Gandhi Faiz-e-Aam College, Shahjahanpur, U.P., India. He received his PhD degree from Aligarh Muslim University (AMU), Aligarh, India, in 2008, prior to conducting postdoctoral research at the Botanical Institute, University of Basel (BIB), Switzerland (2008–2010), and Chonbuk National University (CBNU), Republic of Korea in 2011. He worked as an Assistant Professor at Jimma University in Ethiopia from 2011 to 2014 and as a Fellow researcher at Universiti Putra Malaysia's (UPM) Institute of Tropical Agriculture from 2014 to 2015. He is listed among the top 2% of scientists in the world consecutively for the past three years, according to Stanford University in the United States. Dr. Akhtar is the author and coauthor of more than 150 articles in peer-reviewed journals, conference proceedings, and book chapters and has edited 21 books with international publishers. His numerous honors include the Best Researcher of the Year Award (2016) from the APSF, DST, India; the Outstanding Scientist in Plant Pathology Award (2017) from VIRA, Chennai, India; the Global Teachers Award (2018) from the AKS Education Foundation, New Delhi, India; the Albert Nelson Marquis Award (2018) from the Marquis Who's Who Society, USA; and the Teacher Innovation Award (2019 & 2022) from ZIIEI, Sri Arvindo Society, Puducherry, India. In addition to this, he is a member of the International Society for Development and Sustainability, Japan; Fellow of the Plantae Group, American Society of Plant Biologists; and member of International Society of Environmental Relationship and Sustainability, Chennai, India, and International Service for the Acquisition of Agri-Biotech Applications (ISAAA), Manila, Republic of the Philippines. He serves the scientific community as an editorial board member and reviewer of several high-impact international journals. His current work focuses on molecular biotechnology, bioremediation, biomineralization, nano-fertilizers, and nanobiotechnology of rhizospheric plant-microbe interactions.

Contributors

Ismail Abiola Adebayo Department of Medical Biochemistry, Molecular Biology and Genetics, School of Medicine and Pharmacy, College of Medicine and Health Sciences, University of Rwanda, Butare, Rwanda

Taofeek Tope Adegboyega Department of Medical Microbiology and Parasitology, School of Medicine and Pharmacy, College of Medicine and Health Sciences, University of Rwanda, Butare, Rwanda

Rasheed Omotayo Adeyemo Department of Medical Microbiology and Parasitology, School of Medicine and Pharmacy, College of Medicine and Health Sciences, University of Rwanda, Butare, Rwanda

Basit Akolade Adigun Crop Science Program, Faculty of Fisheries and Food Science, Universiti Malaysia Terengganu, Kuala Nerus, Terengganu, Malaysia

Lukman Ahamad Section of Plant Pathology and Nematology, Department of Botany, Aligarh Muslim University, Aligarh, Uttar Pradesh, India

Mohd. Sayeed Akhtar Department of Botany, Gandhi Faiz-e-Aam College, Shahjahanpur, Uttar Pradesh, India

Saheed Adekunle Akinola Department of Medical Microbiology and Parasitology, School of Medicine and Pharmacy, College of Medicine and Health Sciences, University of Rwanda, Butare, Rwanda

K. Sindhura Bhairavi Department of Entomology, College of Agriculture, Assam Agricultural University, Jorhat, Assam, India

Sanjukta Chakraborty Department of Plant Pathology, College of Agriculture, Assam Agricultural University, Jorhat, Assam, India

Arkadeb Chatterjee Department of Nematology, Assam Agricultural University, Jorhat, Assam, India

Srishti Choudhury Department of Entomology, College of Agriculture, Punjab Agricultural University, Ludhiana, India

Sanhita Chowdhury Department of Entomology, Post Graduate College of Agriculture, Dr. Rajendra Prasad Central Agricultural University, Pusa, Samastipur, Bihar, India

Tolessa Muleta Daba Department of Medical Biochemistry, Molecular Biology and Genetics, School of Medicine and Pharmacy, College of Medicine and Health Sciences, University of Rwanda, Butare, Rwanda

Mohammad Danish Section of Botany, School of Sciences, Maulana Azad National Urdu University, Hyderabad, India

Abhibandana Das Department of Entomology, College of Agriculture, Assam Agricultural University, Jorhat, Assam, India

Vinod Kumar Dubey Department of Entomology, Post Graduate College of Agriculture, Dr. Rajendra Prasad Central Agricultural University, Pusa, Samastipur, Bihar, India

Vaithiyalingam Gopinath Department of Plant Pathology and Nematology, Post Graduate College of Agriculture, Dr. Rajendra Prasad Central Agricultural University, Pusa, Samastipur, Bihar, India

Pranjal Kumar Kaman Department of Plant Pathology, Assam Agricultural University, Jorhat, Assam, India

Anurag Kashyap Department of Plant Pathology, Assam Agricultural University, Jorhat, Assam, India

Deeksha Kashyap Section of Plant Pathology and Nematology, Department of Botany, Aligarh Muslim University, Aligarh, Uttar Pradesh, India

Masudulla Khan Section of Plant Pathology and Nematology, Department of Botany, Aligarh Muslim University, Aligarh, Uttar Pradesh, India

Botany Section, Women's College, Aligarh Muslim University, Aligarh, India

Maneesh Kumar Department of Biotechnology, Magadh University, Bodh Gaya, Bihar, India

Arti Kumari Department of Biotechnology, Patna Women's College, Patna, Bihar, India

Priyanka Kumari Department of Biotechnology, Gautam Buddha University, Greater Noida, Uttar Pradesh, India

Selvaraj Kunjiappan Department of Biotechnology, Kalasalingam Academy of Research and Education, Krishnankoil, Tamil Nadu, India

Waqar Majeed Department of Zoology, Wildlife and Fisheries, University of Agriculture, Faisalabad, Pakistan

Anik Majumdar Division of Plant Pathology, ICAR-Indian Agricultural Research Institute, Pusa, New Delhi, India

Shakirah Binti Mohammad Nahar Department of Biological Sciences and Biotechnology, Faculty of Science and Technology, Universiti Kebangsaan Malaysia, Bangi, Selangor, Malaysia

Yusuf Opeyemi Oyebamiji Department of Biological Sciences and Biotechnology, Faculty of Science and Technology, Universiti Kebangsaan Malaysia, Bangi, Selangor, Malaysia

Aiman Parveen Section of Plant Pathology and Nematology, Department of Botany, Aligarh Muslim University, Aligarh, Uttar Pradesh, India

Subbaiah Revathi Lakshmi Centre for Research and Postgraduate Studies in Botany, Ayya Nadar Janaki Ammal College, Affiliated to Madurai Kamaraj University, Madurai, Sivakasi, Tamil Nadu, India

Merajul Islam Robab Section of Botany, School of Sciences, Maulana Azad National Urdu University, Hyderabad, India

Alex Rugamba Department of Medical Biochemistry, Molecular Biology and Genetics, School of Medicine and Pharmacy, College of Medicine and Health Sciences, University of Rwanda, Butare, Rwanda

Dikshita Saikia Department of Plant Pathology, Punjab Agricultural University, Ludhiana, Punjab, India

Rashmita Saikia Department of Plant Pathology, Punjab Agricultural University, Ludhiana, Punjab, India

Jean Paul Sinumvayo Department of Medical Microbiology and Parasitology, School of Medicine and Pharmacy, College of Medicine and Health Sciences, University of Rwanda, Butare, Rwanda

Madasamy Sundar Department of Biotechnology, Kalasalingam Academy of Research and Education, Krishnankoil, Tamil Nadu, India

Himanshu Thakur Department of Entomology, CSK Himachal Pradesh Krishi Vishvavidyalaya, Palampur, Himachal Pradesh, India

Annette Uwineza Department of Medical Biochemistry, Molecular Biology and Genetics, School of Medicine and Pharmacy, College of Medicine and Health Sciences, University of Rwanda, Butare, Rwanda

Nur Sakinah Binti Mohd Yusri Department of Biological Sciences and Biotechnology, Faculty of Science and Technology, Universiti Kebangsaan Malaysia, Bangi, Selangor, Malaysia

Interactions of Opportunistic Fungi, Plants, and Plant Parasitic Nematodes

1

Mohammad Danish and Merajul Islam Robab

1.1 Introduction

Agriculture output in India has declined for a variety of causes. Diseases are a severe impediment to agricultural cultivation. One of these is plant parasitic nematode (PPN) infestation, which reduces crop quality and production. PPNs drastically impair crop harvest as well as yield, which is a huge financial concern for the world's farming sector (Shakeel et al. 2022). Due to their microscopic size, concealed habitats, and absence of obvious symptoms on the aerial parts of the plants, plant parasitic nematodes are regarded as a hidden enemy of crops. PPNs infiltrate the roots of the plant, where each nematode creates a feeding location (Pervez and Ali 2011). This complex developmental process results in significant modifications to root structure and metabolism, including the development of root galls, syncytia, large cells, lesions, etc. (Danish et al. 2021). Nematodes cause an estimated $100 billion in losses to agriculture every year (Coyne et al. 2018). There is an urgent need to control these plant parasitic nematodes using appropriate treatments, given the monetary losses brought on by root-knot nematodes.

Chemicals, host plant resistance, crop rotation, soil solarization, antagonistic organisms, etc. are only a few of the methods that have been utilized to control them thus far. The most popular method of nematode control is crop rotation, although this method needs enough space to grow crops that are not nematode hosts. Also, the grower must make good revenues from these alternative crops. Chemical

M. Danish (✉)
Section of Botany, School of Sciences, Maulana Azad National Urdu University, Hyderabad, India

M. I. Robab
Section of Botany, School of Sciences, Maulana Azad National Urdu, Hyderabad, India

Section of Botany, School of Sciences, Maulana Azad National Urdu University, Hyderabad, India

M. S. Akhtar (ed.), *Opportunistic Fungi, Nematode and Plant Interactions*, https://doi.org/10.1007/978-981-97-2045-3_1

treatments have been found an effective controlling method against nematodes, but due to their scarcity, high cost, and transient effects, nematodes developed a resistance to them (Chen et al. 2020). Additionally, these treatments can destroy beneficial microorganisms, and prolonged chemical compounds may pose notorious risks to the ecological integrity (Dong et al. 2006). Nematicides may be extremely poisonous to animals and harmful to the developing foetus (Sharma and Rakesh 2009). Thus, it is imperative to create affordable and environmentally acceptable agents due to restrictions on the use of nematicides. To combat and minimize the detrimental effect of nematodes on crop ecosystems, biocontrol agents such as fungi in particular have the potential to be a promising alternative to nematicides and conventional techniques (Danish et al. 2022).

The importance of employing fungi as biological control agents against plant pathogen has substantially increased since fungi reproduce rapidly, selectively, and both sexually and asexually including target specific activity. They also have a short generation time. Also, they can continue to exist sustainably in the absence of the host by transitioning from parasitic infection to saprotrophism (Thambugala et al. 2020). Opportunistic fungi, parasites of nematode eggs, parasites of nematode cysts, and fungi that produce nematode toxic chemicals are examples of nematode opponents. Surprisingly, fungus from quite different orders and families can be found in each of these groups. While fungal infections cause plant physiology to suffer, mutualistic fungi aid hosts in their defence against infection and pests. Fungi produce bioactive compound such as short peptide signalling pathways, enzymes, and bioactive compounds, which encourage invasion and contribute to both symbiotic as well as detrimental connections (Zeilinger et al. 2016). Especially those that can be achieved via induced systemic resistance (ISR) induction are the most important beneficial fungi with biocontrol ability, such as arbuscular mycorrhizas, ectomycorrhizas, endophytes, *Trichoderma* species, and yeasts (Ghorbanpour et al. 2018). Therefore, this chapter focuses on different opportunistic fungi, nematode interactions on host plants and their mechanisms of actions.

1.2 Interaction of Plant Parasitic Nematodes with Different Fungi and Plants

1.2.1 *Trochoderma* Spp.

Numerous species of *Trichoderma* spp. commonly live in the soil, rhizosphere, and endosphere of plants. These fungi have caught our interest because they can inhibit the development and spread of potential infections like nematodes and fungus (Poveda 2021). In compensation for plant-derived sugars, *Trichoderma* spp. have been viewed (mainly) as non-pathogenic and opportunistic plant symbionts that can colonize plant roots and interact favourably with their hosts through *Trichoderma* effector proteins and hormonal crosstalk (Guzmán-Guzmán et al. 2019). As a result of the interaction between *Trichoderma* and plants, the plant gains not only an increase in biomass and general nutrition but also defence against a number of phytopathogens, either directly by acting as a mycoparasite and competing with the pathogen for nutrients or inadvertently by triggering the plant defence system (Mukherjee et al. 2022). Many species of this genus have been investigated under the greenhouse as well as in the field conditions for their biocontrol potentials (Alfiky and Weisskopf 2021). *T. harzianum* inoculation on tomato seeds reduced the level of infection brought on by the nematode *Meloidogyne javanica* in greenhouse conditions, impacting the establishment, reproduction, and nematode development (Sahebani and Hadavi 2008). Also, they observed a significant decline in egg hatching, indicating that *Trichoderma* species has a lot of potential as a biocontrol agent for this nematode. Moreover, *T. harzianum* can promote immunity against nematodes and other diseases. *T. harzianum* increases the gene expression of PR1, PR5, JERF3, and ACO, which are connected to SA- and JA/Et-mediated defence responses in plants, and reduces the symptoms of infection brought on by the nematode *M. incognita* in tomato plants (Leonetti et al. 2017). Furthermore, they noted that second juveniles of *Heterodera avenae* eggs and second-stage juveniles are susceptible to parasitization by *T. longibrachiatum* T6, which decreases the nematode's survival. *T. longibrachiatum* T6 enhanced flavonoid and lignin content, as well as defence-related enzyme activity in wheat roots in a greenhouse experiment, imparting resistance against the nematode *H. avenae* (Zhang et al. 2017). Although, field applications should be handled with caution given that it is the only member of the genus *Trichoderma* to have been identified as an opportunistic human pathogen thus far. According to Pathak et al. (2005), using *T. virens* or *T. harzianum* 1 week before nematode inoculation at 4 and 8 g/kg soil considerably decreased the nematode parameters in soil compared to control. According to Devi et al. (2002), *T. viride* or *T. harzianum* added in soil at 1 g/kg increased plant growth and decreased M. incognita population. According to Goswami and Mittal (2004), *T. viride* culture filtrates were highly toxic against J_2s of *M. incognita* that were infecting tomato plants. They also demonstrated that *T. viride* culture filtrates were also effective at preventing *M. incognita* eggs from hatching. *Trichoderma* species use a variety of strategies to combat plant diseases, including mycoparasitism, competition, inducing host plant resistance, and antibiosis (Zeilinger et al. 2016).

1.2.2 Ovicidal and Larvicidal Fungi

Pochonia chlamydosporia has been discovered in nematode suppressive soils to parasitize eggs and has a global distribution. According to Giné et al. (2016) *P. chlamydosporia* is the primary biotic component causing soil suppressiveness to root-knot nematodes (RKN) in horticultural crops. In the absence of both plant and nematode hosts, the fungus is still able to survive as a saprophyte in soil. The fungus can colonize host plant roots in the rhizosphere, and certain *Pochonia* species have even been noted to exhibit endophytic activity in some plants belong to Poaceae and Solanaceae family—a growth habit that may aid the host plant's defence against soil-borne infections (Maciá-Vicente et al. 2009). The fungus is also a direct and indirect opponent of RKN and cyst nematodes, acting by parasitizing eggs. *P.*

chlamydosporia field populations can include individuals that perform a variety of highly advantageous functions for plants, such as promoting plant growth that increases tolerance of plants, inducing plant defence mechanisms that control infection, growth, and reproduction of RKN, effective egg parasites that suppress the RKN inoculum, or saprophytic behaviour that contributes to the organic matter cycle and plant nutrition (Ghahremani et al. 2019). There have been reports of *P. chlamydosporia* induced plant defensive mechanisms against RKN, including a rise in the activity of defence enzymes such as peroxidases (POX) and polyphenol oxidases (PPO) enzymes at the stage of root nematode invasion within the 2–3 days of nematode inoculation. The effect of local defence mechanisms alone, however, might not be sufficient to achieve significant nematode suppression because *P. chlamydosporia* does not extensively colonize the root (Escudero and Lopez-Llorca 2012), even though this is improved by chitosan irrigation.

1.2.3 *Paecilomyces* Spp.

Different species of genus "*Paecilomyces*" can be discovered in a variety of environments, such as nematodes suppressive soil (Aminuzzamana et al. 2013) and in the root zone of different plants (Kilama et al. 2007). *Paecilomyces* also contributes significantly to the development of many plants as an endophyte by offering a number of benefits. Both directly and indirectly, it has the ability to act as a biostimulant. *Paecilomyce*s or metabolites secrete by them improve agricultural output and plant morphological characteristics when used directly (Moreno-Gavíra et al. 2020.) *Paecilomyces* is a nematophagus fungus that has undergone extensive research and can be found in many different biological formulations that are utilized in agriculture (Dong et al. 2007). By several pathways, the interplay between plants and *Paecilomyces* enhances plants growth and offers defence against plant pathogens (Malhadas et al. 2017). *Paecilomyces* has positive impacts on crop growth when employed against nematodes pathogens by acting as a biological agent (Mezeal et al. 2018). This predatory fungus is common in many types of soils. The root zone can be inoculated with this fungus to encourage plant development and reduce nematode infestation. There are numerous instances when *Paecilomyces* species act against the nematodes, particularly when dealing with *Meloidogyne* species, including other genera such as *Pratylenchus, Globodera, Rotylenchulus, Heterodera* (Lima-Rivera et al. 2016). The treatment of *P. lilacinus* under greenhouse and field conditions has efficiently controlled root-knot and cyst nematode (Moreno-Gavíra et al. 2020). Plants treated with *P. lilacinus* exhibit decreased root galling as well as improved plant development and yield in soils containing root-knot nematodes (Mittal et al. 1995). The use of *P. lilacinus* reduces both the overall population of nematodes in the soil and the hatching of egg masses.

The genus *Paecilomyces* has many species that produce a broad range of secondary metabolites, including nematicidal, with various chemical compositions and biological activity (Lima-Rivera et al. 2016). Eggs containing younger *M. hapla* larvae were far more susceptible to the serine proteases generated by *P. lilacinus* than eggs containing older, more mature larvae. The penetration of the nematode cuticle and subsequent cell breakdown are shown to be facilitated by a variety of hydrolytic proteins, including proteases mainly serine proteases, collagenases, and chitinases (Paul et al. 2013). Similar to this, many secondary metabolites released by *Paecilomyces* are important in biological control of nematodes (Yan et al. 2011). The members of this genus, notably *P. lilacinus*, can enter both the eggshells and structural components of the juvenile and adult stages of various species of nematodes via spore germination, hyphal branching, and appressoria formation (Khan et al. 2006). The synthesis of various enzymes such as amylases, lipases, proteases, and chitinases showing nematicidal effect has been described by (Giné and Sorribas 2017). The virulence and parasitic capacity of *P. lilacinus* against *M. incognita, Panagrellus redivivus*, and *Caenorhabditis elegans* are increased by the overexpression of genes that control the synthesis of all these enzymes (Wang et al. 2014).

1.2.4 Nematode-Trapping Fungi

Nematode-trapping fungi are a distinct and fascinating genus of fungi that capture and kill nematodes using specialized trapping structures. They can develop a variety of trapping methods, including constricting rings, adhesive networks, and adhesive mycelium (Jiang et al. 2017). In a variety of crop species, nematophagous fungi are well-known bio-agents that are effective against a wide range of nematode pathogens (Singh et al. 2013). The ability of *A. oligospora* directly or indirectly encourages plant development (Danish et al. 2022). Reduced nematode pathogens in the rhizosphere and root systems are necessary for the direct effects to occur. The production of metabolites, improvement of other soil microbial flora, and manipulation of root secretion systems are indirect effects that have an impact on plant growth by reducing the growth and reproduction of phytopathogens and other harmful microorganisms. The use of *A. oligospora* appears to favour the promotion of plant development, the management of disease, and the induction of systemic resistance (Singh et al. 2012). The use of *A. oligospora* against *M. incognita* in *Withania somnifera* resulted in a considerable decrease in the nematode population and an increase in plant growth, as demonstrated by Sharma and Pandey (2009). In vitro parasitism of five *A. oligospora* isolates against *M. graminicola* was assessed by Singh et al. (2012) demonstrated that *A. oligospora* isolate VNS-1 killed nematode larvae (J_2) in 8 days. They reported that, when *A. oligospora* was applied to rice plants that had been inoculated with *M. graminicola*, the number of galls decreased and plant growth increased. According to Simon and Anamika (2011), the addition of *A. oligospora* to *M. graminicola* infested soil decreased the frequency of root galls while also promoting rice plant growth. In tomato plants infected with *M. javanica, A. oligospora* dramatically reduced the mean numbers of galls, egg masses, and eggs per egg mass (Singh et al. 2007; Sharma and Pandey 2009; Jamshidnejad et al. 2013). Similarly, Sharf et al. (2014) conducted the research on *P. chlamydosporia* against *M. incognita* in red kidney beans, and the results showed that *P. chlamydosporia* dramatically decreased the number of gall and number of egg masses and improved plant growth.

Following mechanism behind the capturing ability, it was discovered that *A. oligospora* produce two pathogenicity factors: an extracellular serine protease and a carbohydrate-binding protein (lectin) (Yang et al. 2007). Extracellular proteases are crucial in the degradation of nematode structural elements. *A. oligospora* captured nematodes have been reported to be immobilized by proteases (Tunlid et al. 1994). The amount of extracellular proteases produced by nematophagous fungi is associated with their capacity to eradicate nematodes; the greater their output, the more effective they are at eliminating nematodes (Bedelu et al. 1998).

1.2.5 Mycorrhizal Fungi

Arbuscular mycorrhizal fungi are thought to colonize more than 80% of all terrestrial plant species. They are obligate root symbionts. In exchange for the photosynthetic carbon from their host, they boost nutrient intake, which promotes plant development (Smith et al. 2010). Moreover, they can reduce plant stress brought on by both biotic and abiotic causes, such as plant parasitic nematodes (Vos et al. 2012). As previously discussed by Hol and Cook (2005), AMF can also reduce population density of pant parasitic nematodes. AMF possesses anti-PPN properties in plants such as banana, coffee, and tomato, study performed under in vitro, greenhouse, and in field experiments (Koffi et al. 2013). Reduction in root penetration of tomato plant by the false root-knot nematode (*Nacobbus aberrans*) were noted by the application of *Rhizophagus intraradices* and *Funneliformis mosseae* (Marro et al. 2018), and *G. intraradices*, *G. mosseae*, and *G. etunicatum* decrease *M. javanica* population in peach trees (Calvet et al. 2001), have all demonstrated a reduction in nematode population by enhancing the host nutrient status. The effectiveness of AMF on plant nematode populations and their capacity for infection depends heavily on regional environmental factors. Changes in potassium, phosphorus, and moisture are the main factors affecting the positive effect produced by mycorrhizal fungi in this regard (Ferreira et al. 2018). AMF-mediated biocontrol can occur through a variety of mechanisms, including direct effects of AMF on the pathogen, such as competition for resources or habitat, or indirect effects transmitted by plants. The latter group includes the effects of AMF on plant tolerance, plant defence induction, and altered plant exudation, which results in changed rhizosphere associations. The various systems are not completely independent of one another, and biocontrol is most likely the result of a combination of multiple mechanisms (Cameron et al. 2013). Furthermore, the proportional importance of a certain route can fluctuate based on the specific AMF–pathogen–plant relationship. There has been significant advancement in recent years, particularly in the fields of induced systemic resistance (ISR) (Pieterse et al. 2014) and the function of the rhizosphere in biocontrol (Cameron et al. 2013). Its use as biological control agents in the field is relatively uncommon in agriculture, despite several publications on the biocontrol effect of AMF (Salvioli and Bonfante 2013). Performance variations based on the AMF isolate, pathogen, plant type, and environmental conditions are partly to blame for this (Salvioli and Bonfante 2013). The effectiveness of these biocontrol agents will therefore be improved with a better understanding of how they work.

1.3 Conclusions and Future Prospects

A lot of research has been done on the interactions between crop, nematode, and fungal species in both glasshouse and field conditions to gain the comprehensive knowledge in reducing plant parasitic infections, on different crop plant by applying biocontrol fungi. It was most probable that these fungi have a biocontrol effect in conjunction with increased host tolerance, organism competition, and systemic resistance. The inherent mechanisms governing ecosystem stability under field circumstances may be one of the causes of the variations between laboratory-based and field-based trial findings. Before using fungi in the field, case-by-case study will be necessary because the biocontrol effect depends on a number of factors, such as the species involved. It is necessary to conduct more research, including studies into induced systemic resistance and metabolome changes, to better understand the processes underlying how these fungi affect nematode population densities and promote plant growth. Further research on the interactions between fungi and nematodes in crop hosts is necessary because there is a lack of insight about their effects in the rhizosphere. Thus, by comprehending how organismal interactions in native niches are controlled, we may be able to improve applications that take into consideration native agricultural ecosystems and eventually result in sustainable crop protection techniques that preserve biodiversity.

References

Alfiky A, Weisskopf L (2021) Deciphering *Trichoderma* plant–pathogen interactions for better development of biocontrol applications. J Fungi 7:61

Aminuzzamana FM, Xieb HY, Duanc WJ, Sundand BD, Liu XZ (2013) Isolation of nematophagous fungi from eggs and females of *Meloidogyne* spp. and evaluation of their biological control potential. Biocontrol Sci Technol 23:170–182

Bedelu T, Gessesse A, Abate D (1998) Relation of protease production to nematode-degrading ability of two *Arthrobotrys* spp. World J Microbiol Biotechnol 14:731–734

Calvet C, Pinochet J, Hernández-Dorrego A, Estaún V, Camprubí A (2001) Field microplot performance of the peach-almond hybrid GF-677 after inoculation with arbuscular mycorrhizal fungi in a replant soil infested with root-knot nematodes. Mycorrhiza 10:295–300

Cameron D, Neal A, van Wees S, Ton J (2013) Mycorrhiza-induced resistance: more than the sum of its parts. Trends Plant Sci 18:539–545

Chen J, Li QX, Song B (2020) Chemical nematicides: recent research progress and outlook. J Agric Food Chem 68:12175–12188

Coyne DL, Cortada L, Dalzell JJ, Claudius-Cole AO, Haukeland S, Luambano N, Talwana H (2018) Plant-parasitic nematodes and food security in sub-Saharan Africa. Annu Rev Phytopathol 5:381

Danish M, Robab MI, Marraiki N, Shahid M, Zaghloul NS, Nishat Y, Shaikh H (2021) Root-knot nematode *Meloidogyne incognita* induced changes in morpho-anatomy and antioxidant enzymes activities in *Trachyspermum ammi* (L.) plant: a microscopic observation. Physiol Mol Plant Pathol 116:101725

Danish M, Robab MI, Hatamleh AA, Al-Dosary MA, Datta R, Sheikh H (2022) Reduction of root-knot nematode infection with compost and nematode-trapping fungus in greenhouse setting. Rhizosphere 22:100508

Devi LS, Richa S, Sharma R (2002) Effect of *Trichoderma* spp., against root-knot nematode *Meloidgyne incognita* on tomato. Indian J Nematol 32:227–228

Dong JY, Li XP, Li L, Li GH, Liu YJ, Zhang KQ (2006) Preliminary results on nematicidal activity from culture filtrates of basidiomycetes against the pinewood nematode, *Bursaphelenchus xylophilus* (Aphelenchoididae). Ann Microbiol 56:163–166

Dong LQ, Yang JK, Zhang KQ (2007) Cloning and phylogenetic analysis of the chitinase gene from the facultative pathogen *Paecilomyces lilacinus*. J Appl Microbiol 103:2476–2488

Escudero N, Lopez-Llorca LV (2012) Effects on plant growth and root-knot nematode infection of an endophytic GFP transformant of the nematophagous fungus *Pochonia chlamydosporia*. Symbiosis 57(1):33–42

Ferreira BS, Santana MV, Macedo RS, Silva JO, Carneiro MAC, Rocha MR (2018) Co-occurrence patterns between plant-parasitic nematodes and arbuscular mycorrhizal fungi are driven by environmental factors. Agric Ecosyst Environ 265:54–61

Ghahremani-Nahr J, Kian R, Sabet E (2019) A robust fuzzy mathematical programming model for the closed-loop supply chain network design and a whale optimization solution algorithm. Expert Syst Appl 116:454–471

Ghorbanpour M, Omidvari M, Abbaszadeh-Dahaji P, Omidvar R, Kariman K (2018) Mechanisms underlying the protective effects of beneficial fungi against plant diseases. Biol Contr 117:147–157

Giné A, Sorribas FJ (2017) Effect of plant resistance and BioAct WG (*Purpureocillium lilacinum* strain 251) on *Meloidogyne incognita* in a tomato–cucumber rotation in a greenhouse. Pest Manag Sci 73:880–887

Giné A, Carrasquilla M, Martínez-Alonso M, Gaju N, Sorribas FJ (2016) Characterization of soil suppressiveness to root-knot nematodes in organic horticulture in plastic greenhouse. Front Plant Sci 7:179723

Goswami BK, Mittal A (2004) Management of root-knot nematode infecting tomato by *Trichoderma viride* and *Paecilomyces lilacinus*. Indian Phytopathol 57:235–236

Guzmán-Guzmán P, Porras-Troncoso MD, Olmedo-Monfil V, Herrera-Estrella A (2019) *Trichoderma* species: versatile plant symbionts. Phytopathology 109:6–16

Hol WHG, Cook R (2005) An overview of arbuscular mycorrhizal fungi-nematode interactions. Basic Appl Ecol 6:489–503

Jamshidnejad V, Sahebani N, Etebarian H (2013) Potential biocontrol activity of *Arthrobotrys oligospora* and *Trichoderma harzianum* B1 against *Meloidogyne javanica* on tomato in the greenhouse and laboratory studies. Arch Phytopathol Plant Protect 46:1632–1640

Jiang X, Xiang M, Liu X (2017) Nematode-trapping fungi. Microbiol Spectr 5:1–12

Khan A, Williams KL, Nevalainen HK (2006) Infection of plant-parasitic nematodes by *Paecilomyces lilacinus* and *Monacrosporium lysipagum*. BioControl 51:659–678

Kilama P, Dubois T, Coyne D, Niere B, Gold CS, Adipala E (2007) Antagonism of *Paecilomyces* spp. isolated from banana (Musa spp.) roots and rhizosphere against *Radopholus similis*. Nematropica 37:215–226

Koffi MC, Vos C, Draye X, Declerck S (2013) Effects of *Rhizophagus irregularis* MUCL 41833 on the reproduction of *Radopholus similis* in banana plantlets grown under *in vitro* culture conditions. Mycorrhiza 23:279–288

Leonetti P, Zonno MC, Molinari S, Altomare C (2017) Induction of SA-signaling pathway and ethylene biosynthesis in *Trichoderma harzianum* treated tomato plants after infection of the root-knot nematode *Meloidogyne incognita*. Plant Cell Rep 36:621–631

Lima-Rivera DL, Lopez-Lima D, Desgarennes D, Velazquez-Rodriguez AS, Carrion G (2016) Phosphate solubilization by fungi with nematicidal potential. J Soil Sci Plant Nutr 16:507–524

Maciá-Vicente JG, Rosso LC, Ciancio A, Jansson HB, Lopez-Llorca LV (2009) Colonisation of barley roots by endophytic *Fusarium equiseti* and *Pochonia chlamydosporia*: effects on plant growth and disease. Ann Appl Biol 155(3):391–401

Malhadas C, Malheiro R, Pereira JA, de Pinho PG, Baptista P (2017) Antimicrobial activity of endophytic fungi from olive tree leaves. World J Microbiol Biotechnol 33:46

Marro N, Caccia M, Doucet ME, Cabello M, Becerra A, Lax P (2018) Mycorrhizas reduce tomato root penetration by false root-knot nematode *Nacobbus aberrans*. Appl Soil Ecol 124:262–265

Mezeal IA, Mizil SN, Hussin MS (2018) Researching biocontrol of *Trichoderma viride*, *Paecilomyces lilacinus* in contradiction of effectiveness of fungi insulated as of selected therapeutic herbals. Plant Arch 18:1631–1637

Mittal N, Saxena G, Mukerji GK (1995) Integrated control of root-knot disease in three crop plants using chitin and *Paecilomyces lilacinus*. Crop Prot 14:647–651

Moreno-Gavíra A, Diánez F, Sánchez-Montesinos B, Santos M (2020) *Paecilomyces variotii* as a plant-growth promoter in horticulture. Agronomy 10:597

Mukherjee PK, Mendoza-Mendoza A, Zeilinger S, Horwitz BA (2022) Mycoparasitism as a mechanism of *Trichoderma*-mediated suppression of plant diseases. Fungal Biol Rev 39:15–33

Pathak DKN, Ranjan R, Kumar M, Kumar B (2005) Biomanagement of *Meloidogyne graminicola* by *Trichoderma harzianum* and *Trichoderma virens* in rice. Ann Plant Protect Sci 13:438–440

Paul NC, Deng JX, Lee JH, Yu SH (2013) New records of endophytic *Paecilomyces inflatus* and *Bionectria ochroleuca* from chili pepper plants in Korea. Mycobiology 41:18–24

Pervez R, Ali SS (2011) Phytonematodes, a thrust in pulse crops. In: Sharma GL (ed) Phytonematodes in crops: their identification, treatment and management. New India Publishing Agency, New Delhi, pp 25–54

Pieterse CMJ, Zamioudis C, Berendsen RL, Weller DM, van Wees SCM, Bakker PAHM (2014) Induced systemic resistance by beneficial microbes. Annu Rev Phytopathol 52:347–375

Poveda J (2021) *Trichoderma* as biocontrol agent against pests: new uses for a mycoparasite. Biol Control 159:104634

Sahebani N, Hadavi N (2008) Biological control of the root-knot nematode *Meloidogyne javanica* by *Trichoderma harzianum*. Soil Biol Biochem 40:2016–2020

Salvioli A, Bonfante P (2013) Systems biology and "omics" tools: a cooperation for next-generation mycorrhizal studies. Plant Sci 203-204:107–114

Shakeel A, Khan AA, Bhat AH, Sayed S (2022) Nitrogen fertilizer alleviates root-knot nematode stress in beet root by suppressing the pathogen while modulating the antioxidant defence system and cell viability of the host. Physiol Mol Plant Pathol 120:101838

Sharf R, Hisamuddin AM, Akhtar A (2014) Combined effect of biofertilizers and fertilizer in the management of *Meloidogyne incognita* and also on the growth of red kidney bean (*Phaseolus vulgaris*). Int J Plant Pathol 5:1–11

Sharma P, Pandey R (2009) Biological control of root-knot nematode; *Meloidogyne incognita* in the medicinal plant: *Withania somnifera* and the effect of biocontrol agents on plant growth. Afr J Agric Res 4:564–567

Sharma P, Rakesh P (2009) Biological control of root-knot nematode; *Meloidogyne incognita* in the medicinal plant; *Withania somnifera* and the effect of biocontrol agents on plant growth. Afr J Agric Res 4:564–567

Simon S, Anamika (2011) Management of root knot diseases in rice caused by *Meloidogyne graminicola* through nematophagous fungi. J Agric Sci 3:122–127

Singh KP, Jaiswal RK, Kumar N, Kumar D (2007) Nematophagous fungi associated with root galls of rice caused by *Meloidogyne graminicola* and its control by *Arthrobotrys dactyloides* and *Dactylaria brochopaga*. J Phytopathol 155:193–197

Singh UB, Sahu A, Singh RK, Singh DP, Meena KK, Srivastava JS, Manna MC (2012) Evaluation of biocontrol potential of *Arthrobotrys oligospora* against *Meloidogyne graminicola* and *Rhizoctonia solani* in Rice (*Oryza sativa* L.). Biol Control 60:262–270

Singh UB, Sahu A, Sahu N, Singh BP, Singh RK, Singh DP, Jaiswal RK, Sarma BK, Singh HB, Manna MC, Rao AS (2013) Can endophytic *Arthrobotrys oligospora* modulate accumulation of defence related biomolecules and induced systemic resistance in tomato (*Lycopersicon esculentum* mill.) against root knot disease caused by *Meloidogyne incognita*. Appl Soil Ecol 63:45–56

Smith SE, Facelli E, Pope S, Smith FA (2010) Plant performance in stressful environments: interpreting new and established knowledge of the roles of arbuscular mycorrhizas. Plant Soil 326:3–20

Thambugala KM, Daranagama DA, Phillips AJ, Kannangara SD, Promputtha I (2020) Fungi vs fungi in biocontrol: an overview of fungal antagonists applied against fungal plant pathogens. Front Cell Infect Microbiol 10:604923

Tunlid A, Rosen S, Ek BO, Rask L (1994) Purification and characterization of an extracellular serine protease from the nematode-trapping fungus *Arthrobotrys oligospora*. Microbiology 140:1687–1695

Vos CM, Tesfahun AN, Panis B, De Waele D, Elsen A (2012) Arbuscular mycorrhizal fungi induce systemic resistance in tomato against the sedentary nematode *Meloidogyne incognita* and the migratory nematode *Pratylenchus penetrans*. Appl Soil Ecol 61:1–6

Wang Y, Zhang WP, Cao HL, Shek CS, Tian RM, Wong YH, Batang Z, Al-Suwailem A, Qian PY (2014) Diversity and distribution of eukaryotic microbes in and around a brine pool adjacent to the Thuwal cold seeps in the Red Sea. Front Microbiol 5:37

Yan XN, Sikora RA, Zheng JW (2011) Potential use of cucumber (*Cucumis sativus* L.) endophytic fungi as seed treatment agents against root-knot nematode *Meloidogyne incognita*. J Zhejiang Univ Sci B 12:219–225

Yang Y, Yang E, An Z, Liu X (2007) Evolution of nematode-trapping cells of predatory fungi of the Orbiliaceae based on evidence from rRNA-encoding DNA and multiprotein sequences. Proc Natl Acad Sci USA 104:8379–8384

Zeilinger S, Gupta VK, Dahms TES, Silva RN, Singh HB, Upadhyay RS, Gomes EV, Tsui CK-M, Nayak SC (2016) Friends or foes? Emerging insights from fungal interactions with plants. FEMS Microbiol Rev 40:182–207

Zhang S, Gan Y, Ji W, Xu B, Hou B, Liu J (2017) Mechanisms and characterization of *Trichoderma longibrachiatum* T6 in suppressing nematodes (*Heterodera avenae*) in wheat. Front Plant Sci 8:1491

Opportunistic Fungi, Plant, and Nematode Interactions in Agricultural Crops

2

Basit Akolade Adigun, Shakirah Binti Mohammad Nahar, Yusuf Opeyemi Oyebamiji, Tolessa Muleta Daba, Jean Paul Sinumvayo, Taofeek Tope Adegboyega, Saheed Adekunle Akinola, and Ismail Abiola Adebayo

2.1 Introduction

Ecosystem comprises of both the biotic and abiotic components. The abiotic factors influence the survival and functioning of the biotic likewise the living organisms interact, interrelate, and interdepend on one another in the ecosystem. Since, living organisms prefer to live in close association with other organisms in their environment. Ecosystem depends on these several interactions to make it a functional and balance entity. Both the shoot and root system are prone to interactions from various organisms. Naturally, the location of root system seems to favour more interactions than the shoot due to the conduciveness of the rhizosphere serving as a reservoir for various microbes and medium for the interaction. The rhizosphere is the region containing plant roots and numerous soil microorganisms with the microbes surrounding the root. The root and microorganism secrete their metabolites, both of which influences the plant in this region leading to several chemical and biochemicals reactions (Srinivasan et al. 2020). Different crops may have unique microbial

B. A. Adigun
Crop Science Program, Faculty of Fisheries and Food Science, Universiti Malaysia Terengganu, Kuala Nerus, Terengganu, Malaysia

S. B. M. Nahar · Y. O. Oyebamiji
Department of Biological Sciences and Biotechnology, Faculty of Science and Technology, Universiti Kebangsaan Malaysia, Bangi, Selangor, Malaysia

T. M. Daba · I. A. Adebayo (✉)
Department of Medical Biochemistry, Molecular Biology and Genetics, School of Medicine and Pharmacy, College of Medicine and Health Sciences, University of Rwanda, Butare, Rwanda

J. P. Sinumvayo · T. T. Adegboyega · S. A. Akinola
Department of Medical Microbiology and Parasitology, School of Medicine and Pharmacy, College of Medicine and Health Sciences, University of Rwanda, Butare, Rwanda

M. S. Akhtar (ed.), *Opportunistic Fungi, Nematode and Plant Interactions*, https://doi.org/10.1007/978-981-97-2045-3_2

communities closely associated with their organs. According to Velmourougane et al. (2017) root exudates are major factor determining the prevention or accommodation of microbial proliferation in the rhizosphere and these root metabolites contain acids, polysaccharides, amino acids, mucilage, and other cell components.

Cultivated lands contain several eukaryotic and prokaryotic organisms which inhabits the same environment with the planted crops. Some of this organism may be obligate parasite, facultative parasite, pathogenic, opportunistic, and non-pathogenic. Plant–microbe interactions activate several transformations in the rhizosphere, playing their active roles in nutrient cycling, most especially carbon and nitrogen sequestration, which are crucial to the functioning ecosystem (Velmourougane et al. 2017). Health and yield of agricultural crop is influenced by their interactions with pathogenic microbes, beneficial microbes, plants, and animals in their environment. Such interactions may be beneficial or harmful to the cultivated crops. Basically, interactions in functioning ecosystem can be mutualistic, antagonistic or commensalism.

Mutualistic interactions are associations in which all partners benefit. Examples include nitrogen fixing, nutrifying, and denitrifying bacteria, all of which are important components of the nitrogen cycle; insects and flowering plants; plant roots and fungi in mycorrhiza; algae and fungi in lichens. Mutualistic associations provide plants with services and resources which include pollination, nutrient supply, drought resistance, and pathogen protection (Delavaux et al. 2022). Grass and endophytic fungi interaction has been described as mutualistic because the fungi enhanced growth of the host (grass) by increasing tillering, conferring insect resistance and drought tolerance, and toxic reaction in cattle that feeding on the infected grasses (Bacon and Hill 1996).

Antagonistic interactions may include competition between plant populations in a field, parasitic flowering plants and their host, *Trichoderma* spp. against some plant pathogens, *Penicillium* species antimicrobial activity on some microorganism. Most biocontrol agents exhibit antagonism on the organisms they control. Sharangi (2011) have reported the use of some crops against some weed and the advantages of incorporating them in cropping system. However, allelopathy can go both ways (harmful or beneficial) depending on the allelopathic chemical been secreted. On the other hand, commensalism interactions in agronomical plants may include intercropping of banana or plantain (*Musa* spp.) and cocoa (*Theobroma cacao* L.) which does not affect one another because growth stage of both plants differs and the banana tends to shade the cocoa seedlings from excess sunlight, rooted climbers or epiphytes and their host plant, both do not depend on their host for nutrient but only need them to get access to sunlight therefore they pose any harm to their host plant. Therefore, this chapter explores the various physical and biochemical interactions between agricultural plant and opportunistic microorganisms.

2.2 Opportunistic Organisms

Opportunistic organisms are organisms that tend to dominate due to certain circumstances limiting the metabolism of the other organism. Their expression depends on the vulnerability of the host organism. Gerg et al. have described opportunistic pathogens as diseases causing organisms which takes advantage of the immunocompromised condition of the host. Marçais and Bréda (2006) defined opportunistic organism as an organism incapable of colonizing a host unless the host has been weakened in advance by another stress.

2.3 Factors Affecting the Interactions Opportunistic Organisms and Agricultural Crops

Factors leading to impaired physiological state of agricultural crops will lead to the expression of the opportunistic organism and vice versa. Several biotic and abiotic factors can influence nature of interactions among living organisms. Researchers have proved that interactions among organisms are dynamic and can be impacted by abiotic factors and consequently influence both the nature and magnitude of their interactions (Ragozzino and D'Errico 2011; Zhang et al. 2020). Biotic factors influencing interaction may include previous infections on the host plant which might have weaken the defence system of the plant and providing chance for the expression of opportunistic organism. For example, *Fusarium solani* in the ornamental plants (*Kalanchoe blossfeldiana*) requiring prior infection of *Thielaviopsis basicola* (Madriz-Ordeñana et al. 2019).

Growth or developmental stage of agricultural crops: during flowering period some pollinators (e.g., insects) may introduce inoculum of certain opportunistic organism to the host plant, early growth stages, and senescence period of some plant may exert some level of vulnerability or weakness on some plants. Non-pathogenic microbes of harvested fruits and vegetables may influence shelf life of their host by depleting the quality of the produce (Al-Kharousi et al. 2016).

Wounding of agricultural crops can lead to expression of opportunistic infection. Herbivory or damages made on crop plant may expose some plant internal tissues which have high level of susceptibility to opportunistic pathogen. Injuries caused by some vectors or carrier of opportunistic organism can facilitate the introduction of the pathogen, e.g., *Fusarium verticillioides* by *Diatraea saccharalis* caterpillar (Franco et al. 2021) or *Ostrinia nubilalis* (European corn borer) larvae (Gatch and Munkvold 2002). Nutrient deficiency and unfavourable weather conditions tend to stress agronomical plants and render most crop plants vulnerable to various opportunistic pathogens. In maize plant stalk rot fungi infection was promoted by abiotic stresses that render the host vulnerable, especially after anthesis which is characterized with decrease in carbohydrate level (Dodd 1980). Production of aflatoxin commenced only when phytoalexin production stopped in drought peanut stressed immature peanuts (Dorner et al. 1989).

2.4 Opportunistic Fungi and Agricultural Crops Interactions

Plant interaction with opportunistic pathogen with time often led to the detriment of the host plant while interactions involving non-pathogenic fungi could be beneficial by promoting nutrient absorption, growth, disease resistance. Several fungi species have been studied and several reports on various fungi influencing their host plants. The fungal genera *Fusarium* and *Rhizoctonia* contain both phytopathogenic and non-pathogenic species and strains with the non-pathogenic strains serving as bio-control agent and inducing plant resistant to the pathogenic strains (Harman et al. 2004). An interaction involving some opportunistic fungi species is discussed below.

2.4.1 *Trichoderma* Spp.

One of the most utilized fungi by man in agricultural crops is *Trichoderma* spp. due to its ability to serve as a biological control agent. *Trichoderma* spp. has been described as opportunistic, avirulent plant symbionts, and one of the most interactive free-living fungi found in the rhizosphere (Harman et al. 2004). However, a remarkable instance has been described in which supposed non-pathogenic group of fungi such as *Trichoderma* spp., some strains of *Fusarium* and *Rhizoctonia* and, mycorrhizae strains have the potential to infect plant roots and become pathogenic but would limit their infection to superficial root cells (Harman et al. 2004). Bailey and Lumsden (1998) have reported very few cases, where certain strains of *Trichoderma* spp. become pathogenic to plants, producing toxins, and causing diseases in crops like apples, maize, and alfalfa. Some authors have reported the case of *Trichoderma* as ear rot pathogen in maize plant in the USA (Munkvold and White 2016). Similarly, in Europe, the occurrence of *Trichoderma* ear rot infection has been reported (Pfordt et al. 2020). Previous reports showed that *T. afroharzianum* ear rot infection is opportunistic depending on prior mechanical damage or injury caused by insects or birds (Munkvold and White 2016). Conversely, Pfordt et al. (2020) proofed that no mechanical injury is required for *T. afroharzianum* ear rot infection and therefore stated that *T. afroharzianum* is not an opportunistic pathogen. Symptoms and sign of *Trichoderma* ear rot infection include appearance of dark, grey-green layers of conidia on and between the kernels of infected ears causing premature germination of the kernels (Pfordt et al. 2020). *Trichoderma* spp. has also been reported as a causative agent of green mould disease in mushroom cultivation (Sánchez-Montesinos et al. 2021). In Trichoderma penetration is restricted to the intercellular spaces in plant root only and it is superficial in nature, not growing beyond the epidermal layer and colonization of the plant root is promoted by the secretion of a protein called swollen in (Strakowska et al. 2014).

2.4.2 Arbuscular Mycorrhizal Fungi

Agronomical plants interactions involving arbuscular mycorrhizal fungi proved beneficial to the plant. Mycorrhiza associated with tomato roots can secrete exudates serving as substrate to support the normal growth of bacteria, which in turns enhance nutrient availability to both Arbuscular mycorrhizal fungi and the plant (Emmanuel and Babalola 2020). Synergistic effect of a bacteria (*Bradyrhizobium japonicum*) and fungi (*Aspergillus niger*) increased nitrogen fixation in soybean (Sharma and Kumawat 2011). Similarly, the interplay between arbuscular mycorrhizal (AM) and *B. japanicum* stimulated the growth of tomato plants (Sagar et al. 2022). These interactions involving arbuscular mycorrhizal lead to the increase in growth and therefore beneficial to the plants. Farias et al. (2018) investigated the effect of multiple fungi species (*Beauveria bassiana, Metarhizium anisopliae, Pochonia chlamydosporia, Purpureocillium lilacinum,* and *T. asperella*) on the growth response and arbuscular mycorrhizal colonization in sugarcane, corn, and soybean. Increased growth of the roots of the three plants was reported, and they suggested this may be due to the fungi stimulating auxin production in the plants. In the same study, soybean and sugarcane AM colonization of the root were 20% and 28% greater, respectively, when compared with the uninoculated treatment (control) the increment was insinuated to be as a result production of metabolites promoting AM colonization (salicylic acid and abscisic acid) or inhibiting (jasmonic acid) by the fungi treatment. Besides no increment in root colonization was recorded in maize.

2.4.3 *Fusarium* Spp.

Franco et al. (2021) assessed the interplay among *Saccharum* spp. (sugar cane plant), *Fusarium verticillioides* and *Diatraea saccharalis* caterpillar (plant–fungi–insect interaction) in which the fungus secretes volatile organic compounds (VOCs) which attracts the insect caterpillar to feed as a result the caterpillar serve as a carrier for the fungus and when it develop into adult, the fungus is transmitted vertically to their offspring, which continues the cycle by inoculating the fungus into healthy plants. The result revealed that *F. verticillioides* is an opportunistic organism and capable of influencing both the host plant and the insect to be instrumental in its dissemination as it navigates its life cycle stages. Likewise, based on their electroantennography (EAG) result it was suggested that *F. verticillioides* can affect nervous system of the insect ensuring that only uncontaminated insects feeds on the infected plants leading to efficient transmission of the fungus (Franco et al. 2021). *F. verticillioides* infected maize plants have attractant effect on the insect *Eldana saccharina* (stem borer) (Schultess et al. 2002) and *Ostrinia nubilalis* (European corn borer) (Gatch and Munkvold 2002) leading to pest infestation and stalk rot infection, respectively. *F. verticillioides* can lead to abnormal growth characteristics in maize plant affecting the main stem and foliage of the plant (Schultess et al. 2002). Deformation of the shoot by *F. verticilloides* may be severe in maize, decreasing agronomical yield of the plant since maize lack tillering system (Bosque-Pérez and Mareck 1991; Cardwell et al. 1997). Numerous *Fusarium* species were most often isolated from sugar cane in tissues damaged by *E. saccharina* borings, but some species were also isolated endophytically in undamaged and non-diseased stalks of the sugar cane plant (Mcfarlane et al. 2009). *F. solani* isolated from infected *K. blossfeldiana* is not pathogenic based on pathogenicity test and was proven to be opportunistically dependent on *T. basicola* pre-infection (Madriz-Ordeñana et al. 2019). Avirulent strains of *F. oxysporum* opportunistically took advantage of water-stressed (excessive water exposure) to infect *Zinnia hybrida* plants in International Space Station (ISS), signs (aerial mycelia), and leaf margin necrosis were observed (Schuerger et al. 2021).

2.4.4 *Aspergillus flavus*

This is one of the major economically important fungi notorious for its production of aflatoxin. It is an opportunistic pathogen affecting maize and other oil seed crops (Klich 2007; Kelley et al. 2009). *A. flavus* took advantage of heat and drought stress to attack its host plant (Klich 2007; Dolezal et al. 2014). Production and accumulation of aflatoxin in maize and some crops have been associated with the antecedent of heat and drought stress (Payne 1998). Klich (2007) described the mode of entry of *A. flavus* stating that the fungi can invade plants through various opening including insect openings. Besides, there are possibilities of manifestation of *A. flavus* infection in plants without obvious opening or injury. *A flavus* have been described to be unspecialized in its choice of host, having a broad host range in plants and insects (St. Leger et al. 2000). Series of symptoms have been reported from plants infected with *A. flavus*. Healthy pinto bean leaves did not develop infection with the inoculation of *A. flavus* while leaves stressed with heat or dry ice prior to inoculation showed yellowing and necrotic symptoms which initiated from the wound site and proliferated to healthy tissues of the plant (St. Leger et al. 2000). This indicated that the heat and cold stress is necessary for the infection of pinto bean leaves by *A. flavus*. Likewise, in maize kernel very little fungal colonization was reported from the inoculation of unstressed kernels while stressed kernels exhibited varying levels symptoms ranging from formation of conidia on the inoculated kernel to browning and rots in underneath kernels depending on virulent nature of strain of the pathogen (Mellon et al. 2005). Substrate consumption of *A. flavus* in corn and cotton seeds occurs in a sequence beginning with free saccharides followed by lipid reserves and finally utilizing starch, in case of undamaged corn kernels, *A. flavus* preferentially targets the germ tissue which has the highest percentage of saccharides and triglycerides (Mellon et al. 2000; Mellon et al. 2005). *A. flavus* infection in maize kernel can effect metabolic changes, not limited to the production of metabolites for defence and disruption of kernel development (Dolezal et al. 2014). Fragile endosperm, discoloured and wrinkled germ with mycelium growth in the embryo have also been reported in infected maize kernel (Dolezal et al. 2014).

2.4.5 *Penicillium* Spp.

Penicillium has been described as having both epiphytic and endophytic association with plants (Srinivasan et al. 2020). Their interaction is mostly beneficial due to the antagonistic nature of the fungus on some plant pathogens although there are reports of fungus been pathogenic in few cases. In *Solanum melongena* (egg plants) *Penicillium* spp. has been reported as a saprophytic fungus, capable of opportunistically infecting the fruit when physiological state of the fruit is impaired by injuries caused by insects or birds, advanced maturity, solar burn, previous fungi infection, etc. (INRAE 2021). *Penicillium* spp. causing blue mould in egg plant is very rear but often attack ripe fruit, during harvest and in storage. Similarly, saprophytes like *Cladosporium* spp., and *Alternaria alternata* opportunistically caused cladosporium rot and black rot diseases, respectively, in egg plant fruit, taking the advantage of the injuries on the fruit. In *Solanum tuberosum* (potato plants), Taheri and Kakooee (2017) reported the occurrence of *A. tenuissima* as an opportunistic and necrotrophic fungus causing alternaria blight in potato plantations in Iran. *A. tenuissima* is characterized with infection of older or stressed leaves leading to necrosis with concentric rings which expand necrotic lesions on leaves. In the presence of favourable condition Alternaria blight can escalate to major symptoms on the stem and defoliation which can result huge yield loss if not controlled early. *A. tenuissima* is species very close to *A. solani* which causes early blight in the same plant.

2.4.6 Plant Fungi and Nematode Interactions

Most interaction involving crop plants, fungi, and nematodes usually have antagonistic effect on the nematode but beneficial to the plant due to the antagonism of the fungi against the phytonematodes. Several fungi have predatory effects on plant-parasitic nematodes in the rhizosphere which makes them suitable for control against the phytonematodes. *Arthrobotrys oligospora* has been described as the most frequently isolated, sufficiently present in the environment and the first to be recognized nematophagous fungus (Niu and Zhang 2011). Barron (2003) descried *Arthrobotrys* as a predatory fungus of nematode which feed by making use of traps formed from specialized hyphae which comes in various forms to attract, capture, crushed, kill, and digest using the enzyme secreted. The nutrient from the digested nematode is then transported and distributed to other part of the mycelium for fungi metabolism. *Pochonia chlamydosporia* (egg-parasitic fungus) has also been described as predator of nematodes with potentials to be utilized as biological control agent of plant nematodes (Larriba et al. 2015). *Paecilomyces* fungi have been characterized with their antagonistic interaction against and serving as control for pest and diseases in some crops (Moreno-Gavíra et al. 2021). *P. lilacinus* parasitizes some stages of *Meloidogyne* spp. (root knot nematode) in its environment, feeding on them in a way very similar to what was described with *Arthrobotrys* thereby serving as control against the nematode (Senthilkumar et al. 2020). As reported in Larriba et al. (2015), the application of *P. chlamydosporia* as a biological control of nematodes in wide array of crops and growth promoters in many crops including wheat, tomato, lettuce, and barley. In early research on effect of *A. oligospora* and *Pochonia chlamydosporia* (=*Verticilum chlamydosporium*) on some plant roots reported the chemotropic movement of *A. oligospora* towards the root of plants (pea, barley, and white mustard) and they suggested that this could explain the relative abundance of the fungi in the rhizosphere as compared to other part of the soil (Bordallo et al. 2002). In the same study, necrotic regions caused by *A. oligospora* were reported in root epidermal cells of barley plants which was absent in roots colonized by *P. chlamydosporium*. Both fungi proliferated into the root intercellular spaces of barley and tomato via direct penetration or the use of appressoria. Both fungi induce cell appositions in tomato and barley roots. Kassam et al. (2021) studied the effect of *A. thaumasia* on 1-month-old tomato seedlings infected with nematode (*M. incognita*) and observed that fungus increases growth parameters of infected plants even when compared the control (uninfected plants). The result of the study showed that *A. thaumasia* enhanced plant growth in nematode-infected plants indicating the fungi protection on the plant against the nematode and further application of fungal filtrate would reduce the nematode disease burden. Colonization of barley root by *P. chlamydosporium* has also been reported with the fungus modifying the transcriptome of barley leading to the promotion of biosynthesis of auxin, chaperones, and induction of genes related both biotic and abiotic stress responses in the plant (Larriba et al. 2015). Ashraf and Khan (2005) studied the effect of some opportunistic fungi (*Paecilomyces lilacinus*, *Cladosporium oxysporum*, *Gliocladium virens*, and *Talaromyces flavus*) on the life cycle of a nematode (*M. javanica*) on brinjal seedlings (*Solanum melongena*). Their study revealed that each of the four opportunistic fungi implemented have inhibitory effect on the penetration and proliferation of the nematode on the plant. Similarly, the life cycle of the nematode was delayed in all the fungi studied. This study clearly showed the antagonistic effect of some opportunistic fungi species on a nematode.

2.4.7 Opportunistic Fungi in Woody Plants

Cases of opportunistic fungus causing opportunistic diseases have also been reported in woody plants. *Armillaria gallica* is an opportunistic fungus characterized with low level of level of pathogenicity and unable to infect vigorously growing hosts (Wargo and Harrington 1991). *A. gallica* often opportunistically invades oak trees damaged by insect defoliation or drought (Marçais and Bréda 2006). In NW America forests, another *Armillaria* species (*A. ostoyae*) was reported to invade optimally growing tree stumps created by selective logging from quiescent lesions and subsequently colonize and disseminate its inoculum in the vicinity, leading to increased infection in surrounding trees (Cruickshank et al. 1997; Morrison et al. 2001). *Cylindrocarpon* causes stem end rot in kiwifruit (*Actinia* spp.), while *Phomopsis* sp. and *Phoma* sp. *Botrytis* are grouped as opportunistic fungi taking advantage of injured spot on the fruit and causing a disease which they referred to as wound rot (Burdon and Lallu 2011).

2.4.8 Metabolites Produce by some Opportunistic Fungi and Plant Response

When opportunistic organisms become pathogenic, they exhibit properties like pathogens. Pathogens produce several toxins and proteins which help to facilitate the establishment of infection and maintain disease in the host plant (de Jonge et al. 2011; Kale and Tyler 2011). The role of exudates produce by opportunistic organism during interaction with plant may vary depending on the nature of the interaction. Usually, these secondary metabolites have their roles which help in the survival of the microorganism. Dolezal et al. (2013) illustrated that it may be a difficult task to determine which secondary metabolites is produced as a mere response to changes in the fungi environment or for pathogenic purpose. Some scientists are of the view that some of these metabolites serve as a protectant for the fungus against stress agent during its growth (Dolezal et al. 2013). Swollenin produced by *Trichoderma* spp. contains cellulase and hemicellulase which function by influencing the binding of the enzyme to the insoluble substrate and altering the plant cell wall structure (Strakowska et al. 2014). Array of toxins and metabolites including gibberellic acid, moniliformin, fumonisins, fusaric acid, beauvericin, fusaproliferin, fusaproliferin as reported in Mcfarlane et al. (2009). Bartelt and Wicklow (1999) assayed the volatile organic compounds produced *F. verticillioides* and revealed the presence of several low molecular weight alcohols, esters, aldehydes, and other compounds that can entice sap beetles (*Carpophilus humeralis*). Lower dosage of Aflatoxin has been reported to inhibit the development of callus, shoot and root growth of Tobacco plants in vitro culture (McLean et al. 1994a, b). When interactions lead to infection of the host plants, the plant also reacts to combat the infection. These reactions by the plant against infection may include oxidative burst due to accumulation of reactive oxygen species, programmed cell death, and hypersensitivity response against infections which may be noticeable at early stages of the infection (Lamb and Dixon 1997; Dickman and Fluhr 2013; Taheri and Kakooee 2017). Taheri and Kakooee (2017) result also revealed the formation of reactive oxygen species (ROS) in infected and neighbouring cells of the potato plant which was observed as one of the initial histochemically and cytologically measurable pointer of the interaction between the plant and the pathogen.

2.5 Opportunistic Plants and Agricultural Crops Interactions

An opportunistic plant is a plant that uses any available resources to rapidly establish itself in a new environment. These plants are frequently thought to be fast-growing and adaptable, able to flourish in a variety of environments. Some opportunistic species can survive on nearly any available nutrient. Like other organisms, interspecies interactions are common among Plantae. These relationships between different species can be mutualistic, communalistic, or parasitic to the crops grown in the environment (Subrahmaniam et al. 2018). Kuebbing and Nuñez

(2015) discovered that when a plant interacts with a neighbouring plant, it can change the physiology of the plant. This can include changes in gene regulation or the activation of a defence mechanism. Certain plants serve as beneficial neighbours in positive plant–plant interactions by alerting plants of the same or different species to impending danger (Babikova et al. 2014). Plant–plant interactions are important in controlling the communities and ecosystems. They not only affect the composition of plant communities, but they also have far-reaching impacts on ecosystems, such as the availability of resources and the structure of habitats. However, the effects of plant–plant interactions can be changed by external (both natural and anthropogenic) drivers such as climatic circumstances or nutrient availability (Brooker 2006). One good example of opportunistic plants and agricultural crops interactions is fern *Azolla*. *Azolla* is a widespread free-water fern in tropical, subtropical, and mild temperate regions (Raja et al. 2012). The application of *Azolla* as a biofertilizer provides nutrients from natural sources and has the potential to significantly improve soil health and increase yield sustainability. Interaction between fern azolla and paddy fields has been shown to increase rice yield (Akhtar et al. 2020; Thapa and Poudel 2021; Santhiya and Jeeva 2022). *Azolla* is capable of suppressing weeds, which will improve the availability of N, P, K, and other mineral nutrients. *Azolla* biofertilizer has the capacity to maintain an optimal soil pH and to fix organic C and N, thereby enhancing mineralization, microbial activity, and crop yield. A group of researchers discovered that the addition of *Azolla* increases paddy yield by 8–14% (Yao et al. 2018). *Azolla* application at a rate of 10–12 tons per acre increases soil nitrogen content by 50–60 kg/ha and lowers the rice crop's nitrogenous fertilizer requirement by 30–35 kg (Roy et al. 2016).

Invasive weeds are another example of opportunistic plants *Centaurea stoebe*. According to the study of Broz et al. (2010) this fungus could be cultivated with the same and distinct plant species. They concluded that the results *C. stoebe* significantly upregulated its defensive mechanism when grown with the same species plants as opposed to those grown with various species plants. It demonstrates that *C. stoebe* can modify its growth and defensive strategy based on the nature of nearby plants, highlighting the importance of biochemical processes in plant sensing and responding mechanisms. Plants grown with the same species may be more resistant to pathogens than those grown with different species.

When grown in a mixed cropping system, grass can have several beneficial impacts on other crops. This plant is also a model of an opportunistic plant. Grass is a valuable plant for farming systems because it offers a variety of advantages to the crops that are grown around it. The ability to enhance soil structure is one of the most significant benefits (Ball et al. 2020). Grass roots are capable of penetrating compacted soil, allowing other plants to receive more water and nutrients. This enhanced soil structure also improves soil porosity, allowing for enhanced aeration and drainage and resulting in healthier plants (Blanco-Canqui and Jasa 2019). Grass is essential for nutrient cycling in addition to enhancing the soil's structure. When grasses expire, they decompose and return nutrients to the soil, thereby making them accessible to other plants (Dabney et al. 2001). The use of artificial fertilizers, which can be both expensive and harmful to the environment, is minimized as a result of this procedure.

Several characteristics of opportunistic plants cause negative impact to other plant. Although they are capable of preventing erosion and stabilizing the soil, they can also be extremely harmful and risky to ecosystems. Allelopathy is a good example of a negative plant interaction in which one plant secretes chemicals that inhibit or kill the development of another plant, either of the same species (autotoxicity) or of a different species (Zhang et al. 2021). There are multiple methods in which allelopathy may affect other plants. Some allelochemicals can inhibit the growth of competing plants by suppressing their seed germination, growth rate, or nutrient absorption (Wang et al. 2022). Sorghum, black walnut, eucalyptus, and sunflowers are examples of plants that exhibit allelopathic effects. According to Lekberg et al. (2018), allelopathy reduced plant performance by 45%.

The displacement of native plant species, which can contribute to a reduction in biodiversity, is one of the most significant effects. This is due to the fact that opportunistic plants are typically fast-growing and extremely competitive, allowing them to outcompete other plants for resources such as light, water, and nutrients (Oduor 2022). Consequently, native species with weak competitive ability may not be able to endure or reproduce in the presence of these plants, causing their decline or extinction in the ecosystem (Kandori et al. 2009).

Allelopathy can also have a negative impact on the allelopathic plant itself. Some allelochemicals can accumulate in the soil over time, leading to reduced growth and survival of the allelopathic plant due to autotoxicity (Ghimire et al. 2019; Xu et al. 2023). In summary, interaction between opportunistic plants and agricultural crops is a complex phenomenon that can be positive or negative impact. Their potential adverse effects on the environment should be thoughtfully considered and controlled in order to reduce the extent of the damage they cause. For the purpose of sustainable agriculture and ecosystem management, gaining an understanding of its mechanisms and effect is essential.

2.6 Opportunistic Nematodes and Agricultural Crops Interactions

Nematodes are a diverse collection of organisms that are classified in the phylum Nematoda. They are also referred to as roundworms (Kiontke and Fitch 2013). As a result of their ability to thrive in a wide variety of ecosystems, they now boast the greatest diversity of any phylum in the animal world. Free-living and parasitic nematodes can also be found in a wide variety of plant and mammal species. Nematode bodies are typically long, slender, and threadlike, but are not segmented like earthworm bodies. The intestine and gonad are encompassed by the body wall, which consists of dorsal and ventral longitudinal musculature, epidermis, and cuticle (Basyoni and Rizk 2016). There is a pressurized, fluid-filled cavity between the inner and outer cylinders that functions as a hydrostatic skeleton. This structure allows nematodes to move gracefully in sinusoidal waves while lying on their sides.

The stylet is a structure that is shared by all phytoparasitic nematodes. It is found in the stoma, also known as the mouth, and it includes a channel that allows food and secretions to pass through. This structure enables phytoparasitic nematodes to pierce the cell wall of their host and inject enzymes that begin the process of partially decomposing the contents of the cell before sucking those contents into their digestive system (Iqbal and Jones 2017).

Interaction between agricultural products and nematodes is detrimental to the plant. Plant-parasitic nematodes are a major biotic stress that cause significant yield losses to numerous agriculture production (Kumar et al. 2020; Phani et al. 2021; Kantor et al. 2022). To date, nearly 4100 species of plant parasitic nematodes (PPNs) have been identified and are regarded as a significant threat to global food security (Nicol et al. 2011). Their attacks directly reduce the quality and quantity of cultivated crops, thereby diminishing their productivity. Globally, PPNs are considered significant parasites and are responsible for significant crop losses, estimated at USD 173 billion annually (Gamalero and Glick 2020). Plant-parasitic nematodes infect plant organs using a variety of strategies including leaves, roots, stems, seeds, and flowers. They have remarkable capacity to affect the growth, physiology, and defence mechanisms of plants (Eves-van den Akker 2021). The three categories of phytoparasitic nematodes are endoparasites, ectoparasites, and semi-endoparasites, and each group is subdivided into its own subcategory based on the life cycle and feeding pattern of its members (Mesa-Valle et al. 2020).

Endoparasites are the most prevalent nematodes that enter the plant and cause significant growth damage. The parasites infect plant roots and induce the formation of gigantic feeding cells, which reduces plant nutrition and water absorption (El-Sappah et al. 2019). Endoparasitic nematodes are classified into two types: migratory endoparasites and sedentary endoparasites. Endoparasites that migrate, such as *P. penetrans* and *R. similis*, feed on the cytoplasm of cells from soil to plant tissues, whereas sedentary endoparasites, such as root-knot nematodes, *Meloidogyne* spp., establish feeding sites near the vascular region (Sijmons et al. 1994). Root-knot and cyst nematodes are considered as most damaging pests in agriculture (Kiriga et al. 2018; Mukhtar and Hussain 2019; Verdejo-Lucas and Talavera 2019). In Pakistan, the root-knot nematode (*M. javanica*) is posing a significant threat to okra production based on Mukhtar and Hussain (2019) research. Root-knot nematodes have been found to predominate in 85% of okra fields with an average incidence of 39% and of the four most prevalent root-knot species. This parasite caused the greatest reductions in cultivar shoot weight and root length. Recent analyses using various statistical approaches, according to Bandara et al. (2020), estimate yield losses caused by soybean cyst nematode (SCN) in the United States accumulated nearly $32 billion from 1996 to 2016, or more than $1.5 billion annually (Arjoune et al. 2022).

Ectoparasite nematodes, such as *Belonolaimus longicaudatus* Rau and *Tricodorus* spp., perforate and feed on the outer surface of the plant cell wall without entering the host plant root. This disrupts the plant's ability to absorb water and nutrients, which in turn prevents the plant's growth and causes significant damage (Fosu-Nyarko and Jones 2016). The findings of Willett et al. (2020) reveal that ecto-parasites nematode (*B. longicaudatus* Rau) modify global metabolome on amino acid levels in Bermudagrass lines. It reduced nutrient uptake from a compromised root system of the plant. Meanwhile, research conducted by Abu Habib et al. (2020) proved that ectoparasites nematodes *Tricodorus* spp. reduce the productivity of many citrus orchards in Egypt. This parasite was detected in the collected citrus soil samples showed that this nematode is recognized as serious constraints for global food security.

Third categories of phytoparasitic nematodes are semi-endoparasites. Semi-endoparasites generally have a part of their body inside the host while the other part is outside the host. They may enter the host's body to feed, lay eggs, or reproduce, and then come out of the host's body to complete their life cycle (Mesa-Valle et al. 2020). Example of semi-endoparasites are *R. reniformis, Sphaeronema* spp., *Hoplolaimus* spp., and *Helycotylenchus* spp. *R. reniformis*, is a nematode that capable of infecting >300 plant species, including a large number of crops such as cotton, soybean, and pineapple (Wubben et al. 2010). Dyer et al. (2020) reported that *R. reniformis* reduced seed cotton yields by an average of 42% between fields without infestation and those with infestation. In Alabama, an estimated 3% yield loss or approximately 26,000 bales of cotton were lost (Lawrence and McLean 2001). In India, *H. multicinctus* reportedly causes detrimental effects on the banana root system and leads to impaired plant growth and production, resulting in substantial crop yield reductions (Tharani et al. 2021).

Plants have the ability to recognize and respond to microbe or pathogen. Numerous transcriptional changes, including the upregulation or downregulation of particular genes or pathways, are induced in plants in response to nematode invasion. Some alterations are primary host defence responses, which in resistant hosts may prevent further invasion, such as the induction of specific defence genes, whereas in susceptible hosts, alterations in gene expression can facilitate successful plant parasitism (Fosu-Nyarko and Jones 2016). During infection, plant-parasitic nematodes exude pheromones and unknown proteinaceous ligands during infection. Both substances are identified by the immune system of the host plant. Nematode pheromones are altered in the peroxisome to produce shorter chain molecules that are repulsive to nematodes. This process occurs regardless of whether the pheromones are recognized. Both the dorsal gland and the subventral gland of nematodes are responsible for the expression of effectors. These organs secrete effectors that are transported into and around the plant cell in order to perform a variety of manipulations on the plant's development, metabolism, physiology (Eves-van den Akker 2021).

As we all know nematodes are well known for having the ability for damaging crops, but few are still conscious about non-parasitic nematodes, which have many beneficial roles to plant growth. Nonparasitic worm families were classified into five groups. These vary from r-strategists with a short lifecycle, high reproduction rates, high colonization ability, and high stress tolerance (c-p 1) to k-strategists with a long lifecycle, few offspring, low colonization ability, and high stress sensitivity (c-p 2, c-p 2, c-p 5) (Bongers 1990). In addition to being an essential component of soil food webs, some nematodes play an important role in all soil processes and ecological operations. They contribute to the transformation of carbon (C) and the cycling of nutrients from active plant roots and decomposing plant remains. Nematodes are beneficial to plant development and primary productivity, the cycling of carbon and nutrients, and the control of pests and diseases (Campos-Herrera 2015). Successful biocontrol activity against numerous commercially significant insect pests has led to the widespread use of entomopathogenic nematodes for the biological control of insect pests (Wilschut and Geisen 2021). Indirectly, nematodes contribute to the process of decomposition of organic materials in the soil. Process of recycling nutrients by nematodes results in the release of pertinent nutrients for the growth of plants (Luo et al. 2021).

Numerous nematodes may not directly affect plants, but indirect interactions with other organisms in the rhizosphere can influence plant performance. Nematodes that feed on plant roots improve plant productivity. According to Gebremikael et al. (2016), root feeders can implicitly stimulate plant performance by inducing plant exudation, which stimulates microbial activity, resulting in increased N and P availability. Furthermore, when working with bacterivorous protists, these root feeding nematodes also increase root branching, which allows for greater N uptake and has positive impact on plant performance (Trap et al. 2016). From Jiang et al. (2023) case study, nematodes and their bacterial prey enhance phosphorus acquisition in wheat. These findings indicate that nematodes play an important role in P biogeochemical cycling and plant growth, as well as in ecosystem processes.

According to Gebremikael et al. (2016) at modest abundances, root feeders can indirectly stimulate plant performance by inducing increased plant exudation, which in turn stimulates microbial activity and increases N and P availability. In conjunction with bacterivorous protists, these root-feeding nematodes also increase root branching, which facilitates greater N uptake and has positive effects on plant performance (Trap et al. 2016). However, Jiang et al. (2023) reported that the nematodes and their bacterial prey enhance wheat's phosphorus uptake. These findings demonstrate that nematodes contribute to P biogeochemical cycling, plant growth, and ecosystem function.

2.7 Solutions to Minimize the Effect of Opportunistic Organisms on Cultivated Crops

Opportunistic organisms such as fungi, nematodes, and plant can cause substantial damage to cultivated crops, which can lead to lower yields, lower quality, and economic losses. These organisms can rapidly spread and proliferate, resulting in outbreaks that can be devastating to crops. As a result, it is critical to adopt strategies that can help reduce the impact of opportunistic organisms and ensure crop production that is sustainable. One of the most effective strategies for minimizing the impact of opportunistic organisms is crop rotation. Crop rotation is a farming technique that involves growing different crops in a specific sequence over a period of years in the same field. It has been widely used to mitigate weed, insect, and helps to break the life cycle of pests and diseases (Zhao et al. 2020; Neupane et al. 2021).

Pathogens such as bacteria, fungi, and viruses can accumulate in the soil over time and infect plants, causing crop diseases. By rotating crops, it is possible to interrupt the life cycles of these pathogens and reduce their population levels in the soil. According to Leandro et al. (2018) long-term crop rotations on maize and soybean help to supress soybean sudden death syndrome (SDS) caused by *F. virguliforme*. SDS is a fungus that infects the roots and lower stem of soybean plants, producing toxins. This toxin causes severe injury to the plant, resulting in a substantial loss of yield (Bradley et al. 2021). Additionally, crop rotations with more than two crops produced in succession result in increased yield, a greater variety of microorganisms, as well as altered microbial activity and community structure. Venter et al. (2016) reported an average 3% increase in microbial diversity in soils across 17 studies benefit from crop rotation. The use of resistant varieties is another method for reducing the effect of opportunistic organisms. Through the plant breeding programme, plant breeders have created crop varieties that are resistant to various pests and diseases. These varieties can help reduce the impact of disease and pests on crops and may necessitate fewer pesticide applications. Rice blast disease is one of the most notorious phytopathogenic fungi cause by *Magnaporthe oryzae* and responsible for severe economic loss up to 30% of the rice crop annually (Sakulkoo et al. 2018). At any stage of maturation, rice blast can infect the aboveground tissues of rice plants, resulting in crop failure. The pathogen causes lesions on leaves, leaf collars, culm nodes, panicle neck nodes, and panicles. To combat this disease, tolerant or resistant plant varieties have been developed. Pongsuseribu 2 is a variety that is resistant to rice blast (Hasan et al. 2018; Agbowuro et al. 2020). This variety is resistant to *M. oryzae* due to the presence of at least two dominant alleles, ‘Pi-b’ and ‘Pi-kh’ (Tanweer et al. 2015). Rutgers tomato (*L. lycopersicum*) is another resistant variety that can be used to reduce the impact of opportunistic organisms, particularly fungi. This variety is resistant to a number of prevalent tomato diseases, including TMV (Li et al. 2021).

The impact of opportunistic organisms on cultivated crops can also be reduced through the use of biological control. Biological control is the use of natural predators, parasites, or pathogens to control the population of pests and diseases by attacking them directly or through their indirect effects on the plant. BCAs are an attractive alternative to chemical pesticides because they are environmentally benign, sustainable, and economical. The most common methods for controlling plant pathogens involve the use of microorganisms, such as bacteria, fungi, viruses, and other microorganisms. *Trichoderma*, for instance, is used to control soil-borne plant pathogens such as *Fusarium* and *Pythium* (Degani and Dor 2021). Microbial agents function by producing toxic metabolites, competing with pathogens for nutrients, or inducing plant defence responses that increase the plant’s resistance to the pathogen (Köhl et al. 2019). Biological control can be an effective and sustainable alternative to chemical pesticides.

Good agricultural practices, such as proper irrigation, nutrient management, and pruning, can also help minimize the impact of opportunistic organisms on cultivated crops. Nutrient management is the most important component of effective agricultural practice for reducing the occurrence of plant diseases. Proper nutrient management ensures that crops receive the necessary nutrients for growth and development, which can improve plant health, increase resistance to disease, and reduce the likelihood of nutrient deficiencies that can make plants more susceptible to disease (Liu et al. 2021). By promoting healthy plants, the risk of pests and diseases becoming established and spreading throughout the field can be reduced.

Finally, chemical pesticides can be used to control pests and diseases as a last resort. Chemical pesticides are substances designed to eliminate or manage crop-damaging insects, diseases, and vegetation. These pesticides play an essential role in agricultural development given that they can reduce agricultural product losses and improve yield and quality at a reasonable price (Strassemeyer et al. 2017). Crops can be treated with chemical pesticides in a variety of ways, including spraying, dusting, and soil application. The choice of pesticide and method of application will depend on the specific pest, disease, or weed issue to be addressed. However, they should be used sparingly and only when other methods have failed, as excessive use can result in resistance and environmental damage. In a nutshell, the impact of opportunistic organisms on cultivated crops can be minimized by implementing a combination of strategies. Crop rotation, biological control, genetic resistance, cultural practices, and chemical control are all effective methods to reduce the impact of opportunistic organisms on crops. By adopting these strategies, farmers can reduce crop losses, increase yields, and promote sustainable agricultural practices.

2.8 Conclusions and Future Prospects

Interactions among opportunistic fungi, nematode, and plant could be harmful or beneficial to agricultural crops. The beneficial impact includes availability of nutrient leading to growth increase and conferring resistance against plant pathogenic, while the harmful effect on the plant could be as worse as leading to various plant infections. The role of host vulnerability cannot be overemphasized because rarely do optimally growing plants get overtaken by opportunistic infections. To prevent negative impact of opportunistic organism interactions proper measures should be in place to ensure the health of agricultural plant is at its peak. Environmentally friendly methods like crop rotation, biological control, genetic resistance, hygienic cultural practices, eradication of pest and alternative hosts are encouraged in prevention of negative effect of opportunistic organisms.

References

Abu Habib A, Younes H, Ibrahim I, Khalil A (2020) Plant parasitic nematodes associated with citrus trees and reaction of two citrus cultivars to *Tylenchulus semipenetrans* in northern Egypt. J Adv Agric Res 25:166–175

Agbowuro GO, Afolabi MS, Olamiriki EF, Awoyemi SO (2020) Rice blast disease (*Magnaporthe oryzae*): a menace to rice production and humanity. Int J Pathog Res 4:32–39

Akhtar M, Sarwar N, Ashraf A, Ejaz A, Ali S, Rizwan M (2020) Beneficial role of *Azolla* sp. in paddy soils and their use as bioremediators in polluted aqueous environments: implications and future perspectives. Arch Agron Soil Sci 67:1242–1255

Al-Kharousi ZS, Guizani N, Al-Sadi AM, Al-Bulushi IM, Shaharoona B (2016) Hiding in fresh fruits and vegetables: opportunistic pathogens may cross geographical barriers. Int J Microbiol 2016:4292417

Arjoune Y, Sugunaraj N, Peri S, Nair SV, Skurdal A, Ranganathan P, Johnson B (2022) Soybean cyst nematode detection and management: a review. Plant Methods 18:110

Ashraf MS, Khan TA (2005) Effect of opportunistic fungi on the life cycle of the root-knot nematode (*Meloidogyne javanica*) on brinjal. Arch Phytopathol Plant Protect 38:227–233

Babikova Z, Johnson D, Bruce T, Pickett J, Gilbert L (2014) Undergroundallies: how and why do mycelial networks help plants defend them-selves? What are thefitness, regulatory, and practical implicationsof defence-related signalling between plants via common mycelialnetworks. BioEssays 36:21–26

Bacon CW, Hill NS (1996) Symptomless grass endophytes: products of coevolutionary symbioses and their role in the ecological adaptations of grasses. In: Redlin SC, Carris LM (eds) Entophytic fungi in grasses and Woody plants: systematics, ecology, and evolution. The American Phytopathological Society, St. Paul, MN, pp 155–178

Bailey BA, Lumsden RD (1998) Direct effect of *Trichoderma* and *Gliocladium* on plant growth and resistance to pathogen. In: Harman GE, Kubicek CP (eds) *Trichoderma* and *Gliocladium*, vol 2. Taylor and Francis, London, pp 185–204

Ball KR, Baldock JA, Penfold C, Power SA, Woodin SJ, Smith P, Pendall E (2020) Soil organic carbon and nitrogen pools are increased by mixed grass and legume cover crops in vineyard agroecosystems: detecting short-term management effects using infrared spectroscopy. Geoderma 379:114619

Bandara AY, Weerasooriya DK, Bradley CA, Allen TW, Esker PD (2020) Dissecting the economic impact of soybean diseases in the United States over two decades. PLoS One 15:1–28

Barron GL (2003) Predatory fungi, wood decay, and the carbon cycle. Biodiversity 4:3–9

Bartelt RJ, Wicklow DT (1999) Volatiles from *Fusarium verticillioides* (Sacc.) Nirenb. and their attractiveness to Nitidulid beetles. J Agric Food Chem 47:2447–2454

Basyoni MMA, Rizk EMA (2016) Nematodes ultrastructure: complex systems and processes. J Parasit Dis 40:1130–1140

Blanco-Canqui H, Jasa PJ (2019) Do grass and legume cover crops improve soil properties in the long term. Soil Sci Soc Am J 83:1181–1187

Bongers T (1990) The maturity index: an ecological measure of environmental disturbance based on nematode species composition. Oecologia 83:14–19

Bordallo JJ, Lopez-Llorca LV, Jansson HB, Salinas J, Persmark L, Asensio L (2002) Colonization of plant roots by egg parasitic and nematode trapping fungi. New Phytol 154:491–499

Bosque-Pérez NA, Mareck JH (1991) Effect of the stemborer *Eldana saccharina* (Lepidoptera: Pyralidae) on the yield of maize. Bull Entomol Res 81:243–247

Bradley CA, Allen TW, Sisson AJ, Bergstrom GC, Bissonnette KM, Bond J, Byamukama E, Chilvers MI, Collins AA, Damicone JP, Dorrance AE, Dufault NS, Esker PD, Faske TR, Fiorellino NM, Giesler LJ, Hartman GL, Hollier CA, Isakeit T, Jackson-Ziems TA, Jardine DJ, Kelly HM, Kemerait RC, Kleczewski NM, Koehler AM, Kratochvil RJ, Kurle JE, Malvick DK, Markell SG, Mathew FM, Mehl HL, Mehl KM, Mueller DS, Mueller JD, Nelson BD, Overstreet C, Padgett GB, Price PP, Sikora EJ, Small I, Smith DL, Spurlock TN, Tande CA, Telenko DEP, Tenuta AU, Thiessen LD, Warner F, Wiebold WJ, Wise KA (2021) Soybean yield loss estimates due to diseases in the United States and Ontario, Canada, from 2015 to 2019. Plant Health Prog 22:483–495

Brooker RW (2006) Plant–plant interactions and environmental change. New Phytol 171:271–284

Broz AK, Broeckling CD, De-la-Peña C, Lewis MR, Greene E, Callaway RM, Sumner LW, Vivanco JM (2010) Plant neighbour identity influences plant biochemistry and physiology related to defence. BMC Plant Biol 10:115–119

Burdon J, Lallu N (2011) Kiwifruit (*Actinidia* spp.). In: Yahia EM (ed) Postharvest biology and technology of tropical and subtropical fruits. Cocona to Mango, Woodhead Publishing Ltd, Cambridge, London, pp 326–362

Campos-Herrera R (2015) Nematodes pathogenesis of insect and other Pest: ecology and applied Technologies for Sustainable Plant and Crop Protection. Springer, Cham, pp 285–508

Cardwell KF, Schulthess F, Ndemah R, Ngoko Z (1997) A systems approach to assess crop health and maize yield losses due to pests and diseases in Cameroon. Agric Ecosyst Environ 65:33–47

Cruickshank MG, Morrison DJ, Punja ZK (1997) Incidence of *Armillaria* species in precommercial thinning stumps and spread of *Armillaria ostoyae* to adjacent Douglas fir trees. Can J For Res 27:481–490

Dabney SM, Delgado JA, Reeves DW (2001) Using winter cover crops to improve soil and water quality. Commun Soil Sci Plant Anal 32:1221–1250

de Jonge R, Bolton MD, Thomma B (2011) How filamentous pathogens co-opt plants: the ins and outs of fungal effectors. Curr Opin Plant Biol 14:400–406

Degani O, Dor S (2021) *Trichoderma* biological control to protect sensitive maize hybrids against late wilt disease in the field. J Fungi 7:315

Delavaux CS, Weigelt P, Magnoli SM, Kreft H, Crowther TW, Bever JD (2022) Nitrogen-fixing symbiotic bacteria act as a global filter for plant establishment on islands. Commun Biol 5:1209

Dickman MB, Fluhr R (2013) Centrality of host cell death in plant-microbe interactions. Annu Rev Phytopathol 5:543–570

Dodd JL (1980) The role of plant stresses in development of corn stalk rots. Plant Dis 64:533–537

Dolezal AL, Obrian GR, Nielsen DM, Woloshuk CP, Boston RS, Payne GA (2013) Localization, morphology and transcriptional profile of *Aspergillus flavus* during seed colonization. Mol Plant Pathol 14:1–12

Dolezal AL, Shu X, GR OB, Nielsen DM, Woloshuk CP, Boston RS, Payne GA (2014) *Aspergillus flavus* infection induces transcriptional and physical changes in developing maize kernels. Front Microbiol 5:384

Dorner JW, Cole RJ, Sanders TH, Blankenship PD (1989) Interrelationship of kernel water activity, soil temperature, maturity, and phytoalexin production in preharvest aflatoxin contamination of drought stressed peanuts. Mycopathologia 105:117–128

Dyer DR, Groover W, Lawrence KS (2020) Yield loss of cotton cultivars due to *Rotylenchulus reniformis* and the added benefit of a nematicide. Plant Health Prog 21:113–118

El-Sappah AH, Islam MM, El-Awady HH, Yan S, Qi S, Liu J, Cheng GT, Liang Y (2019) Tomato natural resistance genes in controlling the root-knot nematode. Genes 10:925

Emmanuel OC, Babalola OO (2020) Productivity and quality of horticultural crops through co-inoculation of arbuscular mycorrhizal fungi and plant growth promoting bacteria. Microbiol Res 239:126569

Eves-van den Akker S (2021) Plant-nematode interactions. Curr Opin Plant Biol 62:102035

Farias CP, De Carvalho RC, Resende FML, Azevedo LCB (2018) Consortium of five fungal isolates conditioning root growth and arbuscular mycorrhiza in soybean, corn, and sugarcane. Ann Braz Acad Sci 90:3649–3660

Fosu-Nyarko J, Jones MGK (2016) Advances in understanding the molecular mechanisms of root lesion nematode host interactions. Annu Rev Phytopathol 54:253–278

Franco FP, Túler AC, Gallan DZ, Gonçalves FG, Favaris AP, Peñaflor MFGV, Leal WS, Moura DS, Bento JMS, Silva-Filho MC (2021) Fungal phytopathogen modulates plant and insect responses to promote its dissemination. ISME J 15:3522–3533

Gamalero E, Glick BR (2020) The use of plant growth-promoting bacteria to prevent nematode damage to plants. Biology 9:381

Gatch EW, Munkvold GP (2002) Fungal species composition in maize stalks in relation to European corn borer injury and transgenic insect protection. Plant Dis 86:1156–1162

Gebremikael MT, Steel H, Buchan D, Bert W, De Neve S (2016) Nematodes enhance plant growth and nutrient uptake under C and N-rich conditions. Sci Rep 6:32862

Ghimire BK, Ghimire B, Yu CY, Chung IM (2019) Allelopathic and autotoxic effects of *Medicago sativa*-derived allelochemicals. Plan Theory 8:233

Harman GE, Howell CR, Viterbo A, Chet I, Lorito M (2004) *Trichoderma* species opportunistic, avirulent plant symbionts. Nat Rev Microbiolol 2:43–56

Hasan MM, Rafii MY, Ismail MR, Mahmood M, Rahim HA, Latif MA, Amirul Alam M, Ahmad F, Malek MA (2018) Genetic analysis of the resistance to rice blast in the BC2F1 population derived from MR263 × Pongsu seribu 1. Biotech Biotechnol Equip 32:1134–1140

INRAE (2021) Fungi reported on eggplant. http://ephytia.inra.fr/en/C/26948/Eggplant-Penicillium-spp. Accessed 28 Oct 2023

Iqbal S, Jones MGK (2017) Nematodes. In: Thomas B, Murray BG, Murphy DJ (eds) Encyclopedia of applied plant sciences, vol 3. Academic Press, Waltham, MA, pp 113–119

Jiang Y, Wang Z, Liu Y, Han Y, Wang Y, Wang Q, Liu T (2023) Nematodes and their bacterial prey improve phosphorus acquisition by wheat. New Phytol 237:974–986

Kale SD, Tyler BM (2011) Entry of oomycete and fungal effectors into plant and animal host cells. Cell Microbiol 13:1839–1848

Kandori I, Hirao T, Matsunaga S, Kurosaki T (2009) An invasive dandelion unilaterally reduces the reproduction of a native congener through competition for pollination. Oecologia 159:559–569

Kantor M, Handoo Z, Kantor C, Carta L (2022) Top ten most important US regulated and emerging plant-parasitic nematodes. Horticulture 8:208

Kassam R, Yadav J, Chawla G, Kundu A, Hada A, Jaiswal N, Bollinedi H, Kamil D, Devi P, Rao U (2021) Identification, characterization, and evaluation of nematophagous fungal species of *Arthrobotrys* and *Tolypocladium* for the management of *Meloidogyne incognita*. Front Microbiol 12:790223

Kelley RY, Williams WP, Mylroie JE, Boykin DL, Hawkins LK, Windham GL, Wilkinson JR (2009) Genomic profile of maize response to *Aspergillus flavus* infection. Toxin Rev 28:129–141

Kiontke K, Fitch DHA (2013) Nematodes. Curr Biol 23:862–864

Kiriga AW, Haukeland S, Kariuki GM, Coyne DL, Beek NV (2018) Effect of *Trichoderma* spp. and *Purpureocillium lilacinum* on *Meloidogyne javanica* in commercial pineapple production in Kenya. Biol Control 119:27–32

Klich MA (2007) *Aspergillus flavus*: the major producer of aflatoxin. Mol Plant Pathol 8:713–722

Köhl J, Kolnaar R, Ravensberg WJ (2019) Mode of action of microbial biological control agents against plant diseases: relevance beyond efficacy. Front Plant Sci 10:845

Kuebbing SE, Nuñez MA (2015) Negative, neutral, and positive inter-actions among nonnative plants: patterns, processes, and manage-ment implications. Glob Change Biol 21:926–934

Kumar V, Khan MR, Walia RK (2020) Crop loss estimations due to plant-parasitic nematodes in major crops in India. Natl Acad Sci Lett 43:409–412

Lamb C, Dixon R (1997) The oxidative burst in plant disease resistance. Annu Rev Plant Physiol Plant Mol Biol 48:251–275

Larriba E, Jaime MD, Nislow C, Martın-Nieto J, Lopez-Llorca LV (2015) Endophytic colonization of barley (*Hordeum vulgare*) roots by the nematophagous fungus *Pochonia chlamydosporia* reveals plant growth promotion and a general defense and stress transcriptomic response. J Plant Res 128:665–678

Lawrence GW, McLean KS (2001) Reniform nematodes. In: Kirkpatrick TL, Rothrock C.S (eds), Compendium of cotton diseases. APS Press, St. Paul, MN, pp. 42–43

Leandro LFS, Eggenberger S, Chen C, Williams J, Beattie GA, Liebman M (2018) Cropping system diversification reduces severity and incidence of soybean sudden death syndrome caused by *Fusarium virguliforme*. Plant Dis 102:1748–1758

Lekberg Y, Bever JD, Bunn RA, Callaway RM, Hart MM, Kivlin SN, Klironomos J, Larkin BG, Maron JL, Reinhart KO, Remke M, van der Putten WH (2018) Relative importance of competition and plant-soil feedback, their synergy, context dependency and implications for coexistence. Ecol Lett 21:1268–1281

Li S, Zhang Z, Zhou C, Li S (2021) RNA-dependent RNA polymerase 1 delays the accumulation of viroids in infected plants. Mol Plant Pathol 22:1195–1208

Liu G, Simonne EH, Morgan KT, Hochmuth GJ, Agehara S, Mylavarapu R, Williams P (2021) Fertilizer management for vegetable production in Florida. In: Dittmar P, Agehara S, Dufault

NS (eds) Vegetable production handbook. UF-IFAS Extension, University of Florida, Gainesville, pp 1–12
Luo J, Zhang X, Kou X, Xie H, Bao X, Mahamood M, Liang W (2021) Effects of residue mulching amounts on metabolic footprints based on production and respiration of soil nematodes in a long-term no-tillage system. Land Degrad Devel 32:2383–2392
Madriz-Ordeñana K, Jørgensen HJL, Balan A, Sørensen DM, Nielsen KL, Thordal-Christensen H (2019) Prevalence of soil-borne diseases in *Kalanchoe blossfeldiana* reveals a complex of pathogenic and opportunistic fungi. Plant Dis 103:2634–2644
Marçais B, Bréda N (2006) Role of an opportunistic pathogen in the decline of stressed oak trees. J Ecol 94:1214–1223
McFarlane SA, Govender P, Rutherford RS (2009) Interactions between *Fusarium* species from sugarcane and thestalk borer, *Eldana saccharina* (Lepidoptera: Pyralidae). Ann Appl Biol 155:349–359
McLean M, Watt MP, Berjak P, Dutton MF (1994a) Effects of aflatoxin B1 on *in vitro* cultures of *Nicotiana tabacum* vat. Samsun 1: callus growth and differentiation. Mycopathologia 125:93–105
McLean M, Watt MP, Berjak P, Dutton MF, Snyman C (1994b) Effects of aflatoxin B1 on *in vitro* cultures of *Nicotiana tabacum* var. Samsun. II. Root and shoot development in tobacco plantlets. Mycopathologia 125:107–117
Mellon JE, Cotty PJ, Dowd MK (2000) Influence of lipids with and without other cotton seed reserve materials on aflatoxin B1 production by *Aspergillus flavus*. J Agric Food Chem 48:3611–3615
Mellon JE, Dowd MK, Cotty PJ (2005) Substrate utilization by *Aspergillus flavus* in inoculated whole corn kernels and isolated tissues. J Agric Food Chem 53:2351–2357
Mesa-Valle CM, Garrido-Cardenas JA, Cebrian-Carmona J, Talavera M, Manzano-Agugliaro F (2020) Global research on plant nematodes. Agronomy 10:1148
Moreno-Gavíra A, Diánez F, Sánchez-Montesinos B, Santos M (2021) Biocontrol effects of *Paecilomyces variotii* against fungal plant diseases. J Fungi 7:415
Morrison DJ, Pellow KW, Nemec AFL, Norris DJ (2001) Effects of selective cutting on the epidemiology of *Armillaria* root disease in the southern interior of British Columbia. Can J For Res 31:59–70
Mukhtar T, Hussain MA (2019) Pathogenic potential of javanese root-knot nematode on susceptible and resistant okra cultivars. Pak J Zool 51:1891–1897
Munkvold GP, White DG (2016) Compendium of corn diseases, 4th edn. APS Press, St. Paul, MN, p 165
Neupane A, Bulbul I, Wang Z, Lehman RM, Nafziger E, Marzano SYL (2021) Long term crop rotation effect on subsequent soybean yield explained by soil and root-associated microbiomes and soil health indicators. Sci Rep 11:9200
Nicol JM, Turner SJ, Coyne DL, Den NL, Hockland S, Mafi ZT (2011) Current nematode threats to world agriculture. In: Jones J, Gheysen G, Fenoll C (eds) Genomics and molecular genetics of plant-nematode interactions. Springer, Dordretch, pp 21–43
Niu XM, Zhang KQ (2011) *Arthrobotrys oligospora*: a model organism for understanding the interaction between fungi and nematodes. Mycology 2(2):59–78
Oduor AMO (2022) Native plant species show evolutionary responses to invasion by *Parthenium hysterophorus* in an African savanna. New Phytol 233:983–994
Payne GA (1998) Process of contamination by aflatoxin-producing fungi and their impact on crops. In: Bhatnagar D, Sinha KK (eds) Mycotoxins in agriculture and food safety. Marcel Dekker, New York, pp 279–306
Pfordt A, Schiwek S, Karlovsky P, von Tiedemann A (2020) *Trichoderma afroharzianum* ear rot-a new disease on maize in Europe. Front Agron 2:547758
Phani V, Khan MR, Dutta TK (2021) Plant-parasitic nematodes as a potential threat to protected agriculture: current status and management options. Crop Protect 144:105573
Ragozzino A, D'Errico G (2011) Interactions between nematodes and fungi: a concise review. J Zool Red 94:123–125

Raja W, Rathaur P, John SA, Ramteke PW (2012) Azolla: an aquatic pteridophyte with great potential. Int J Res Biol Sci 2(2):68–72

Roy DC, Pakhira MC, Bera S (2016) A review on biology, cultivation and utilization of *Azolla*. Adv Life Sci 5:11–15

Sagar A, Rai S, Ilyas N, Sayyed RZ, Al-Turki AI, El Enshasy HA, Simarmata T (2022) Halotolerant rhizobacteria for salinity-stress mitigation: diversity, mechanisms and molecular approaches. Sustain For 14:490

Sakulkoo W, Osés-Ruiz M, Garcia EO, Soanes DM, Littlejohn GR, Hacker C, Correia A, Valent B, Talbot NJ (2018) A single fungal MAP kinase controls plant cell-to-cell invasion by the rice blast fungus. Science 359:1399–1403

Sánchez-Montesinos B, Santos M, Moreno-Gavíra A, Marín-Rodulfo T, Gea FJ, Diánez F (2021) Biological control of fungal diseases by *Trichoderma aggressivum* f. *europaeum* and its compatibility with fungicides. J Fungi 7:598

Santhiya B, Jeeva S (2022) *Azolla* as a source of biofertilizer for sustainable crop production-a review. J Xi'an Shiy Univer 18:598–605

Schuerger AC, Amaradasa BS, Dufault NS, Hummerick ME, Richards JT, Khodadad CL, Smith TM, Massa GD (2021) *Fusarium oxysporum* as an opportunistic fungal pathogen on *Zinnia hybrida* plants grown on board the international space station. Astrobiology 21:1029–1048

Schultess F, Cardwell KF, Gounou S (2002) The effect of endophytic *Fusarium verticillioides* on infestation of two maize varieties by lepidopterous stemborers and coleopteran grain feeders. Phytopathology 92:120–128

Senthilkumar M, Anandham R, Krishnamoorthy R (2020) Paecilomyces. In: Amaresan N, Senthil Kumar M, Annapurna K, Kumar K, Sankaranarayanan A (eds) Beneficial microbes in agro-ecology. Academic Press, Boca Raton, pp 793–808

Sharangi AB (2011) In search of allelopathy from common Alliaceae crops in managing weeds in coriander: an overview. Int J Agric Res 6:209–217

Sharma MK, Kumawat DM (2011) Co-inoculation study of *Bradyrhizobium japonicum* and *Aspergillus niger* in soybean for nitrogen fixation. J Microbiol Biotechnol Food Sci 1:383–394

Sijmons PC, Atkinson HJ, Wyss U (1994) Parasitic strategies of root nematodes and associated host cell responses. Annu Rev Phytopathol 32:235–259

Srinivasan R, Prabhu G, Prasad M, Mishra M, Chaudhary M, Srivastava R (2020) Penicillium. In: Amaresan N, Senthil Kumar M, Annapurna K, Kumar K, Sankaranarayanan A (eds) Beneficial microbes in agro-ecology. Academic, Boca Raton, pp 651–667

St. Leger RJ, Screen SE, Shams-Pirzadeh B (2000) Lack of host specialization in *Aspergillus flavus*. Appl Environ Microbiol 66:320–324

Strakowska J, Błaszczyk L, Chełkowski J (2014) The significance of cellulolytic enzymes produced by *Trichoderma* in opportunistic lifestyle of this fungus. J Basic Microbiol 54:2–13

Strassemeyer J, Daehmlow D, Dominic AR, Lorenz S, Golla B (2017) SYNOPS-WEB, an online tool for environmental risk assessment to evaluate pesticide strategies on field level. Crop Protect 97:28–44

Subrahmaniam HJ, Libourel C, Journet EP, Morel JB, Munos S, Niebel A, Raffaele S, Roux F (2018) The genetics underlying natural variation of plant-plant interactions, a beloved but forgotten member of the family of bioticinteractions. The Plant J 93:747–770

Taheri P, Kakooee T (2017) Reactive oxygen species accumulation and homeostasis are involved in plant immunity to an opportunistic fungal pathogen. J Plant Physiol 216:152–163

Tanweer FA, Rafii MY, Sijam K, Rahim HA, Ahmed F, Latif MA, Ashkani S (2015) Cloning and characterization of two major blast resistance genes pi-b and pi-kh from Malaysian rice variety Pongsu Seribu 2. Plant Omics 8:257–263

Thapa P, Poudel K (2021) *Azolla*: potential biofertilizer for increasing rice productivity, and government policy for implementation. J Wastes Biomass Manage 3:62–68

Tharani G, Alagesan A, Jawahar S, Saranya S, Balakrishnan A, Padmanaban B, Manivannan S (2021) Distribution and molecular characterization of *Helicotylenchus multicinctus* (Tylenchida: Hoplolaimidae) from banana roots in Namakkal region, Tamil Nadu. Int J Bot Stud 6:965–971

Trap J, Bonkowski M, Plassard C, Villenave C, Blanchart E (2016) Ecological importance of soil bacterivores for ecosystem functions. Plant Soil 398:1–24

Velmourougane K, Saxena G, Prasanna R (2017) Plant-microbe interactions in the rhizosphere: mechanisms and their ecological benefits. In: Singh D, Singh H, Prabha R (eds) Plant-microbe interactions in agro-ecological perspectives. Springer, Singapore, pp 193–219

Venter ZS, Jacobs K, Hawkins HJ (2016) The impact of crop rotation on soil microbial diversity: a meta-analysis. Pedobiologia 59:215–223

Verdejo-Lucas S, Talavera M (2019) Root-knot nematodes on zucchini (*Cucurbita pepo* sub sp. pepo): pathogenicity and management. Crop Protect 126:10494

Wang C, Liu Z, Wang Z, Pang W, Zhang L, Wen Z, Zhao Y, Sun J, Wang ZY, Yang C (2022) Effects of autotoxicity and allelopathy on seed germination and seedling growth in *Medicago truncatula*. Front Plant Sci 13:908426

Wargo PM, Harrington TC (1991) Host stress and susceptibility. In: Shaw CG, Kile GA (eds) *Armillaria root disease*, agriculture handbook no. 691. USDA Forest Service, Washington, DC, pp 88–101

Willett DS, Filgueiras CC, Benda ND, Zhang J, Kenworthy KE (2020) Sting nematodes modify metabolomic profiles of host plants. Sci Rep 10:2212

Wilschut RA, Geisen S (2021) Nematodes as drivers of plant performance in natural systems. Trends Plant Sci 26:237–247

Wubben MJ, Callahan FE, Scheffler BS (2010) Transcript analysis of parasitic females of the sedentary semi-endoparasitic nematode *Rotylenchulus reniformis*. Mol Biochem Parasitol 172:31–40

Xu Y, Chen X, Ding L, Kong C-H (2023) Allelopathy and allelochemicals in grasslands and forests. Forests 14:562

Yao Y, Zhang M, Tian Y, Zhao M, Zeng K, Zhang B, Yin B (2018) Azolla biofertilizer for improving low nitrogen use efficiency in an intensive rice cropping system. Field Crop Res 216:158–164

Zhang Y, Li S, Li H, Wang R, Zhang K-Q, Xu J (2020) Fungi–nematode interactions: diversity, ecology, and biocontrol prospects in agriculture. J Fungi 6:206

Zhang Z, Liu Y, Yuan L, Weber E, van Kleunen M (2021) Effect of allelopathy on plant performance: a meta-analysis. Ecol Lett 24:348–362

Zhao J, Yang Y, Zhang K, Jeong J, Zeng Z, Zang H (2020) Does crop rotation yield more in China? A meta-analysis. Field Crops Res 245:107659

Opportunistic Fungi, Plant, and Nematode Interactions in Vegetable Crops

3

Deeksha Kashyap, Aiman Parveen, Masudulla Khan, and Lukman Ahamad

3.1 Introduction

Vegetables are a vital source of vitamins, minerals, and antioxidants that are beneficial to human health (Asaduzzaman and Asao 2018). In India, vegetable production has increased several times with gradual increase in the productivity and area of cultivation over the years as well as maintains its second position after China in terms of vegetable production (Sahni and Kumari 2017). In addition, the vegetable crops attacked by various plant pathogens and limit their productivity throughout the world. Among them, nematodes parasitic on plants provide the biggest hazard to agriculture. PPNs, or plant-parasitic nematodes, are the unseen adversaries and cause major constraint on crop productivity around the world (Stirling 2014). Between $US80 billion and $US157 billion in yield losses annually occur due to plant infections caused by about 4100 different species of nematodes (Khan and Khan 2021). Various nematodes associated with vegetables crops are cyst nematodes (*Heterodera* spp.), root-knot nematodes (*Meloidogyne* spp.), reniform nematodes (*Rotylenchulus* sp.), stem and bulb nematode (*Ditylenchus* spp.), lesion nematodes (*Pratylenchus* sp.), lance nematodes (*Hoplolaimus* spp.), etc. However, root-knot nematodes and *Meloidogyne* spp. are important pests of vegetables, including cruciferous (cabbage, cauliflower, broccoli, sprouts, and brussels), solanaceous (tomato, brinjal and chilli), leguminous (pea, cowpea, bean), cucurbitaceous (bottle gourd, cucumber, biter ground, pumpkin), okra, and many other root

D. Kashyap · A. Parveen · L. Ahamad (✉)
Section of Plant Pathology and Nematology, Department of Botany, Aligarh Muslim University, Aligarh, Uttar Pradesh, India

M. Khan
Section of Plant Pathology and Nematology, Department of Botany, Aligarh Muslim University, Aligarh, Uttar Pradesh, India

Botany Section, Women's College, Aligarh Muslim University, Aligarh, India

M. S. Akhtar (ed.), *Opportunistic Fungi, Nematode and Plant Interactions*,
https://doi.org/10.1007/978-981-97-2045-3_3

and bulb crops (lettuce, carrot, onion, celery, garlic and radish). Onions, garlic, potatoes, peas, carrots, and other vegetables are frequently attacked by stem and bulb nematodes (Ditylenchus spp.) (Bahadur 2021). One of the most prevalent and significant nematode species for root knots is *Meloidogyne incognita,* which can seriously harm a variety of vegetable crops, including cucumber, tomato, eggplant, okra, carrot, pea, beetroot (Hussey and Barker 1973; Agaba et al. 2015). After the nematode infects the host plant's roots, it forms distinctive galls or knots that obstruct the passage of nutrients and water, resulting in stunted growth, wilting, and a lower yield.

Through parasitism, predation, or antagonism, opportunistic fungi can inhibit the growth of nematodes (Siddiqui and Mahmood 1996). Opportunistic fungi have the ability to negatively impact nematode reproductive structures by colonizing them. Heteroderid group nematodes that are in the sedentary stages of their life cycle are susceptible to fungal attacks from within the host roots, on the root surface, or in the surrounding soil. It is well known that these fungi grow more quickly in the rhizosphere. Compared to the two earlier categories, these fungi often play a more significant role in the biological control of nematodes (Park et al. 2012). Numerous interactions exist between fungi and plants, and each connection results in unique changes in both partners. While fungal diseases negatively impact plant physiology, mutualistic fungi enhance host defences against pathogens and/or enhance plant uptake of nutrients (Zeilinger et al. 2016). However, some fungi are also detrimental to nematodes which belong to different groups such as endoparasitic, opportunistic, predacious, mycorrhizal, and plant pathogenic fungi. They may impede the growth of nematodes by parasitism, predation, or antagonism (Siddiqui and Mahmood 1996). Furthermore, due to their ongoing association with nematodes in the rhizosphere, these fungi consistently kill nematodes in nearly all soil types. Thus, the aim of this chapter is to focus on the function of fungi in both beneficial and harmful interactions with nematodes on various vegetable crops.

3.2 Interactions Between Nematodes and Opportunistic Fungi

Plant-parasitic nematodes are endoparasitic, semi-endoparasitic, and ectoparasitic in nature that a feed on various plant parts (roots, seeds, stems, leaves, and flowers). Root-knot and cyst nematodes are the most destructive parasites throughout the world (Jones et al. 2013; Smant et al. 2018). However, root-knot nematode *Meloidogyne* species have broad host range, which can infect almost all crops (Khan and Khan 2021). The cyst nematodes usually have narrow host range in infecting crop plants. Among cyst nematodes, *Globodera rostochiensis* and *G. pallida* reproduce primarily on potato and some other solanaceous crop species, while *H. glycines* causes damage to the soybeans and remains mostly limited to this crop. The potato cyst nematodes cause an estimated yield loss of 9% in the crops worldwide (Khan and Khan 2021; Zheng et al. 2022). However, juveniles (J2s) of *Meloidogyne* species are infective in nature which cause root perforation followed by the initiation of giant cells formation. In addition, juveniles of cyst nematodes also have ability to puncture the plant roots leading to development of feeding sites known as syncytia (Phillips et al. 2017; Khan and Khan 2021). In many cases, root-knot nematodes (*Meloidogyne* spp.), cyst nematodes (*Globodera* spp. and *Heterodera* spp.), reniform nematodes (*Rotylenchulus* spp.), root-lesion nematodes (*Pratylenchus* spp.), and other nematodes (*Xiphinema* spp. and *Longidorus* spp.) have involved in the formation of disease complexes with fungal pathogens. The associations of nematodes and fungi in formation of disease complexes have been reviewed previously by several authors (Back et al. 2002; Zhang et al. 2020).

The multi-pathogenic concept of plant diseases came when Melendez and Powell (1967) observed that plant-parasitic nematodes onset the disease in association with some soil fungi which are not normally pathogenic to cause disease. Though during the last few decades research studies have been done on multi-pathogenic diseases, yet there is a great scope of conducting research on the complex interactions between plant-parasitic nematodes, fungi, and host roots. The plants grown in such soils are continuously influenced by the physical, chemical structure of the soil and also the flora and fauna present therein. The roots of plants are continuously exposed to the soil environment, developing an association between plant and microorganisms such as fungi and nematodes under field conditions. These microorganisms interact among themselves having an impact on plant growth and development.

3.2.1 Synergistic Interaction of Nematodes and Rhizoctonia on Vegetable Crops

The genus *Rhizoctonia* is a group of fungi that comes under Basidiomycota, which includes more than 100 species. It is a soil-borne fungus that can cause many diseases in vegetable crops. Firstly, the fungus *R. solani* Kühn was observed on potato tubers in the year 1885 (Wilson et al. 2008; Maymon et al. 2015). This fungus causes two major diseases in potato crops such as black scurf and stem canker, which in turn lower the quality and number of tubers (Brewer and Larkin 2005). It also has the ability to cause damping-off, crown and root-root disease in sugar beet (Khan and Liu 2016). In addition, *R. solani* (AG 2–2 IIIB) can cause damping-off, root rot in broccoli and cauliflower (Türkkan et al. 2020). However, *R. solani* affects the root, and crown of pumpkin, cucumber, watermelon, and melon plants. The fungus can spread through contaminated soil, water, and infected plant material (Anjum et al. 2011). The disease severity varied according to their isolates from 8.3 to 91.6% (Huang et al. 2020; Aiello et al. 2012; Mirmajlessi et al. 2012; Zhang et al. 2016). The symptoms caused by *Rhizoctonia* species can be severe, leading to stunted growth, reduced yield, and even crop failure on different vegetable crops.

Nematodes can affect plants by feeding on their roots, leading to a reduction in plant growth and productivity. This can make the plant more susceptible to fungal infection, including infection caused by *Rhizoctonia* fungi that can exacerbate the root rot disease severity in vegetable crops (Abuzar 2013; Al-Hazmi and Al-Nadary 2015). The interaction between *Meloidogyne* nematodes and *Rhizoctonia* fungi can vary depending on the specific nematode and fungus species and the vegetable crop involved. However, *Meloidogyne incognita* and *R. solani* are important pathogens of tomato crop that produce symptoms such as root-knot and damping-off, root rot, stem canker, respectively (Kumar and Haseeb 2009). This study shows that *R. solani* can increase the infection rate and development in *M. incognita* infected tomato plants. Björsell et al. (2017) reported that *Pratylenchus* spp., stubby-root nematodes and *Globodera* spp. with *R. solani* interaction shown that the degree of stem canker was higher on tomato plants. There was no regional relationship between *R. solani* and potato cyst nematodes or stubby-root nematodes, in contrast to root-lesion nematodes. In addition, there is evidence that plant growth is impaired when co-infected with *R. solani* and *Globodera pallida* in potato plants (Back et al. 2006). Other researchers also hypothesized that plant roots that have already been exposed to fungal infections are more susceptible to nematode penetration. In addition, sedentary endoparasitic nematodes have great potential in favouring infection for plant pathogenic fungi in various vegetable crops (Back et al. 2002). However, the relationship between nematodes and *Rhizoctonia* fungi can therefore be considered to be opportunistic, in that each organism can take advantage of the conditions created by the other to enhance its own survival and spread. The presence of nematodes can create wounds or entry points on the roots of vegetable crops that can make them more susceptible to infection by *Rhizoctonia* fungi. At the same time, the effects of *Rhizoctonia* fungi on plant roots can create conditions that favour nematode infection and reproduction. The disease was more severe when both pathogens were present (Safiuddin and Shahab 2012). Additionally, secondary infections can exacerbate the damage caused by root-lesion nematodes, further reducing yield and quality of crop (Edin et al. 2019). Moreover, the damage produced by *Rhizoctonia* fungi can increase the production of exudates from the roots of the plant, which in turn recruit the nematodes (Huang et al. 2014; Table 3.1).

3.2.2 Synergistic Interaction of Nematodes and Alternaria on Vegetable Crops

Alternaria is one of the most widespread genera in the world. It causes pre- and post-harvest damage to grains, fruits, and vegetables. In vegetables, the fungus *Alternaria* had ability to cause a variety of diseases and reduce the yield by making it a significant threat to food security and economic stability (Neergaard 1945; Garg and Singh 2016; Meena et al. 2016). There are some species of *Alternaria* which includes *A. solani* a pathogen of early blight in tomato and potato plants. The early blight disease typically starts from the lower leaves of the plant and moves upward as plant reaches to maturity. The symptoms caused *A. solani* are brown, irregularly shaped lesions on the leaves, stems, and fruit of the plant. As the disease progresses, the leaves may turn yellow and die, and the fruit may rot. The fungus overwinters in plant debris and can survive in the soil for up to 2 years (Locke 1949; Mirkarimi et al. 2013). However, *A. alternata* is a common pathogen that affects a wide range of vegetable crops such as cucumber, pepper, and leafy greens, etc. The symptom

Table 3.1 Synergistic interactions of nematodes and fungi on different vegetable crops

Host Plant	Nematode	Fungus	Effect on plant	References
Tomato	*M. incognita*	*A. solani*	The interaction between the nematode and the fungus appeared to be the cause of the yield drop in plants infected with *M. incognita* over controls	Barker (1972)
Potato	*P. penetrans* *P. crenatus*	*V. dahliae*	Roots were colonized by *V. dahliae* to a significantly greater extent when grown in soil infested with *V. dahliae* and *P. penetrans* than in soil infested with *V. dahliae* alone	Bowers et al. (1996)
Potato	*P. neglectus*	*V. dahliae*	More destructive for potato when both pathogens present together	Hafez et al. (1999)
Potato	*P. penetrans*	*V. dahliae*	Roots by *P. penetrans* systemically affect disease physiology associated with stem vascular infection by *V. dahliae*	Rotenberg et al. (2004)
Potato	*Globodera*	*R. solani*	Six weeks after planting, a correlation between the density of *G. rostochiensis* juveniles in potato roots and the frequency of *R. solani*-infected stolon's was discovered	Back et al. (2006)
Potato	*Pratylenchus* spp.	*V. dahliae*	Causing significant yield losses in potato fields when infected with both pathogens	Harding and Wicks (2007)

(continued)

Table 3.1 (continued)

Host Plant	Nematode	Fungus	Effect on plant	References
Potato	*G. pallida G. rostochiensis*	*R. solani*	*R. solani* disease in potatoes is more common when nematodes are present than when simply *R. solani* is present	Bhattarai et al. (2009)
Tomato	*Meloidogyne incognita*	*R. solani*	When *M. incognita* was present, tomato growth and fungal root colonization were dramatically reduced compared to when the fungus was present alone	Kumar and Haseeb (2009)
Sugar beet	*H. schachtii*	*R. solani*	Plant damage was most severe when both species were treated together	Hillnhütter et al. (2011)
Okra	*Meloidogyne incognita*	*R. solani*	When both species were treated at the same time, plant damage was at its worst	Safiuddin and Shahab (2012)
Chilli	*Meloidogyne incognita*	*R. solani* and *F. oxysporum*	The presence of nematodes increased the severity of root rot and the root system compared to fungus alone	Abuzar (2013)
Potato	*G. rostochiensis*	*R. solani*	*R. solani* inoculation caused serious harm, especially when it was combined with large densities of PCN	Kiani et al. (2013)

Host Plant	Nematode	Fungus	Effect on plant	References
Potato	*Pratylenchus* spp. *Stubby-root nematodes and Globodera* spp.	*R. solani*	On plants evaluated in the middle of the patches as opposed to those rated on the boundaries, the degree of stem canker was greater. In contrast to root-lesion nematodes, there was no regional association between *R. solani* and stubby-root nematodes or potato cyst nematodes	Björsell et al. (2017)
Cabbage	*M. incognita*	*Sclerotinia sclerotiorum*	Sequential inoculation of nematodes followed by fungus increased the white rot disease severity	Kalaivanan et al. (2017)
Eggplant	*Meloidogyne incognita*	*Phomopsis vexans*	The fungus adversely affected nematode multiplication. More reduction in plant growth was observed in 20 days prior inoculation of nematode to fungus as compared to prior inoculation of fungus to nematode	Khan and Siddiqui (2017)
Eggplant	*M. incognita*	*Phomopsis vexans*	Simultaneous inoculation of *M. incognita* and *P. vexans* increased the severity of leaf blight in eggplant	Khan and Siddiqui (2018)

(continued)

Table 3.1 (continued)

Host Plant	Nematode	Fungus	Effect on plant	References
Cucumber	*M. incognita*	*F. oxysporum* f. sp. *cucumerinum*	Nematode inoculation caused significantly more reduction in plant growth parameter of cucumber than fungus alone	Patil et al. (2018)
Tomato	*Meloidogyne* spp.	*F. oxysporum* f. sp. *lycopersici*	80-100% yield loss was reported due to Fusarium wilt-root-knot nematode complex	Wanjohi et al. (2018)
Carrot	*M. incognita*	*Alternaria dauci* and *R. solani*	When nematodes were introduced into the fungus, the disease's severity was quite high in plants	Ahmad et al. (2019)
Potato	*Nematodes* sp.	*R. solani*	Inoculation with *R. solani* proved extremely harmful, especially when paired with high nematode densities.	Edin et al. (2019)
Tomato	*M. incognita*	*Fusarium oxysporum* f. sp. *lycopersici*	Simultaneous infestation of both increased crop losses than singly	Mwangi et al. (2019)
Beetroot	*Meloidogyne incognita*	*Rhizoctonia solani*	Sequential and simultaneous inoculation of *M. incognita* and *R. solani* increased the severity of root-rot in plants	Khan and Siddiqui (2020)

caused by *A. alternata* includes small or circular lesions on the leaves that may be surrounded by a yellow halo. The lesions may also have a concentric ring pattern on leaves. In severe cases, the leaves may turn yellow and die (Vaughn and Eller 1996). In addition, *A. brassicicola* is a pathogen that affects cruciferous vegetables, including cabbage, broccoli, and cauliflower. The symptom produced by *A. brassicicola* infection includes small, dark lesions on the leaves that may enlarge over time. The lesions may also have a concentric ring pattern on leaves. In severe cases, the leaves may become yellow and die, and the plant may produce fewer and smaller heads (Michereff et al. 2012). *A. cucumerina* is a pathogen that affects cucurbits, including cucumber, squash, and melon (Gannibal 2011). Recent study shows that *A. dauci* and *A. porri* fungal pathogens caused infections on carrot and onion respectively (Ahmad and Siddiqui 2019; Islam et al. 2020).

The interaction between nematodes and *Alternaria* fungi as a disease complex on vegetable crops can have significant impacts on crop yield and quality. Several studies have investigated that interaction between these two pathogens can increase the severity of *Alternaria* infection in the presence of nematodes or vice versa (Table 3.1). One study conducted by Ahmad et al. (2019) found that the infection of carrot plants with root-knot nematodes *M. incognita* and *A. solani* significantly reduced plant growth parameters. The authors reported that the nematode-infected plants showed higher disease severity with fungus than the non-infected plants. Parasitic nematodes exacerbated the plant disease by injuring the root system and allowing fungi to colonize the host. Fungus establishment on the other hand had a suppressive impact on the nematode population and was unable to worsen the damage. In addition, the interaction between root-knot nematode, *M. incognita* and *A. solani* was observed on tomato plants (Barker 1972). The interaction between nematodes and fungi had negative impact on vegetable crops. The nematodes can create wounds in the root system that provide entry points for the fungi leading to an increase in disease severity in plants. In addition, competitive interactions were noticed between nematodes, *Aphelenchus avenae, Aphelenchoides saprophilus,* and *A. besseyi* on potato dextrose agar in the presence of fungus *A. tenuis* at different temperatures. The study shows that number of nematodes was decreased when they grow in mixed cultures (Migunova and Shesteperov 2011).

3.2.3 Synergistic Interaction of Nematodes and Fusarium on Vegetable Crops

The genus *Fusarium* is a very frequently distributed soil-borne fungus which lives as saprophytes on decaying organic matter while few are important plant pathogens. These fungi are necrotrophs and produce toxins that kill plant cells. *Fusarium* pathogens on vegetables produce characteristic spores such as macroconidia, microconidia, and small resting spores (chlamydospores and sclerotia). However, soil inhabiting fungus *F. oxysporum* can cause wilt infection in tomatoes and sweet potatoes (Table 3.1). In sweet potatoes, wilt disease caused by the *F. oxysporum* f. sp. *batatas* induces plant wilting and yellowing of the leaves (Muimba-Kankolongo

2018). However, other fungal pathogen *F. oxysporum* f. sp. *lycopersici* causes the vascular wilt disease in tomato plants. The initial symptoms caused by the fungus appear on lower leaves followed by wilting in plants (Srinivas et al. 2019).

The first report published by Atkinson (1892) which showed that interaction of root-knot nematode *Meloidogyne* spp. and fungal pathogen *F. oxysporum* involved in serious yield losses of cotton plants as compared to wilt disease caused by fungal pathogen alone. The disease complex caused by the *Fusarium* fungus and *Meloidogyne* species were distributed widely and greatly damaging to the vegetables grown in India and other countries (Khan and Sharma 2020). In addition, various other studies also showed the formation of disease complex between root-knot nematodes and *F. oxysporum* f. sp. *lycopersici* in tomato and many other crops (Mai and Abawi 1987; Back et al. 2002). Similarly, simultaneous inoculation of *M. incognita* and *F. oxysporum* in tomato plants decreased the growth followed by increase in disease intensity over inoculation of single pathogen (Yitayih et al. 2020). However, reproduction in nematodes was reduced in the presence of *F. oxysporum* f. sp. *lycopersici* and damage caused by both pathogens was greater in tomato plants (Regmi et al. 2022).

3.2.4 Synergistic Interactions of Nematodes and Other Fungi on Vegetable Crops

Other fungal pathogens such as *Verticillium* spp., *Phomopsis vexans,* and *Sclerotinia sclerotiorum* caused significant damage in vegetable crops. The *Verticillium* wilt pathogens particularly *V. albo-atrum* or *V. dahliae* caused characteristic symptoms appear first on oldest leaves and spread fast to the entire younger leaves. The infected plant leaves have become yellowish brown, wilt, and die (Muimba-Kankolongo 2018). However, Phomopsis blight caused by *P. vexans* is also a devastating disease and caused significant yield and quality loss in eggplant (Bhanushree et al. 2022). Diseases caused by *Sclerotinia* are the primary contributors in losses of horticultural crops including leafy and root vegetables as well as vegetable brassicas (Saharan and Mehta 2008).

Interaction between nematodes and other fungi were also noticed in various vegetable crops (Table 3.1). In this regard, root-knot pathogen *M. javanica* with wilt fungus *V. dahliae* were involved in the decrease of fungal colonization, number of galls, and egg masses in olive plants (Saeedizadeh et al. 2009). However, synergistic interaction of *V. dahliae* and *P. penetrans* were also involved in development of early dying disease in potato (Rowe et al. 1985). Moreover, root-lesion nematodes *P. penetrans* and *P. scribneri* and wilt fungus *V. dahliae* were studied and found that nematode reproduction is negatively affected by the fungus *V. dahliae* in combined application of these organisms in potato early dying complex (Wheeler and Riedel 1994). Other studies also showed that disease severity was high in eggplants when fungus *Phomopsis vexans* was inoculated with *M. incognita*. These pathogens may have direct or indirect effect on disease severity in plants. The direct effect means physical interactions in the rhizosphere by pathogens and their occupancy in the same infection site inside root. In addition, sequential inoculation of both pathogens caused a significant reduction in growth parameters of eggplants (Khan and Siddiqui 2018). Similarly, sequential and concomitant inoculation of *M. incognita* with *S. sclerotiorum* increases white rot disease severity in cabbage plants. So, it is established that nematode predisposes the cabbage plants for fungal pathogen (Kalaivanan et al. (2017).

3.3 Antagonistic Interactions of Other Fungi with Nematodes on Vegetable Crops

Biological control of nematodes is described as regulation of nematode population and mitigation of nematode damage through naturally occurring antagonistic organisms or through environment manipulation or introduction of antagonists. The antagonistic organisms can interact directly with pathogenic nematodes through various mechanisms such as antibiosis, competition for nutrients and space while indirect interaction with nematodes involves the induction of systemic resistance (SAR, ISR) in plants (Xiang et al. 2018; Stirling 2018). A large number of microorganisms including bacteria, fungi, and others have ability to parasitize and prey on phyto-parasitic nematodes (Stirling 1991). Among them, fungal organisms hold an important position but some of the fungi have shown greater efficacy as biocontrol agents (Stirling 1991; Jatala 1986). In addition, antagonistic fungi have been interacted with plant-parasitic nematodes through different ways in crops. The antagonistic fungi have ability to suppress the population of nematodes by using trapping structures or by releasing enzymes and metabolic compounds. They also regulate the behaviour of nematodes by interfering in plant nematode recognition and improve plant growth (Devi 2018). The biocontrol fungal organisms have capability to supress the phyto-parasitic nematodes either through antagonism (parasitize, kill, and prey on nematodes) or through plant growth promotion and induction of plant defence towards nematodes. In addition, opportunistic fungi also have to potential to attack on non-motile stages of life cycle including cysts, eggs, and female of nematodes (Liu et al. 2009).

Various other fungal species have potential to reduce the nematodes population through different mechanisms such as predation or antagonism and parasitism in vegetable crops (Table 3.2). The genus *Paecilomyces* is primarily a saprophyte and infects the *M. incognita* eggs and destroys the embryos in 5 days (Jatala 1986). In addition, other nematophagous fungi such as *Arthrobotrys oligospora* also have ability to prey on nematodes and parasitize their eggs. The eggs of *Meloidogyne* species have been penetrated by the *A. oligospora* far more quickly than *Globodera* and *Nacobbus*. In addition, biocontrol fungus *Purpureocillium lilacinum* (formally known as *Paecilomyces lilacinus*) also have potential to parasitize the nematode eggs and reduce the egg hatching with mortality in the juveniles of *M. incognita*. The fungus *P. lilacinum* also found effective in field conditions to control root-knot nematode in tomato plants (Singh et al. 2013). However, strains of *P. chlamydosporia* caused significant reduction in egg masses of nematodes and root galling in

Table 3.2 Antagonistic interactions between different fungi and nematodes on different vegetable crops

Antagonistic Fungi	Nematodes	Host Plant	Effect on plant	References
P. lilacinus	*M. incognita*	Chilli	*P. lilacinus* with organic additives increased plant growth and reduced population of nematodes and root galling in plants	Ahmad and Khan (2004)
P. lilacinus *M. lysipagum*	*M. javanica*	Tomato	Combined application of *P. lilacinus* and *M. lysipagum* reduced the galls and nematode juveniles on plants	Khan et al. (2006)
Beauveria bassiana	*M. hapla*	Tomato	Culture filtrates of *B. bassiana* caused % mortality and inhibition of hatching of nematode and Soil drenching significantly reduced nematode population densities in soil and in the roots and subsequent gall formation and egg-mass production	Liu et al. (2008)
P. penetrans *P. chlamydosporia* *P. lilacinus* *T. harzianum*	*M. incognita*	Okra	Application of antagonists suppressed galling and reproduction of *M. incognita* resulted in increased of plant growth	Mukhtar et al. (2013)
P. lilacinus	*M. incognita* race I	Tomato	Application of *P. lilacinus* suppressed the severity of root galling and egg mass production in tomato	Udo et al. (2014)
P. lilacinus	*M. incognita*	Eggplant	*P. lilacinus* possessed larvicidal and ovipocidal properties against *M. incognita*. However, *P. lilacinus* reduced populations of *M. incognita* and improved plant growth parameters	Abbas et al. (2016)
P. lilacinus *A. niger*	*M. javanica*	Carrot	Both bio-agents together resulted in a greater increase in plant dry weight and caused reduction in galling and nematode multiplication	Nesha and Siddiqui (2017)

(continued)

Table 3.2 (continued)

Antagonistic Fungi	Nematodes	Host Plant	Effect on plant	References
P. lilacinum PLSAU-1	*M. incognita*	Eggplant	Application of *P. lilacinum* PLSAU-11 reduced gall index and egg masses when the crop was challenged by 800 and 400 eggs/100 cm^3 soil	Sarven et al. (2019)
Trichoderma citrinoviride Snef 1910	*M. incognita*	Tomato	Significantly inhibited egg hatching and Metabolites of *T. citrinoviride* significantly decreased the number of root galls, J2s, and nematode egg masses and J2s population density in soil and significantly promoted the growth of plants	Fan et al. (2020)
Rhizophagus intraradices B1, *Rhizophagus intraradices* A2 and *Funneliformis mosseae*	*Nacobbus aberrans*	*Capsicum annuum*	Use of AMF in pepper production to reduce the population of *N. aberrans* and *R. intraradices* B1 as the most promising candidate in suppressing nematode	Bernardo et al. (2021)
Trichoderma harzianum Pochonia chlamydosporium	*Meloidogyne incognita*	Eggplant	*T. harzianum* significantly caused egg inhibition, followed by P. chlamydosporium after 72 h. soil drench treatment of fungal bio-control agents significantly suppressed *M. incognita* parameters and upraised the plant growth	Uddin et al. (2023)
Chaetomium globosum KPC3	*Globodera rostochiensis*	Potato	Fungus mycelium fully parasitized the cyst after 72 h and also capable of parasitizing the eggs inside the cysts. The culture filtrate of fungus caused mortality in J_2s of *G. rostochiensis* after 72 h combined application of fungus as a tuber treatment and a soil application of farmyard manure resulted in significantly lesser reproduction of *G. rostochiensis*	Bairwa et al. (2023)

tomato (Yang et al. 2012). Some other fungi such as *Trichoderma*, endophytic and mycorrhizal fungi have also been used as biocontrol agents towards phyto-parasitic nematodes by inducing resistance in host plants. They also have ability to decrease the damage in plants posed by nematodes through various antagonistic mechanisms (Poveda et al. 2020).

3.4 Mechanisms of Interactions of Opportunistic Fungi with Nematodes

All the interactions between phyto-parasitic nematodes and fungi involve three components, namely, nematode, host, and other pathogen. The nematodes have ability to interact with other organisms in a mixed community. The interaction mechanism mainly have role of the phyto-parasitic nematodes and other fungal pathogens in plant disease development and their relationship with host plant (Khan 1993). The phyto-parasitic nematodes can facilitate the entry of fungal pathogens through physical injuries or wounding in host plants. In addition, the nematodes can also predispose the hosts or modify the expression of their resistance towards pathogens by changing the physiology of host plant (Riedel 1988; Back et al. 2002).

3.5 Conclusions and Future Prospects

Nematodes and fungi are the most dominant organisms of soil ecosystem. They interact each other through direct or indirect ways to the plants. However, plant-parasitic nematodes and fungi interacted to each other through synergistic or antagonistic mechanisms. Various fungi have potential to decrease the nematodes density in soil but few of them are nematode parasites. However, nematodes act as predisposing factor for other pathogens in forming disease complex in vegetable crops. So, the disease severity in these plants increased in combined application of one or more pathogens and impacted plant growth and development. The application of fungal biocontrol agents are also promising strategy towards nematode diseases in agricultural crops. Therefore, potential applications of synergistic and antagonistic fungi were needed to control phyto-parasitic nematodes in agriculture.

References

Abbas H, Javed N, Khan SA, Kamran M, Atiq M (2016) Exploitation of nematicidal potential of *Paecilomyces lilacinus* against root-knot nematode on eggplant. Int J Veg Sci 22:85–90

Abuzar S (2013) Antagonistic effects of some fluorescent *pseudomonas* strains against root rot fungi (*Rhizoctonia solani* and *Fusarium oxysporum*) and root-knot nematodes (*Meloidogyne incognita*) on chilli (*Capsicum annum*). World Appl Sci J 27:1455–1460

Agaba TA, Fawole B, Claudius-cole B (2015) Pathogenicity of *Meloidogyne incognita* (Chitwood) on some pepper (*capsicum* spp.) cultivars. J Biol Agric Health 5:91–96

Ahmad SF, Khan TA (2004) Management of Root-Knot Nematode, *Meloidogyne incognita,* by integration of *Paecilomyces lilacinus* with organic materials in Chilli. Arch Phytopathol Plant Protect 37:35–40

Ahmad L, Siddiqui ZA (2019) Effect of different inoculum levels of *Meloidogyne incognita, Alternaria dauci* and *Rhizoctonia solani* on the growth, chlorophyll and carotenoid content and disease progression of carrot (*Daucus carota* L.). Acta Phytopathol Entomol Hung 54:211–220

Ahmad L, Siddiqui ZA, Abd Allah EF (2019) Effects of interaction of *Meloidogyne incognita, Alternaria dauci* and *Rhizoctonia solani* on the growth, chlorophyll, carotenoid and proline contents of carrot in three types of soil. Acta Agric Scand B-Soil Plant Sci 69:324–331

Aiello D, Vitale A, Hyakumachi M, Polizzi G (2012) Molecular characterization and pathogenicity of binucleate *Rhizoctonia* AG-F associated to the watermelon vine decline in Italy. Eur J Plant Pathol 134:161–165

Al-Hazmi AS, Al-Nadary SN (2015) Interaction between *Meloidogyne incognita* and *Rhizoctonia solani* on green beans. Saudi J Biol Sci 22:570–574

Anjum T, Chilvers MI, Grau CR (2011) *Rhizoctonia solani* AG-2-2 IIIB as the primary causal agent of soybean seedling blight in Michigan. Plant Dis 95:714–720

Asaduzzaman M, Asao T (2018) Quality vegetable production and human health benefits. In: Asaduzzaman M, Asao T (eds) Vegetables: importance of quality vegetables to human health. InTech, London, p 79430

Atkinson GF (1892) Some diseases of cotton. Ala Agric Exp Stn Auburn Univ Bull 41:5–61

Back MA, Haydock PP, Jenkinson P (2002) Disease complexes involving plant parasitic nematodes and soilborne pathogens. Plant Pathol 51:683–697

Back MA, Haydock PPJ, Jenkinson P (2006) Interactions between the potato cyst nematode *Globodera rostochiensis* and diseases caused by *Rhizoctonia solani* AG3 in potatoes under field conditions. Eur J Plant Pathol 114:215–223

Bahadur A (2021) Nematodes diseases of fruits and vegetables crops in India. IntechOpen, London, pp 1–18

Bairwa A, Dipta B, Mhatre PH, Venkatasalam EP, Sharma S, Tiwari R, Singh B, Thakur D, Naga KC, Maharana C, Sharma AK (2023) *Chaetomium globosum* KPC3: an antagonistic fungus against the potato cyst nematode, *Globodera rostochiensis*. Curr Microbiol 80:125

Barker KR (1972) Correlation of initial densities of *Meloidogyne incognita* on tomato with yield and incidence of early blight. Phytopathology 62:801

Bernardo VF, Garita SA, Arango MC, Ripodas JI, Saparrat MCN, Ruscitti MF (2021) Arbuscular mycorrhizal fungi against the false root-knot nematode activity in *Capsicum annuum*: physiological responses in plants. Biocontr Sci Technol 31:119–131

Bhanushree N, Saha P, Tomar BS, Munshi AD (2022) Phomopsis blight in eggplant and strategies to manage through resistance breeding. J Hortic Sci Biotechnol 97:34–45

Bhattarai S, Haydock PPJ, Back MA, Hare MC, Lankford WT (2009) Interactions between the potato cyst nematodes, *Globodera pallida*, *G. Rostochiensis*, and soil-borne fungus, *Rhizoctonia solani* (AG3), diseases of potatoes in the glasshouse and the field. Nematology 11:631–640

Björsell P, Edin E, Viketoft M (2017) Interactions between some plant-parasitic nematodes and *Rhizoctonia solani* in potato fields. Appl Soil Ecol 113:151–154

Bowers JH, Nameth ST, Riedel RM, Rowe RC (1996) Infection and colonization of potato roots by *Verticillium dahliae* as affected by *Pratylenchus penetrans* and *P. Crenatus*. Phytopathology 86:614–621

Brewer M, Larkin RP (2005) Efficacy of several potential biocontrol organisms against *Rhizoctonia solani* on potato. Crop Protect 24:939–950

Devi G (2018) Utilization of nematode destroying fungi for management of plant-parasitic nematodes-a review. Biosci Biotechnol Res Asia 15:377–396

Edin E, Gulsher M, Andersson Franko M, Englund JE, Flöhr A, Kardell J, Viketoft M (2019) Temporal interactions between root-lesion nematodes and the fungus *Rhizoctonia solani* lead to reduced potato yield. Agronomy 9:361

Fan H, Yao M, Wang H, Zhao D, Zhu X, Wang Y, Liu X, Duan Y, Chen L (2020) Isolation and effect of *Trichoderma citrinoviride* Snef1910 for the biological control of root-knot nematode, *Meloidogyne incognita*. BMC Microbiol 20:299

Gannibal PB (2011) *Alternaria cucumerina* causing leaf spot of pumpkin newly reported in North Caucasus. New Dis Rep 23:2044–0588

Garg A, Singh S (2016) *Alternaria* species in aerospora of vegetable and fruit market at Agra and their mycotoxigenic potential. Asian J Agric Life Sci 1:4–7

Hafez SL, Al-Rehiayani S, Thornton M, Sundararaj P (1999) Differentiation of two geographically isolated populations of *Pratylenchus neglectus* based on their parasitism of potato and interaction with *Verticillium dahliae*. Nematropica 29:25–36

Harding RB, Wicks TJ (2007) *Verticillium dahliae* and *Pratylenchus* spp: populations in potato soils and plants in Australia. Austral Plant Pathol 36:62–67

Hillnhütter C, Sikora RA, Oerke EC (2011) Influence of different levels of resistance or tolerance in sugar beet cultivars on complex interactions between *Heterodera schachtii* and *Rhizoctonia solani*. Nematology 13:319–332

Huang X, Lu X, Wang M (2014) Effects of soil temperature on the activity of root-knot nematodes in tomato. Nematology 16:153–160

Huang LQ, Niu YC, Su L, Deng H, Lyu H (2020) The potential of endophytic fungi isolated from cucurbit plants for biocontrol of soil-borne fungal diseases of cucumber. Microbiol Res 231:126369

Hussey RS, Barker KR (1973) A comparison of methods of collecting inocula of *Meloidogyne* spp., including a new technique. Plant Dis Rep 57:1025–1028

Islam MM, Begum F, Nahar N, Habiba UA, Fakruzzaman KM (2020) *In-vivo* management of purple blotch of onion caused by *Alternaria porri* (Ellis) Cif. Through fungicides. Am J Plant Sci 11:1847–1859

Jatala P (1986) Biological control of plant parasitic nematodes. Ann Rev Pathopathol 24:453–489

Jones JT, Haegeman A, Danchin EG, Gaur HS, Helder J, Jones MG, Kikuchi T, Manzanilla-López R, Palomares-Rius JE, Wesemael WM, Perry RN (2013) Top 10 plant-parasitic nematodes in molecular plant pathology. Mol Plant Pathol 14:946–961

Kalaivanan R, Dhivya M, Karthikeyan G, Devrajan K, Manonmani K (2017) Interaction of white rot fungus [*Sclerotinia sclerotiorum* (lib.) de Bary] and root-knot nematode [*Meloidogyne incognita* (Kofoid and white) chit.] cabbage (*Brassica oleracea* L.). Int J Curr Microbiol App Sci 6:641–647

Khan MW (1993) Mechanism of interactions between nematodes and other plant pathogens. In: Khan MW (ed) Nematode interactions. Chapman Hall, London, pp 55–78

Khan M, Khan AU (2021) Plant parasitic nematodes effectors and their crosstalk with defense response of host plants: a battle underground. Rhizosphere 17:100288

Khan MFR, Liu Y (2016) Penthiopyrad applied in close proximity to *Rhizoctonia solani* provided effective disease control in sugar beet. Crop Protect 85:33–37

Khan MR, Sharma RK (2020) Fusarium nematode wilt disease complexes, etiology and mechanism of development. Indian Phytopathol 73:615–628

Khan M, Siddiqui ZA (2017) Interactions of *Meloidogyne incognita, Ralstonia solanacearum* and *Phomopsis vexans* on eggplant in sand mix and fly ash mix soils. Sci Hortic 225:177–184

Khan M, Siddiqui ZA (2018) Zinc oxide nanoparticles for the management of *Ralstonia solanacearum, Phomopsis vexans* and *Meloidogyne incognita* incited disease complex of eggplant. Indian Phytopathol 71:355–364

Khan MR, Siddiqui ZA (2020) Effects of *Meloidogyne incognita*, *Pectobacterium betavasculorum* and *Rhizoctonia solani* interactions on growth, physiological and biochemical changes of beetroot. Israel J Plant Sci 67:194–211

Khan A, Williams KL, Nevalainen HKM (2006) Control of plant-parasitic nematodes by *Paecilomyces lilacinus* and *Monacrosporium lysipagum* in pot trials. Biocontr 51:643–658

Kiani Z, Zafari D, Rezaee S, Arjmandian A, Gitti M, Struik PC (2013) Co-limitation of potato growth by potato cyst nematode (*Globodera rostochiensis*) and *Rhizoctonia solani*. Arch Phytopathol Plant Protect 46:2401–2408

Kumar V, Haseeb A (2009) Interactive effect of *Meloidogyne incognita* and *Rhizoctonia solani* on the growth and yield of tomato. Indian J Nematol 39:178–181

Liu T, Wang L, Duan YX, Wang X (2008) Nematicidal activity of culture filtrate of *Beauveria bassiana* against *Meloidogyne hapla*. World J Microbiol Biotechnol 24:113–118

Liu X, Xiang M, Che Y (2009) The living strategy of nematophagous fungi. Mycoscience 50:20–25

Locke SB (1949) Resistance to early blight and septoria leaf spot in the genus *Lycopersicon*. Phytopathology 39:829–836

Mai WF, Abawi GS (1987) Interactions among root-knot nematodes and *Fusarium* wilt fungi on host plants. Annu Rev Phytopathol 25:317–338

Maymon M, Freeman S, Nitzani Y (2015) Control of *Rhizoctonia* crown and root rot of tomato by *Trichoderma asperellum* in soilless growing media. Biol Control 81:8–14

Meena M, Zehra A, Dubey MK, Aamir M, Gupta VK, Upadhyay RS (2016) Comparative evaluation of biochemical changes in tomato (*Lycopersicon esculentum* mill.) infected by *Alternaria alternata* and its toxic metabolites (TeA, AOH, and AME). Front Plant Sci 7:1408

Melendez PL, Powell NT (1967) Histological aspect of *Fusarium* with-root-knot complex in flue-cured tobacco. Phytopathology 57:286–292

Michereff SJ, Noronha MA, Xavier Filha MS, Câmara MP, Reis A (2012) Survey and prevalence of species causing *Alternaria* leaf spots on brassica species in Pernambuco. Hortic Bras 30:345–348

Migunova V, Shesteperov A (2011) Competitive interactions between the nematodes *Aphelenchus avenae, Aphelenchoides saprophilus* and *Aphelenchoides besseyi* by cultivation on the fungus *Alternaria tenuis* at different temperatures. IOBS-WPRS Bull 63:49–52

Mirkarimi HR, Abasi-Moghadam A, Mozafari J (2013) Assessment on early blight of potato in order to compare the two methods in vitro using pathogenic fungi *Alternaria solani*. Nat Sci 5:1189–1192

Mirmajlessi SM, Safaie N, Mostafavi HA, Mansouripour SM, Mahmoudy SB (2012) Genetic diversity among crown and root rot isolates of *Rhizoctonia solani* isolated from cucurbits using PCR based techniques. Afr J Agric Res 7:583–590

Muimba-Kankolongo A (2018) Root and tuber crops. In: Muimba-Kankolongo A (ed) Food crop production by smallholder farmers in southern Africa. Academic Press, Boca Raton, USA, pp 123–172

Mukhtar T, Hussain MA, Kayani MZ (2013) Biocontrol potential of *Pasteuria penetrans, Pochonia chlamydosporia, Paecilomyces lilacinus* and *Trichoderma harzianum* against *Meloidogyne incognita* in okra. Phytopathol Mediterr 52:66–76

Mwangi MW, Muiru WM, Narla RD, Kimenju JW, Kariuki GM (2019) Management of *Fusarium oxysporum* f. sp. *lycopersici* and root-knot nematode disease complex in tomato by use of antagonistic fungi, plant resistance and neem. Biocontr Sci Technol 29:229–238

Neergaard P (1945) Danish species of *Alternaria* and *Stemphylium*: taxonomy, parasitism, economic significance. Oxford University Press, London, pp 260–287

Nesha R, Siddiqui ZA (2017) Effects of *Paecilomyces lilacinus* and *aspergillus Niger* alone and in combination on the growth, chlorophyll contents and soft rot disease complex of carrot. Sci Hortic 218:258–264

Park JH, Kim SJ, Choi JH, Yoon MH, Chung DY, Kim HJ (2012) Biological control by nematophagous fungi for plant-parasitic nematodes in soils. Korean J Soil Sci Fertil 45:74–78

Patil J, Goel SR, Yadav S (2018) Effect of *Meloidogyne incognita* and *Fusarium oxysporum* f. sp. *cucumerinum* on cucumber grown under protected cultivation. J Entom Zoo Stud 6:1004–1110

Phillips WS, Howe DK, Brown AM, Eves-Van Den Akker S, Dettwyler L, Peetz AB, Zasada IA (2017) The draft genome of *Globodera ellingtonae*. J Nematol 49:127–128

Poveda J, Abril-Urias P, Escobar C (2020) Biological control of plant-parasitic nematodes by filamentous fungi inducers of resistance: *Trichoderma*, mycorrhizal and endophytic fungi. Front Microbiol 11:992

Regmi H, Vallad GE, Hutton SF, Desaeger J (2022) Interaction studies between *Meloidogyne javanica* and *Fusarium oxysporum* f. sp. *lycopersici* (Fol) race 3 on different isolines of tomato cv. Tasti lee. J Nematol 54:20220018

Riedel RM (1988) Interactions of plant-parasitic nematodes with soil-borne plant pathogens. Agric Ecosyst Environ 24:281–292

Rotenberg D, MacGuidwin AE, Saeed IAM, Rouse DI (2004) Interaction of spatially separated *Pratylenchus penetrans* and *Verticillium dahliae* on potato measured by impaired photosynthesis. Plant Pathol 53:294–302

Rowe RC, Riedel RM, Martin MJ (1985) Synergistic interactions between *Verticillium dahliae* and *Pratylenchus penetrans* in potato early dying disease. Phytopathology 75:412–418

Saeedizadeh A, Kheiri A, Zad J, Etebarian HR, Bandani AR, Nasiri MB (2009) A study of interaction between Verticillium wilt *Verticillium dahliae* and root-knot nematode *Meloidogyne javanica* in olive cultivars. Commun Agric Appl Biol Sci 74:567–572

Safiuddin, Shahab S (2012) Interactive effect of root-knot nematode, *Meloidogyne incognita* and root-rot fungus, *Rhizoctonia solani* on okra (*Abelmoschus esculentus* L.). Arch Phytopathol Plant Protect 45:660–666

Saharan GS, Mehta N (2008) Sclerotinia diseases of crop plants: biology, Ecology and disease management. Springer, Dordrecht, p 486

Sahni RK, Kumari S (2017) Current status of vegetables in India. https://biotecharticles.com/Agriculture-Article/Current-Status-of-Vegetables-in-India 3839.html

Sarven MS, Aminuzzaman FM, Huq ME (2019) Dose-response relations between *Purpureocillium lilacinum* PLSAU-1 and *Meloidogyne incognita* infecting brinjal plant on plant growth and nematode management: a greenhouse study. Egypt J Biol Pest Contr 29:26

Siddiqui ZA, Mahmood I (1996) Biological control of plant parasitic nematodes by fungi: a review. Bioresour Technol 58:229–239

Singh S, Pandey RK, Goswami BK (2013) Bio-control activity of *Purpureocillium lilacinum* strains in managing root-knot disease of tomato caused by *Meloidogyne incognita*. Biocontrol Sci Tech 23:1469–1489

Smant G, Helder J, Goverse A (2018) Parallel adaptations and common host cell responses enabling feeding of obligate and facultative plant parasitic nematodes. Plant J 93:686–702

Srinivas C, Devi DN, Murthy KN, Mohan CD, Lakshmeesha TR, Singh B, Kalagatur NK, Niranjana SR, Hashem A, Alqarawi AA, Tabassum B (2019) *Fusarium oxysporum* f. sp. *lycopersici* causal agent of vascular wilt disease of tomato: biology to diversity-a review. Saudi J Biol Sci 26:1315–1324

Stirling GR (1991) Biological control of plant parasitic nematodes. In: Progress, problems and prospects. CAB International, Wallingford

Stirling GR (2014) Biological control as a component of integrated nematode management: the way forward. In: Stirling GR (ed) Biological control of plant Parsitic nematodes: soil ecosystem Management in Sustainable Agriculture, 2nd edn. CABI, Boston, MA, pp 393–407

Stirling GR (2018) Biological control of plant-parasitic nematodes. In: Poinar GO, Jansson HB (eds) Diseases of nematodes. CRC Press, Boca Raton, FL, pp 103–150

Türkkan M, Kılıçoğlu MÇ, Erper I (2020) Characterization and pathogenicity of *Rhizoctonia* isolates collected from *Brassica oleracea* var. *acephala* in Ordu, Turkey. Phytoparasitica 48:273–286

Uddin J, Ullah F, Naz I, Ahmad S, Saljoqi AU, Khan SS, Salim M (2023) Root-knot nematode pathogen suppression in eggplant using antagonistic fungi. Egypt J Biol Pest Contr 33:1–8

Udo IA, Osai EO, Ukeh DA (2014) Management of root-knot disease on tomato with bioformulated *Paecilomyces lilacinus* and leaf extract of *Lantana camara*. Braz Arch Biol Technol 57:486–492

Vaughn SF, Eller FJ (1996) Pepper (*capsicum* spp.) wound volatiles suppress the growth of *Alternaria alternata*, the causal agent of internal mold of pepper. Hort Sci 31:678–678

Wanjohi WJ, Wafula GO, Macharia CM (2018) Integrated management of *Fusarium* wilt-root-knot nematode complex on tomato in central high lands of Kenya. Sustain Agri Res 7:8–18

Wheeler TA, Riedel RM (1994) Interactions among *Pratylenchus penetrans, P. Scribneri*, and *Verticillium dahliae* in the potato early dying disease complex. J Nematol 26:228–234

Wilson SP, Ahvenniemi MP, Lehtonen JM, Kukkonen M, Rita H, Valkonen PTJ (2008) Biological and chemical control and their combined use to control different stages of the *Rhizoctonia* disease complex on potato through the growing season. Ann Appl Biol 153:307–320

Xiang N, Lawrence KS, Donald PA (2018) Biological control potential of plant growth-promoting rhizobacteria suppression of *Meloidogyne incognita* on cotton and *Heterodera glycines* on soybean: a review. J Phytopathol 166:449–458

Yang JI, Loffredo A, Borneman J, Becker JO (2012) Biocontrol efficacy among strains of *Pochonia chlamydosporia* obtained from a root-knot nematode suppressive soil. J Nematol 44:67–71

Yitayih GK, Mohamed Y, Awol SE (2020) Interaction effect between *Meloidogyne incognita* and *Fusarium oxysporum* f. sp. *lycopersici* on selected tomato (*Solanum lycopersicum* L.) genotypes. Afr J Agric Res 15:330–342

Zeilinger S, Gupta VK, Dahms TE, Silva RN, Singh HB, Upadhyay RS, Gomes EV, Tsui CK, Nayak SC (2016) Friends or foes? Emerging insights from fungal interactions with plants. FEMS Microbiol Rev 40:182–207

Zhang Y, Zhao J, Ding G, Yang L, Chen W (2016) Identification and characterization of *Rhizoctonia solani* causing damping-off on cucumber in China. Plant Dis 100:633–639

Zhang Y, Li S, Li H, Wang R, Zhang KQ, Xu J (2020) Fungi-nematode interactions: diversity, ecology, and biocontrol prospects in agriculture. J Fungi 6:206

Zheng Q, Bertran A, Brand A, van Schaik CC, van de Ruitenbeek SJS, Smant G, Goverse A, Sterken MG (2022) Comparative transcriptome analysis reveals the specific activation of defense pathways against *Globodera pallida* in Gpa2 resistant potato roots. Front Plant Sci 13:909593

4 Biochemical Changes in Host Plant Invaded by Opportunistic Fungi and Nematodes

Rashmita Saikia, Dikshita Saikia, Arkadeb Chatterjee, Himanshu Thakur, and Mohd. Sayeed Akhtar

4.1 Introduction

In nature, every plant is affected by different biotic agents such as fungi, bacteria, viruses, nematodes, etc. Although these plants may be damaged to varying degrees by such biotic agents, many among them survive all of these attacks and continue to grow well and produce appreciable yields (Agrios 2005). The biotic stress caused by numerous phytopathogens such as viruses, mycoplasmas, bacteria, and fungi is continually influencing crop health and yield. Pathogens disrupt physiological, biochemical, and metabolic processes and pathways in plants, resulting in a loss of production and quality (Kumar and Verma 2018). Plant defence mechanisms mainly constitutes of preformed or constitutive defence responses and induced defence responses which appear on subsequent attack by the phytopathogens. Preformed structural barriers like cell walls, bark and cuticles provide rigidity and strength to the growing plant in addition to protecting them from pathogen invasion. Apart from having such physical or structural barriers, plants also have the ability to protect themselves against invading pathogen attack by responding with induced defences that includes certain biochemical compounds produced inside the cells and tissues of the plant which are either toxic to the invading pathogen or lead to conditions which are unfavourable for the growth and multiplication of the

R. Saikia (✉) · D. Saikia
Department of Plant Pathology, Punjab Agricultural University, Ludhiana, Punjab, India

A. Chatterjee
Department of Nematology, Assam Agricultural University, Jorhat, Assam, India

H. Thakur
Department of Entomology, CSK Himachal Pradesh Krishi Vishvavidyalaya, Palampur, Himachal Pradesh, India

M. S. Akhtar
Department of Botany, Gandhi Faiz-e-Aam College, Shahjahanpur, Uttar Pradesh, India

M. S. Akhtar (ed.), *Opportunistic Fungi, Nematode and Plant Interactions*, https://doi.org/10.1007/978-981-97-2045-3_4

phytopathogen inside the plant. Induced defences against phytopathogens in plants include production and release of phenolic compounds, flavonoids, lignins, and enzymes for phenol metabolism such as phenylalanine ammonia-lyase (PAL), polyphenol oxidase (PPO), different antioxidant enzymes such as peroxidases (POX), superoxide dismutase (SOD), catalase (CAT), accumulation of tannins, phytoalexins, etc. (Akter et al. 2015; Debona et al. 2012). The biochemical changes mainly occurring in host plants due to invasion by opportunistic fungi or nematodes play a vital role in safeguarding the plants against the harmful effects of phytopathogens. Until they detect the invading pathogen attack, plants usually wait before producing proteins or toxic substances involved in their defence because the amount of energy and nutritional requirements associated with their production and maintenance is quite high. Plants often respond to microbial invasion by causing a variety of metabolic or biochemical changes linked with stress signalling, thereby activating defence pathways. As a result, they are well-equipped with biochemical compounds that help them in combatting plant diseases. The chapter aims to discuss the various plant biochemical changes that occur in host plants as a result of opportunistic fungus and nematode invasion.

4.2 Induction of Biochemical Changes in Host Plants

The process of inducing various biochemical changes in the plants leading to defence responses against phytopathogens involves certain steps starting from the identification of the pathogen elicitors by the host plant to signal transduction for various biochemical responses as follows:

4.2.1 Recognition of Microbial Elicitors by the Host Plant Receptors

Early recognition of the invading microbe or pathogen by the host plant is important for mobilizing the various defensive biochemical reactions and their products within the host plant system to protect itself from further infections. As soon as the invading pathogen makes contact with the plant surface, the plant starts receiving signal molecules indicating the presence of the pathogen. Various phytopathogens such as fungi, bacteria, and nematodes, secrete different chemicals in their immediate surrounding environment that serve as nonspecific elicitors for pathogen recognition by the host plant. Toxins, glycoproteins, peptides, and extracellular microbial enzymes (such as proteases and pectinases) are some of the pathogen elicitors recognized by the plants and in addition to these non-specific elicitors, certain Avr gene products, hrp gene products secreted by the pathogen along with suppressor molecules serve as specific pathogen elicitors for identification by the host plant. Additionally, the breakdown of certain polysaccharides on the pathogen surface by the host plant enzymes, or the breakdown of a portion of the plant surface polysaccharides by the invading pathogen enzymes releases certain oligomers or monomers of the polysaccharides which serve as recognition elicitors for the host plant.

The receptors which identify the incoming pathogen elicitors are usually situated on the host cell membrane, while several of them may present intracellularly.

4.2.2 Defence Mobilization

Once the plant's receptors recognize and react to the incoming pathogen elicitors, a sequence of biochemical reactions and structural changes are initiated in an attempt to dissolve the pathogen, including its enzymes. The time taken by the plant to recognize a specific pathogen followed by mobilization of its defences to prevent further infection by that pathogen determines whether infection will occur at all (hypersensitive response) or severity of the symptoms (leaf spots, stem, fruit, or root lesions, etc.) before the host defences finally get activated to further stop the development of the invading pathogen.

4.2.3 Signal Transduction

When the host plant recognizes pathogen-derived elicitors, it sends out a certain alarm signals that activate host nuclear genes and proteins to produce inhibitory compounds, and to mobilize them or their products to tissues targeted by the invading pathogen. Some signal transductions occur exclusively intracellularly, but in certain situations, the signal is carried to many adjacent cells, and the signal is sent systemically within plant cells or tissues.

A number of molecules have been identified to be involved in intracellular signal transduction in plants. Some of the common signal transducers appear to be calcium ions, protein kinases, phosphorylases, phospholipases, ATPases, ethylene, etc. Additionally, salicylic acid, oligogalacturonides (derived from plant cell walls), jasmonic acid, fatty acids, ethylene, systemin, and certain other compounds, contribute to systemic signal transmission, resulting in induced systemic resistance. Some natural or synthetic chemicals, such as salicylic acid and dichloroisonicotinic acid, also activate the signalling pathway, resulting in systemic acquired resistance against phytopathogenic viruses, bacteria, nematodes, and fungi.

4.3 Biochemical Defence Responses in Host Plants Against Opportunistic Fungi and Nematodes

The biochemical defence responses set out against invading pathogens are of two types, namely: Pre-existing and Induced.

4.3.1 Pre-existing Biochemical Defence Responses Against Opportunistic Fungi and Nematodes

Physiological or biochemical defence is much more important and common than the prevailing structural defences resisting invasion by plant pathogens (Singh 2017). The pre-existing biochemical defence mechanisms are present before the attack of the pathogen but mostly they appear by the activation of host defences upon pathogen attack. Some of the pre-existing biochemical defences in the plants include:

4.3.1.1 Antimicrobial Compounds Emitted by the Plant in Its Environment

During their growth, plants release a range of biochemical compounds in their surrounding environment. Some of these compounds released into the phyllosphere or rhizosphere may have beneficial properties which promotes the growth of beneficial microbes around the rhizosphere or phyllosphere whereas some are found to be toxic to certain groups of pathogens. Certain leaf and root exudates released by plants are found to be toxic to the incoming pathogen. For example, red scale onions are found to produce protocatechuic acid and catechol on their scales which imparts resistance to *Colletotrichum circinans* causing onion smudge disease. Similarly, powdery mildew resistant varieties of apple produce certain toxic chemicals which prevent conidial germination of the powdery mildew pathogen *Podosphaera leucotricha*. Root exudates of certain varieties of Linseed are resistant to wilt due to the release of hydrocyanides (HCN) in their root exudates.

4.3.1.2 Antimicrobials Present in the Plant Cells

Plants may naturally synthesize thousands of compounds which are unique to specific taxonomic groups and are found to be toxic to animals including insects and microorganisms (Webster 1974) such as cyanogenic glycosides, phenols and phenolic glycosides, unsaturated lactones, saponines, such as tomatine, solanine, avenacin, etc.

4.3.1.3 Lack of Essential Nutrients or Growth Factors for the Growth of Pathogen

Many facultative saprophytes and obligate parasites require certain essential nutrients to establish successful infection if their respective hosts. Absence of such factors or nutrients may hamper the growth or development of those pathogen on or inside the host. Moreover, these nutrients must be in available form for the pathogens to utilize them. If these pre-existing substances are not available to them or cannot be utilized due to certain reactions taking place in the host after infection, then the pathogen will fail to grow. For example, the successful infection of *Rhizoctonia solani* relies upon the formation of infection cushion in the seedlings of radish and lettuce and cushion formation in susceptible hosts depends on the presence of certain essential nutrients present inside the hosts. In resistant varieties, no such substances are produced and hence no mycelial cushions are being formed by the pathogen (Singh 2017). Similarly, for reproduction of certain nematodes Indole acetic acid and its precursor tryptophan are required which when not present in the hosts, stops the reproduction of the nematode.

4.3.2 Induced Biochemical Defence Responses in Host Plants Against Opportunistic Fungi and Nematodes

This kind of defence involves the production of various chemical substances at the onset of signalling pathways in response to the pathogen attack (Fig. 4.1). These substances may either have direct toxic effect on pathogens or may suppress the effect of pathogen toxins. The various induced biochemical defence mechanisms are as discussed as follows:

4.3.2.1 Detoxification of Pathogen Toxins

In resistant plant varieties, the toxins released by the invading pathogen are usually detoxified by metabolic processes of the host plant. For example, blast of rice is caused by the fungus *Magnaporthe grisea* by releasing pyricularin and picolinic acid inside the plant cells. The resistant varieties are able to convert about 40–60% of picolinic acid into its methylester and *N*-methylpicolinic acid, and pyriculain into other compounds which are non-toxic to the host plants (Singh 2005). Additionally, the toxins produced by phytopathogenic bacteria *Xanthomonas albilineans* causing leaf scald disease in sugarcane block chloroplast development in susceptible sugarcane plants. In light of this, Zhang et al. (1999) generated transgenic sugarcane plants expressing a detoxification enzyme, AlbD, and showed that the gene could confer resistance to pathogen multiplication in the host. In certain situations, resistant plants appear to metabolize these toxins more quickly, or generate less toxic or non-toxic compounds by interacting with other molecules. Usually, the quantity of non-toxic compound produced is proportionate to the variety's disease resistance.

4.3.2.2 Induction of Hypersensitive Response (HR)

Hypersensitive response is known as localized cell defence induced at the point of pathogen invasion. It is the result of a sequence of defence responses set out by the infected and nearby cells and the successive release of harmful or toxic substances which kill both the infected and neighbouring cells along with the pathogen (Agrios 2005). The response is mostly observed in host–pathogen interactions that are incompatible and occurs as a part of post infection biochemical defence. It is often triggered in the plants when the products of a resistant gene (R) in the plant cell recognize the presence of specific effector molecules (products of an avirulent gene) of the phytopathogen. Many phytopathogenic bacteria, viruses, fungi and nematodes are capable of inducing hypersensitive response in plants (Hutcheson 1998). Moreover, oxidative burst is considered as an important feature associated with the hypersensitive response during resistance (Harvey et al. 2008).

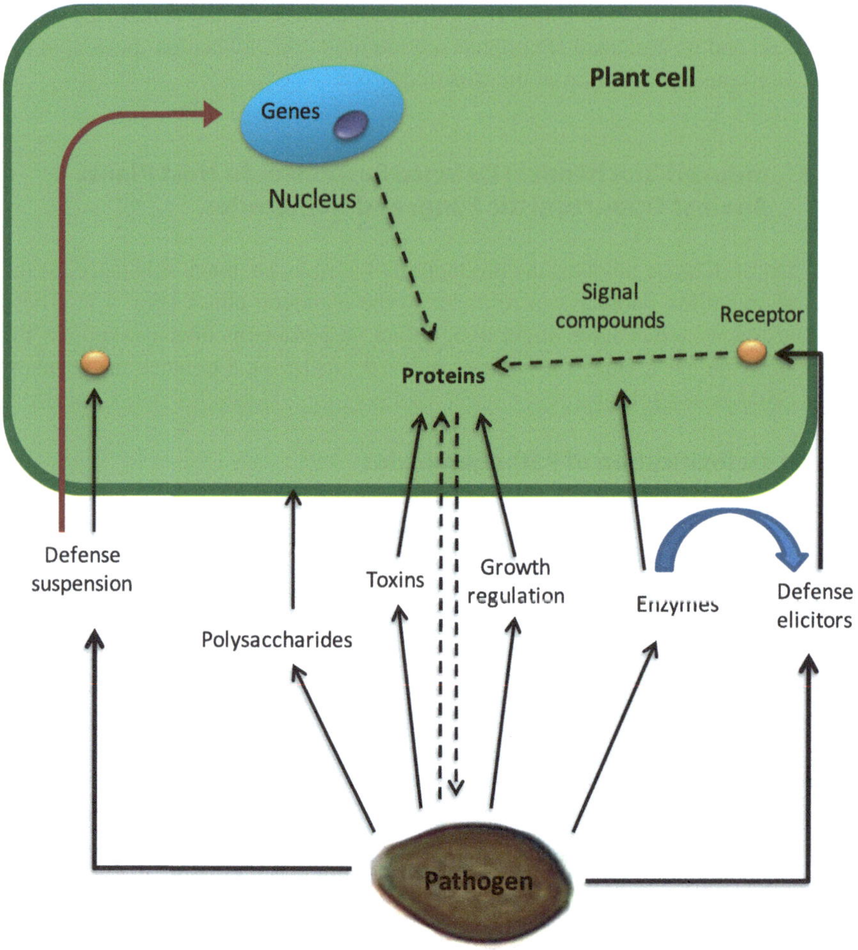

Fig. 4.1 Schematic representation of pathogen interactions with host plant cells. On interaction with a pathogen the plant cell may react with numerous defences, which may include cell wall structural defences (waxes, cutin, suberin, lignin, phenolics, cellulose, callose, cell wall proteins) or biochemical defences which may involve bursts of oxidative reactions, production of elicitors, hypersensitive cell death, ethylene, phytoalexins, pathogenesis-related proteins (hydrolytic enzymes, b-1,3-glucanases, chitinases), inhibitors (thionins, proteinase inhibitors, thaumatin-like proteins), and so on. (Modified from Agrios 2005)

4.3.2.3 Production of Pathogenesis-Related (PR) Proteins

Pathogenesis-related proteins are a wide category of plant proteins which are toxic to invading phytopathogens. Normally, they are found in trace amounts throughout plant system, but are produced in considerably higher concentrations upon pathogen infection or any other stress. Majority of PR proteins exist in intercellular locations, whereas primary PR proteins are seen to be present inside vacuoles (Arabi et al. 2019). Different plant components, such as leaves, seeds, and roots, may produce varying levels of PR proteins. These PR proteins are classified based on their function, amino acid sequence, serological relationship, molecular weight, and various other characteristics and are usually very acidic or basic, making them highly soluble and reactive. Liu et al. (2010) divided PR-proteins into 17 families such as β-1,3-glucanases, chitinases, peroxidases, thaumatin-like proteins, ribosome inactivating proteins, thionins, non-specific lipid switch proteins, oxalate oxidase, and oxalate oxidase-like proteins (Table 4.1). The PR proteins show strong antimicrobial activity such as inhibiting spore release and germination, strengthening the host cell wall and its outgrowths and papillae. For example, β-1,3-glucanase and chitinase break down the chitinous structure of fungal cell walls, while lysozymes break down glucosamine and muramic acid components of bacterial cell walls. Some such as lipoxygenases and lipid peroxidases generate antimicrobial metabolites as well as secondary signal molecules such as jasmonic acid which stops the pathogen invasion inside the cells by inducing resistance in plants. Additionally, it has also been shown that overexpression of PR proteins in transgenic plants results in an enhanced resistance to the incoming pathogens. For example, in tobacco the overexpression of PR1a gene showed increased tolerance to *Peronospora tabacina* and *Phytophthora parasitica* var. *nicotianae* (Alexander

Table 4.1 Classification of various PR proteins involved in plant defence (Moosa et al. 2018)

Family	Size (kDa)	Member	Properties
PR-1	15	Tobacco PR-1a	Antifungal
PR-2	30	Tobacco PR-2	β-1,3-glucanase
PR-3	25–30	Tobacco P, Q	Chitinase type I, II, IV, V, VI, VII
PR-4	15–20	Tobacco "R"	Chitinase type I, II
PR-5	25	Tobacco S	Thaumatin-like
PR-6	8	Tomato inhibitor I	Proteinase inhibitor
PR-7	75	Tomato P69	Endoproteinase
PR-8	28	Cucumber chitinase	Chitinase type III
PR-9	35	Tobacco "lignin-forming peroxidase"	Peroxidase
PR-10	17	Parsley "PR1"	Ribonuclease-like
PR-11	40	Tobacco "class V" chitinase	Chitinase, type I
PR-12	5	Radish Rs-AFP3	Defensin
PR-13	5	Radish Rs-AFP3	Thionin
PR-14	9	Barley LTP4	Lipid transfer protein
PR-15	20	Barley OxOa (germin)	Oxalate oxidase
PR-16	20	Barley OxOLP	Oxalate oxidase-like
PR-17	27	Tobacco PRp27	Antiviral and antifungal

et al. 1993). Similarly, overexpression of thaumatin-like PR-5 in transgenic rice and orange plants showed increased tolerance to *Rhizoctonias solani* and *Phytophthora citrophthora*, respectively (Datta et al. 2001; Fagoaga et al. 2001). Moreover, Hammond-Kosack et al. (1989) demonstrated the accumulation of PR proteins in the apoplast of potato leaves due to infection with nematode *Globodera rostochiensis*. Also, accumulation of chitinases and β-l,3-glucanases have been seen in the leaves and roots of different potato cultivars infected with *Globodera pallida* (Rahimi et al. 1995). The activity of chitinases increased significantly in infected roots, while the activity of β-l,3-glucanases increased in the leaves. Unlike other plant proteins, pathogenesis-related proteins are stabilized mostly by disulphide bonds, making them resistant to proteolysis and higher temperatures of their surrounding environment (Gorjanovic 2009).

4.3.2.4 Phenolics

These are secondary metabolites generated by plant cells or tissues in response to pathogen infection or injury. These compounds have one or more benzene rings and hydroxyl groups (Bergey et al. 1996). Plants synthesize phenols using either the shikimic acid or acetic acid pathways (Görlach et al. 1995).

Simple phenolic chemicals already exist in plants, but their synthesis is further enhanced when the plants are attacked by phytopathogens. Their concentration in the plant system before pathogen attack is insufficient to prevent infection; nevertheless, once infected, the plant rapidly accumulates these chemicals in the tissues surrounding the infection site or leads to the movement of these compounds from healthy to diseased tissues. Examples of simple phenolic compounds include caffeic acid, chlorogenic acid, ferulic acid, scopoletin, and isocumarin.

Unlike simple phenolics, which occur in trace levels in healthy plants, toxic phenolic compounds, such as phytoalexins, are formed only in response to pathogen attack.

4.3.2.5 Phytoalexins

Phytoalexins are toxic antimicrobial substances whose production is stimulated by different phytopathogenic microorganisms or by any type of chemical or mechanical injury to the plants and are produced by healthy cells near localized injured cells in reaction to substances that diffuse out from the diseased cells. More than 300 phytoalexin-like substances have been identified from plants spanning more than 30 families (Agrios 2005). The production of phytoalexins by different plants is family-specific. For example, sulphur-containing phytoalexins such as Sanguinarine, Brassinin, Cyclobrassinin, and others are most common in all Brassicaceae, but the *Poaceae* family contains a variety of phytoalexins such as oryzalexins, zealexins, kauralexins, sakuranetin, phenyl amides, and others. Similarly, Solanaceae has phenylpropanoid-related substances, steroid glycoalkaloids, sesquiterpenoids, and coumarins, and Leguminosae contains isoflavones, coumestans, phaseolin, and stilbenes (Kaur et al. 2022). Some of the commonly known phytoalexins include phaseolin in bean, pisatin in pea, glyceollin in soybean, rishitin in potato, gossypol in cotton, and capsidiol in pepper. Furthermore, phytoalexins have been shown to confer protection against phytopathogenic nematodes. Rich et al. (1977) conducted the first experiment to identify phytoalexins as anti-nematode chemicals, and they observed phytoalexin buildup in host tissues. Similar studies indicated that soybean and potato resistance to endoparasitic nematodes is due to the synthesis of glyceollin (Kaplan and Keen 1980; Huang and Barker 1991). It is observed that phytoalexin extracts treated with an abiotic elicitor can also elicit phytoalexin accumulation, implying that nematode stylet secretions either directly elicit phytoalexin accumulation or indirectly releases an endogenous compound, which in turn elicits phytoalexin accumulation (Zacheo et al. 1997). Although most of the elicitors of phytoalexin production are pathogen-derived, some, such as oligomers of galacturonic acid are released by plant cells during pathogen invasion or are released from the cell walls after being partially broken down by the invading phytopathogen's cell wall degrading enzyme. The buildup of phytoalexin inside plant cells is controlled by increased expression of related biosynthetic enzyme-coding genes, which are activated by different biotic and abiotic stressors. Some of the important phytoalexins produced to resist the incoming pathogen attack are given in Table 4.2.

Table 4.2 Different phytoalexins identified in plants against various pathogens or nematodes

Phytoalexin	Crop	Pathogen/nematode	References
Gossypol (terpenoid aldehydes)	*Gossypium hirsutum*	*Meloidogyne incognita*	Veech (1982)
Glyceollin	*Glycine max*	*Heterodera glycines*	Huang and Barker (1991)
Medicarpin	*Medicago sativa*	*Pratylenchus penetrans*	Baldridge et al. (1998)
Momilactone A and B, phytocassane A, phytocassane E, and sakuranein	Rice	*Magnaporthe oryzae*	Hasegawa et al, Ahuja et al. (2012)
3-Deoxyantho-cynidin	Sorghum	*Fusarium*	Nida et al. (2021)
Medicarpin and 7,40-dihydroxy-flavone	Alfa-alfa	*Fusarium oxysporum* f. sp. *medicaginis*	Gill et al. (2018)
Glyceollins	Soybean	*Fusarium solani*	Abdul and Al-Muwayhi (2020)
3-Deoxy anthocyanidin	Sugarcane	*C. falcatum*	Nandakumar et al. (2021)
Kauralexins and zealexins	Maize	*Rhizopus microsporus, Colletotrichum graminicola, F. graminearum, Cochliobolus heterostrophus,* and *Aspergillus flavus*	Schmelz et al. (2011)
Methoxylchalcones	Barley	*B. sorokiniana* and *F. graminearum*	Ube et al. (2021)

4.3.2.6 Flavonoids

Flavonoids have a diphenyl-propane backbone (C3-C6-C3) that serves as the foundation for flavonoid subgroup classification. They are produced via the shikimate and acetate routes by a cytosolic multienzyme complex that is tethered to the endoplasmic reticulum. Several subgroups of flavonoids have been characterized based on structural features, including anthocyanins, aurones, condensed tannins, chalcones, flavones, flavonols, flavandiols, isoflavonoids, and pterocarpans (Chin et al. 2018). Studies have found increased flavonoid accumulation at infection sites of both endo and ectoparasitic plant pathogenic nematodes, and production has been found to be higher in resistant compared with susceptible host varieties. Flavonoids are commonly implicated as defensive compounds against plant pathogenic nematodes which are mostly from the isoflavonoids and pterocarpan classes (e.g., glyceollin, formononetin, and medicarpin) as well as the flavonols (e.g., kaempferol and quercetin).

Flavonoids are transported to the diseased site which induces hypersensitive reaction in the cells, leading to programmed cell death and they are be incorporated into the cell walls of necrotic and adjacent cells contributing to tightening of the plant tissues by modulating auxin (IAA) activity, leading to the differentiation of tissues, callus and tylose formation, closure of the vascular system to prevent pathogen infection and also are found to inhibit the activity of the pathogen's enzymes by chelating the metals required for their activity. The antifungal activity of flavonoids is often based on inhibiting the development of spores in pathogen and elongation of fungal hyphae or mycelium. Flavonoids are involved in inhibiting a number of root pathogens, hence are generally acknowledged as efficient anti-microbial agents (Mierziak et al. 2014). For example, the invasion of *Verticillium dahlia* in red cotton cultivar was suppressed due to the enhanced levels of flavonoids (Long et al. 2019). Now-a-days numerous scientists has recruited the metabolomics approach to elucidate the secondary metabolite profile of plants to find out the various secondary metabolites involved in pathogen inhibition. For example, metabolomics study revealed that plants can resist infection caused by *Colletotrichum gloeosporioides* through flavonoids production, indicating possible targets for resistance breeding programmes (Jiang et al. 2021).

After a disease or nematode attack, flavonoids accumulate and serve an important role in activating plant defences. For example, in plant pathogenic nematodes (PPNs), flavonoids can reduce egg hatching, as demonstrated by Wuyts et al. (2006), who found that kaempferol inhibited *Radopholus similis* egg hatching. Flavonoids can promote quiescence in juveniles by reducing their mobility, modifying their migration to the roots by repelling them, and even killing them. For example, patuletin, patulitrin, quercetin, and rutin have been shown to kill *Heterodera zeae* juveniles at varying concentrations and exposure times (Faizi et al. 2011). Flavonoid accumulation in plants against plant pathogenic nematodes is species-specific. For example, similar concentrations of flavonols, myricetin, kaempferol, and quercetin repelled the juveniles of *Meloidogyne incognita* and *Radophilus similis*, but not *Pratylenchus penetrans*, whereas the flavonols inhibited the motility of *M. incognita* juveniles but not *R. similis* and *P. penetrans* juveniles (Wuyts et al. 2006).

4.3.2.7 Generation of Reactive Oxygen Species (ROS), Lipoxygenases, and Cell Membrane Disruption of Pathogen

The production, accumulation, and release of reactive oxygen species (ROS) and lipoxygenase enzymes due to host plant invasion by pathogens is a critical membrane-associated defensive response. Plants can commit suicide due to pathogen invasion by producing ROS, followed by hypersensitive response that inhibits pathogen penetration in living tissues and kills the pathogen at the site of disease. One of the first events detected in host cells during host–fungus interactions is the rapid production of superoxide, hydrogen peroxide, and hydroxyl radical, all of which are harmful to invading pathogens. According to Grundler et al. (1997), ROS production is one of the early processes in molecular plant-nematode interactions. Pathogenesis-related proteins are found to be involved in the upregulation of genes involved in ROS production which results in hypersensitive response in plants. Kandoth et al. (2011) discovered that soybean lines with the Rhg1 gene had higher expressions of ROS-related genes and were more resistant to the soybean cyst nematode, *Heterodera glycines*. The emitted radicals may also harm the host plant. For example, hydrogen peroxide is a free radical that has antibacterial properties, and its accumulation in plants is harmful to plant cells. Hence, plants possess enzymes such as peroxidases that can scavenge excess ROS in plant tissue to prevent self-damage. Similarly, PPN secretomes contain superoxide dismutases (SODs), which detoxify ROS produced by plant cells and suppress resistance responses. PPNs can scavenge excess ROS with enzymes and antioxidants such glutathione, peroxidases, thioredoxins, cytochrome C-peroxidases, ascorbate peroxidases, and catalases, which protect nematode cells from ROS-mediated damage. For example, overexpression of a peroxiredoxin-family gene from *Globodera rostochiensis* (potato cyst nematode), shielded it from potato plant defence responses via a variety of redox reactions (Ali et al. 2018).

4.3.2.8 Strengthening of Plant Cell Walls

On encounter with a pathogen, the plant cell walls produce or accumulate defence-related chemicals that strengthen the cell wall's resistance to pathogen attack. These defence-related chemicals include phenolic substances like lignin and suberin, glycoproteins with a high hydroxyproline content like extensin, and minerals like calcium and silicon (Shittu et al. 2021). These chemicals react in response to a pathogen attack and form insoluble cell wall structures by cross-linking with each other which contain the invading pathogen and prevent further disease progression (Agrios 2005).

4.3.2.9 Defence Enzymes

Various defence enzymes such as peroxidase, polyphenol oxidase (PPO), phenylalanine ammonia lyase (PAL), chitinase, and β-1,3-glucanase are produced against the invading phytopathogens in plants. Peroxidases are known for their wide range of defence related processes, including the hypersensitive response, lignification, cross-linking of phenolics and glycoproteins, suberization, and phytoalexin production. Polyphenol oxidase (PPO) catalyses the oxidation of hydroxy phenols to quinone derivatives that have antimicrobial properties. Moreover, the intensification of production of phenolic compounds against pathogens is indicated by an increase in phenylalanine ammonia lyase (PAL) activity in wounded plant tissues. Similarly, chitinases and β-1,3-glucanases are involved in the hydrolysis of cell wall components of invading fungal pathogens such as chitin and β-1,3-glucans (Ebrahim et al. 2011). Some of the examples of enhanced defence enzyme production against invading pathogen are as follows:

(a) Increased activity of polyphenol oxidase (PPO) was indicated in roots of banana plants treated with fungus *Fusarium oxysporum* by Thakker et al. (2007) and also in roots treated with suspension of *Psuedomonas fluorescens* against fusarium wilt (Sarvanan et al. 2004).
(b) After infection with *Ralstonia solanacearum*, potato plants resistant to bacterial wilt showed increased expression of PPO as compared to susceptible plants (El-Argawy and Adss 2016).
(c) In transgenic tomato plants, Khodadadi et al. (2020) observed that the over-expression of PPO provided enhanced resistance to *Pseudomonas syringae* where the disease severity was effectively reduced with a reduced population of the bacteria.
(d) In resistant barley cultivars enhanced activity of PAL was indicated against *Bipolaris sorokiniana* (Kaur et al. 2021), owing to enhanced lignin accumulation and cell wall-bound phenolic compounds by cell walls (Jun et al. 2018).
(e) In transgenic tobacco, PAL over-expression has led to reduced lesion size after attack with *Pseudomonas syringae*.
(f) Similarly, the gene AevPAL1 renders resistance to cereal cyst nematode in bread wheat by altering salicylic acid production and downstream secondary metabolites (Zhang et al. 2021).

4.3.2.10 Induction of Systemic Acquired Resistance (SAR)

SAR is a mechanism of induced defensive responses (Gajanayaka et al. 2014). In SAR, a mobile signal is formed at the point of induction and translocated throughout the plant, resulting in an induced state in tissues far from the elicitor's exposure site. It provides a long-lasting protection from a wide range of microorganisms or pathogens. SAR requires the signal molecule salicylic acid and is commonly coupled with the buildup of PR proteins, which contributes to resistance. SAR is primarily involved with a variety of cellular defence responses, including the PR proteins production, phytoalexins production, changes in the cell wall, and increased activity of defence-related enzymes. SAR is generated systemically upon inoculation with necrotizing pathogens or the application of certain chemicals, such as salicylic acid. Certain plant growth-promoting microorganisms may activate defence activities and increase plant resistance to soil-borne diseases (Prasannath 2017).

4.4 Conclusion and Future Prospects

The knowledge of the different biochemical changes occurring in plants upon incoming pathogen attack is very much important in understanding plants' response to them. Naturally, plants protect themselves against pathogen attack and continue to grow and produce appreciable yields by employing a combination of structural features and biochemical reactions throughout their lifetime. The various defence compounds or mechanisms produced within the plant may occur at any stage of pathogen invasion but they are generally affected by the plant's age, the plant organ affected, the nutritional status of the plant and environmental conditions prevailing during pathogen invasion. Therefore, proper understanding of various constitutive and induced plant defence mechanisms against the invading phytopathogens is of utmost importance improving plant productivity and subsequent development of new and improved diseased resistant varieties. The growing omics approach, specifically metabolomics, has opened up avenues for investigating the interplay of various metabolites in conferring disease resistance. The role of constitutive and induced metabolites in providing resistance against phytopathogens as well as abiotic stress tolerance could be an attractive research domain to investigate in the near future. Research studies should also focus on the use of recently improvised genome editing techniques such as CRISPR-Cas9 to develop crop plants for enhanced innate defence responses against invading phytopathogens, thereby ensuring food security.

References

Abdul M, Al-Muwayhi R (2020) Biotic stress as a defense mechanism in soybean (*Glycine max* L.) toward microbial pathogen: biochemical and physiological pathways study. Life Sci J 17:24–36

Agrios GN (2005) Plant pathology, 5th edn. Elsevier, San Diego, p 922

Ahuja I, Kissen R, Bones AM (2012) Phytoalexins in defense against pathogens. Trends Plant Sci 17:73–90

Akter S, Huang J, Waszczak C, Jacques S, Gevaert K, Breusegem FV (2015) Cysteines under ROS attack in plants: a proteomics view. J Exp Bot 66:2935–2944

Alexander D, Goodman RM, Gut-Rella M, Glascock C, Weymann K, Friedrich L, Maddox D, Ahl-Goy P, Luntz T, Ward ES (1993) Increased tolerance to two oomycete pathogens in transgenic tobacco expressing pathogenesis-related protein 1a. Proc Natl Acad Sci U S A 90:7327–7331

Ali MA, Anjam MS, Nawaz MA, Lam HM, Chung G (2018) Signal transduction in plant-nematode interactions. Int J Mol Sci 19:1648

Arabi E, Jawahar M, Al-daoude A, Al-Shehadah E, Shoaib A, Imad M (2019) Differential expression of defense related genes in susceptible versus resistant barley genotypes challenged with Pyrenophorateres. Mycopathologia 15:61–66

Baldridge GD, O'Neill NR, Samac DA (1998) Alfalfa (*Medicago sativa* L.) resistance to the root-lesion nematode, *Pratylenchus penetrans*: defense-response gene mRNA and isoflavonoid phytoalexin levels in roots. Plant Mol Biol 38:999–1010

Bergey DR, Howe GA, Ryan CA (1996) Polypeptide signaling for plant defensive genes exhibits analogies to defense signaling in animals. Proc Natl Acad Sci U S A 93:12053–12058

Chin S, Behm CA, Mathesius U (2018) Functions of flavonoids in plant-nematode interactions. Plants 7:85

Datta K, Tu J, Oliva N, Ona I, Velazhahan R, Mew TW, Muthukrishnan S, Datta SK (2001) Enhanced resistance to sheath blight by constitutive expression of infection related rice chitinase in transgenic elite indica rice cultivars. Plant Sci 160:405–414
Debona D, Rodrigues FA, Rios JA, Nascimento KJT (2012) Biochemical changes in the leaves of wheat plants infected by *Pyricularia oryzae*. Phytopathology 102:1121–1129
Ebrahim S, Usha K, Singh B (2011) Pathogenesis related (PR) proteins in plant defense mechanism. Microb Pathog 2:1043–1054
El-Argawy E, Adss I (2016) Quantitative gene expression of peroxidase, polyphenoloxidase and catalase as molecular markers for resistance against *Ralstonia solanacearum*. Am J Mol Biol 6:88–100
Fagoaga C, Rodrigo I, Conejero V, Hinarejos C, Tuset JJ, Arnau J, Pina JA, Navarro L, Peña L (2001) Increased tolerance to *Phytophthora citrophthora* in transgenic orange plants constitutively expressing a tomato pathogenesis related protein PR-5. Mol Breed 7:175–185
Faizi S, Fayyaz S, Bano S, Yawar Iqbal E, Siddiqi H, Naz A (2011) Isolation of nematicidal compounds from *Tagetes patula* L. yellow flowers: structure-activity relationship studies against cyst nematode *Heterodera zeae* infective stage larvae. J Agric Food Chem 59:9080–9093
Gajanayaka GM, Prasannath K, De Costa DM (2014) Variation of chitinase and β-1,3 glucanase activities in tomato and chilli tissues grown under different crop management practices and agroecological regions. In: Proceedings of the Peradeniya University International Research sessions, vol 18, Sri Lanka, p 519
Gill US, Uppalapati SR, Gallego-Giraldo L, Ishiga Y, Dixon RA, Mysore KS (2018) Metabolic flux towards the (iso)flavonoid pathway in lignin modified alfalfa lines induces resistance against *Fusarium oxysporum* f. sp. medicaginis. Plant Cell Environ 41:1997–2007
Gorjanovic S (2009) A review: biological and technological functions of barley seed pathogenesis-related proteins (PRs). J Inst Brew 115:334–360
Görlach J, Raesecke HR, Rentsch D, Regenass M, Roy PH, Zala M, Keel C, Boller T, Amrhein N, Schmid J (1995) Temporally distinct accumulation of transcripts encoding enzymes of the pre-chorismate pathway in elicitor-treated, cultured tomato cells. Proc Natl Acad Sci 92:3166–3170
Grundler FM, Sobczak M, Lange S (1997) Defence responses of *Arabidopsis thaliana* during invasion and feeding site induction by the plant-parasitic nematode *Heterodera glycines*. Physiol Mol Plant Pathol 50:419–429
Hammond-Kosack KE, Atkinson HJ, Bowles DJ (1989) Systemic accumulation of novel proteins in the apoplast of the leaves of potato plants following root invasion by the cyst-nematode *Globodera rostochiensis*. Physiol Mol Plant Pathol 35:495–506
Harvey JJ, Lincoln JE, Gilchrist DG (2008) Programmed cell death suppression in transformed plant tissue by tomato cDNAs identified from an *Agrobacterium* rhizogenes-based functional screen. Mol Genet Genomics 279:509–521
Huang JS, Barker KR (1991) Glyceollin I in soybean-cyst nematode interactions: spatial and temporal distribution in roots of resistant and susceptible soybeans. Plant Physiol 96:1302–1327
Hutcheson SW (1998) Current concepts of active defense in plants. Annu Rev Phytopathol 36:59–90
Jiang L, Wu P, Yang L, Liu C, Guo P, Wang H, Wang S, Xu F, Zhuang Q, Tong X, Liu P, Luo L (2021) Transcriptomics and metabolomics reveal the induction of flavonoid biosynthesis pathway in the interaction of Stylosanthes-*Colletotrichum gloeosporioides*. Genomics 113:2702–2716
Jun S-Y, Sattler SA, Cortez GS, Vermerris W, Sattler SE, Kang C (2018) Biochemical and structural analysis of substrate specificity of a phenylalanine ammonia-lyase. Plant Physiol 176(2):1452–1468
Kandoth PK, Ithal N, Recknor J, Maier T, Nettleton D, Baum TJ, Mitchum MG (2011) The soybean Rhg1 locus for resistance to the soybean cyst nematode *Heterodera glycines* regulates the expression of a large number of stress- and defense-related genes in degenerating feeding cells. Plant Physiol 155:1960–1975
Kaplan DT, Keen NT (1980) Mechanism conferring plant incompatibility to nematodes. Rev Nematol 3:123–134

Kaur S, Bhardwaj RD, Kaur J, Kaur S (2021) Induction of defense related enzymes and pathogenesis-related proteins imparts resistance to barley genotypes against spot blotch disease. J Plant Growth Regul 41:682–696

Kaur S, Samota MK, Choudhary M, Choudhary M, Pandey AK, Sharma A, Thakur J (2022) How do plants defend themselves against pathogens: biochemical mechanisms and genetic interventions. Physiol Mol Biol Plants 28:485–504

Khodadadi F, Tohidfar M, Vahdati K, Dandekar AM, Leslie CA (2020) Functional analysis of walnut polyphenol oxidase gene (JrPPO1) in transgenic tobacco plants and PPO induction in response to walnut bacterial blight. Plant Pathol 69(4):756–764

Kumar A, Verma JP (2018) Does plant-microbe interaction confer stress tolerance in plants: a review. Microbiol Res 207:41–52

Liu H, Du Y, Chu H, Shih CH, Wong YW, Wang M, Chu IK, Tao Y, Lo C (2010) Molecular dissection of the pathogen-inducible 3-deoxyanthocyanidin biosynthesis pathway in sorghum. Plant Cell Physiol 51:1173–1185

Long L, Liu J, Gao Y, Xu FC, Zhao JR, Li B, Gao W (2019) Flavonoid accumulation in spontaneous cotton mutant results in red coloration and enhanced disease resistance. Plant Physiol Biochem 143:40–49

Mierziak J, Kostyn K, Kulma A (2014) Flavonoids as important molecules of plant interactions with the environment. Molecules 19:16240–16265

Moosa A, Farzand A, Sahi ST, Khan SA (2018) Transgenic expression of antifungal pathogenesis-related proteins against phytopathogenic fungi-15 years of success. Israel J Plant Sci 65:38–54

Nandakumar M, Malathi P, Sundar AR, Viswanathan R (2021) Host pathogen interaction in sugarcane and red rot pathogen: exploring expression of phytoalexin biosynthesis pathway genes. Indian Phytopathol 74:529–535

Nida H, Lee S, Li Y, Mengiste T (2021) Transcriptome analysis of early stages of sorghum grain mold disease reveals defense regulators and metabolic pathways associated with resistance. BMC Genomics 22:1–7

Prasannath K (2017) Plant defense-related enzymes against pathogens: a review. AGRIEAST: J Agric Sci 11:38–48

Rahimi S, Perry RN, Wright OJ (1995) Studies on pathogenesis-related proteins in potato induced by potato cyst nematode, *Globodera pallida* and several chemicals. Nematologica 41:332

Rich JR, Keen NT, Thomason IJ (1977) Association of coumestanss with the hypersensitivity of Lima bean roots to *Pratylenchus scribneri*. Physiol Plant Pathol 10:105–116

Sarvanan T, Bhaskaran R, Muthuswamy M (2004) *Pseudomonas fluorescence* induced enzymological changes in banana roots (cv. *Rasthali*) against fusarium wilt disease. Plant Pathol J 3:72–70

Schmelz EA, Kaplan F, Huffaker A, Dafoe NJ, Vaughan MM, Ni X, Rocca JR, Alborn HT, Teal PE (2011) Identity, regulation, and activity of inducible diterpenoid phytoalexins in maize. Proc Natl Acad Sci U S A 108:5455–5460

Shittu HO, Aisagbonhi E, Obiazikwor OH (2021) Plants innate defence mechanisms against phytopathogens. J Microbiol Biotechnol Food Sci 2021:314–319

Singh RP (2005) Plant pathology. Kaylani Publishers, New Delhi, p 724

Singh RS (2017) Introduction to principles of plant pathology. Oxford and IBH Publishing, Delhi

Thakker JN, Patel N, Kothari IL (2007) *Fusarium oxysporum* derived elicitor-induced changes in enzymes of banana leaves against wilt disease. J Mycol Plant Pathol 37:510–513

Ube N, Katsuyama Y, Kariya K, Tebayashi SI, Sue M, Tohnooka T, Ueno K, Taketa S, Ishihara A (2021) Identification of methoxylchalcones produced in response to $CuCl_2$ treatment and pathogen infection in barley. Phytochemistry 184:112650

Veech JA (1982) Phytoalexins and their role in the resistance of plants to nematodes. J Nematol 14:2–9

Webster RK (1974) Recent advances in the genetics of plant pathogenic fungi. Annu Rev Phytopathol 12:331

Wuyts N, Swennen R, de Waele D (2006) Effects of plant phenylpropanoid pathway products and selected terpenoids and alkaloids on the behaviour of the plant-parasitic nematodes *Radopholus similis*, *Pratylenchus penetrans* and *Meloidogyne incognita*. Nematology 8:89–101

Zacheo G, Bleve-Zacheo T, Melillo MT (1997) Biochemistry of plant defence responses to nematode infection. In: Fenoll C, Grundler FMW, Ohl SA (eds) Cellular molecular aspects of plant-nematode interactions, Development in plant pathology, vol 10. Springer, Drodrecht, pp 201–213

Zhang L, Xu J, Birch R (1999) Engineered detoxification confers resistance against a pathogenic bacterium. Nat Biotechnol 17:1021–1024

Zhang H, Huang Q, Yi L, Song X, Li L, Deng G, Liang J, Chen F, Yu M, Long H (2021) PAL-mediated SA biosynthesis pathway contributes to nematode resistance in wheat. Plant J 107(3):698–712

Response of ROS in Host Plant Assaulted by Opportunistic Fungi and Nematodes

5

Yusuf Opeyemi Oyebamiji, Nur Sakinah Binti Mohd Yusri, Saheed Adekunle Akinola, Tolessa Muleta Daba, Jean Paul Sinumvayo, Rasheed Omotayo Adeyemo, Alex Rugamba, Annette Uwineza, and Ismail Abiola Adebayo

5.1 Introduction

Globally, the amount of crop yield lost due to pathogens and pests is about 28% for wheat, 41% for rice, 41% for maize, 21% for potatoes, and 32% for soybeans (Gonzalez-Chang et al. 2019). Biotic stress is a real devastation, affecting huge amounts of crop yield reduction (Mardiana-Jansar et al. 2019). Plants' genetic potential constricted under various biotic stresses during either vegetative growth or reproductive growth including pathogenic microbes attack, invasion by weeds, or destruction by herbivorous animals (Ashraf et al. 2012) especially pathogenic nematodes and opportunistic fungi. The most prevalent species in soil ecosystems are nematodes and fungi. They perform vital ecosystem functions and are crucial in promoting nutrient cycling and preserving the sustainability of food webs (Topalovic and Heuer 2019).

Based on the research conducted by Zhang et al. (2020a, b), fungi play a crucial role as primary components in the process of decomposing decaying organic materials. Despite their morphological resemblance to plants, fungi are closely related to animals from an evolutionary perspective. As heterotrophic organisms, fungi and nematodes are frequently found co-existing in various ecosystems, including the

Y. O. Oyebamiji · Nur Sakinah Binti Mohd Yusri
Department of Biological Sciences and Biotechnology, Faculty of Science and Technology, Universiti Kebangsaan Malaysia, Bangi, Selangor, Malaysia

S. A. Akinola · J. P. Sinumvayo · R. O. Adeyemo
Department of Medical Microbiology and Parasitology, School of Medicine and Pharmacy, College of Medicine and Health Sciences, University of Rwanda, Butare, Rwanda

T. M. Daba · A. Rugamba · A. Uwineza · I. A. Adebayo (✉)
Department of Medical Biochemistry, Molecular Biology and Genetics, School of Medicine and Pharmacy, College of Medicine and Health Sciences, University of Rwanda, Butare, Rwanda

M. S. Akhtar (ed.), *Opportunistic Fungi, Nematode and Plant Interactions*,
https://doi.org/10.1007/978-981-97-2045-3_5

rhizosphere of crops, and can cause substantial damage to plants, ultimately resulting in significant harm to the agricultural industry. The manifestation of plant diseases and visible damages observed above the soil surface can be attributed to the destruction of root systems caused by nematode attacks. This results in malnutrition, changes in the color of leaves to yellowish and wilting, stunted growth, and ultimately, death of the plant (Maisarah and Azmi 2015). Around 30% of crop illnesses are caused by pathogenic fungus. The various key structures investigated from the host's leaf descriptions have been the basis for the present procedures for an earlier discovery of leaf illnesses within the host.

Around 90% of the 500,000 species that comprise the second largest phylum in the animal kingdom, namely the Nematoda, are from a smaller group called terrestrial nematodes, which live 15 cm under the soil (Bongers and Bongers 1998; Burros 1996). They are microscopic parasitic invertebrates-worms with a long, circular shape-prevalently metazoans that consume living things (Burros 1996; Maisarah et al. 2018). According to Wilkinson (2014) nematode disease symptoms that are visible above ground may be difficult to spot or misinterpreted as nutritional deficiency. Parasitic nematodes usually use their stylets to get into the tissue of their host plants. They then release cellulase-like substances from their glands to break down the plant's cell wall, which lets them get into the host tissue (Davis et al. 2008). Based on how they feed, parasitic nematodes can be put into three groups: migratory ectoparasites, which feed on the surface of the root, endoparasites, which feed inside the root, and semi-endoparasites. Plant parasitic nematodes eat underground parts of plants like roots, rhizomes, tubers, and bulbs. Their symptoms show up on the plant's leaves, and they are often confused with those caused by things like a lack of nitrogen and water. Other species of nematodes eat stems, leaves, flowers, fruits, and seeds. Some plant parasitic nematodes, such as *Meloidogyne* spp. and *Pratylenchus* spp., are very polyphagous in nature, which means they can infect many different kinds of plants (Palomares-Rius et al. 2012).

For instance, nematode infestations lower plants' resistance to pathogens like fungi and increase their susceptibility to be infested by pathogenic fungi (Atkinson 1892). In his studies at the time, Atkinson (1892) was able to observe the progression of the severity level of cotton wilt in the presence of *Fusarium oxysporum* in combination with root-knot nematodes, *Meloidogyne* spp. This interrelation between fungus and nematode causing a rise in disease intensity in plants was mentioned by Atkinson (1892). Numerous studies have since discovered symbiotic relationships between those two organisms, implicating sessile endoparasitic root-knot fungi and cyst nematodes. As a result, the likelihood of fungi-causing diseases such as *Fusarium* or *Verticillium*, also known as wilt fungi, is increasing, and the presence of cyst nematodes will make the disease even worse.

Meloidogyne spp. or root-knot nematodes are obligatory parasites that often affect crops in tropical and subtropical regions species include *M. incognita*, *M. javanica*, and *M. arenaria* (Swarup and Sosa-Moss 1990; Elling 2013). *Meloidogyne hapla* has brought more potato root gall multiplication causing the symptoms visible at level above ground which showed the plants withering even in moist soil, yellowed leaves, and stunting growth (Wesemael et al. 2014). Ginger root-knot infection symptoms include gall that appears as brown to black lesions with collected fluid on the rhizome (Okorocha et al. 2014). Cyst nematodes feed on plant roots, causing a gradual decline in the capacity of plants to absorb nutrients and water. Eventually, the symptoms appear aboveground as nutritional deficiencies, poor growth, stunting, yellowing, and wilting. The cereal cyst nematode (*Heterodera avenae*) has been found in India (Sharma and Swarup 1984) has lemon-shaped females retains their eggs inside the body until it dead. However, the body wall namely brown cyst, eventually hardens and protecting the alive eggs and develop into parasites that attack the roots of plants (Kort 1972).

Lesion nematodes are a root invader, via plants' cortex, semi-endoparasites that harm roots and tubers in various crop plants, and causing various size of circle in shape of black patches (it enter and surround the tubers, tissue via lenticels), necrosis, soften the tubers, broken and dried the outer skin and eventually, the plant will die (Bridge and Starr 2007; Pinheiro et al. 2011). Attacks by *Pratylenchus* on broadleaf plants, particularly lettuce, slow the plant growth, the leaves become yellow, and the head become tiny (Pinheiro et al. 2010).

Melillo et al. (2011) found that when plants see pathogenic fungi and nematodes, they quickly make ROS and toxic molecules from secondary metabolism. This happens during the early stages of an infection. The host plant creates a toxic environment that stresses the animal through oxidative stress. Reactive oxygen species (ROS) in plants are created when a plant's mitochondria, chloroplast, endoplasmic reticulum, apoplast, peroxisomes, plasma membrane, and cellular walls are under stress from pathogenic fungus and nematode attack (Singla et al. 2019). Major macromolecules, including DNA, lipids, and proteins, may suffer oxidative damage as a result of excessive ROS generation in the cellular environment (Das and Roychoudhury 2014; Mittler 2017).

Pathogen invasion in host root tissues has been extensively studied, according to Goverse and Smant (2014), when the nematode becomes sedentary, the oxidative stress response is immediate from the moment the pathogen penetrated the host, explaining the molecular interaction between the nematode and host plant at the beginning of parasitism. The plant creates an unfavorable oxidative environment, which causes ROS buildup through host immunity (Torres et al. 2006). By triggering the antioxidant system, the host plant detects and fine-tunes its cellular redox state. Pathogen entrance into root tissues, on the other hand, is promoted by mechanical force or the release of enzymes that may break down the cell wall (Bellincampi et al. 2014). The host plant senses and adjusts its cellular redox status by activating the antioxidant pathway and allowing the expression of genes that code for ROS scavenging enzymes. This limits intracellular oxidative damage to the host while causing oxidative stress in the pathogen inside the host apoplast (Foyer and Noctor 2005; Delaunois et al. 2014).

Antioxidants are present everywhere because ROS that surpass the threshold limit in plant cells pose a major hazard by damaging vital cellular entities and producing dysfunctions (Jajic et al. 2015). Antioxidants function as electron donors, converting ROS into molecules that are less dangerous; the oxidized products that are produced as a result are not very reactive or damaging. Plants have a vast array of defensive mechanisms that enable them to identify certain challenges or stressors and respond by creating network signaling that causes an alternation of biochemical and physiological processes to regulate the stress (Lamers et al. 2020). Although only a small quantity of ROS is used as a signal molecule, it plays a vital function in providing a defensive response to thwart pathogen invasion, which stresses the plants. Nevertheless, it must be maintained by their synthesis and scavenging processes. This might prevent the devastating oxidative effects that led to the plant's demise (Waszczak et al. 2015). When pathogens interact with the plants, the plants respond by causing biochemical changes linked to stress signaling and activating their defense pathways, which include different ROS scavenging systems to protect plants from these harms through enzymatic antioxidants and non-enzymatic antioxidants (Noctor 2005). Both types of antioxidants cooperate with one another and are generated by the infected host plant. These systems help plants maintain a non-toxic level of ROS while stabilizing its intensity at either intracellular or extracellular redox homeostasis. As a result, ROS concentration may be controlled and plant damage can be avoided (Vellosillo et al. 2010). This chapter is aimed at discussing biochemical responses in host plants assaulted by opportunistic fungi and nematodes. The role of antioxidant in scavenging ROS in plant and defense mechanisms against pathogen was also discussed

5.2 Biochemical Changes in Plants Induced by Opportunistic Fungi and Nematodes

5.2.1 Sources and Production of ROS

During the plant developmental process, several wide ranges of environmental stresses including pathogen invasion undermine plant overall productivity by altering processes such as the biochemical, physiological, and metabolic pathways. Among the result of these stresses is the accumulation of ROS which is considered an important signaling response against pest and pathogen invasion. The ROS serves as a signal to the plant, directing its metabolic pathways toward preventing damage and defending its machinery (Sahu et al. 2022). The history of ROS discovery can be traced back to 1983 when Doke discovered that the incompatible interaction of oomycete *Phytophthora infestans* with potatoes produced O_2. This discovery sparked a significant wave of research on the generation of ROS in suspension cells or whole plants when they are exposed to live pathogens or other elicitors. It is now clear that ROS plays a crucial role in the plant's reaction to biotic and abiotic stress (Lehmann et al. 2015). The word ROS is used to describe the products of the sequential reduction of molecular oxygen: superoxide radical, hydrogen peroxide and hydroxyl radical, species mostly occur during plant pathogen interactions. Generally, under normal conditions, plant routinely produces ROS but at a low level in chloroplasts and mitochondria, and also by cytoplasmic, membrane-bound or exocellular enzymes involved in redox reactions (Wojtaszek 1997). However, under stress conditions, ROS is produced from secondary sites such as the cell membrane, cell wall, endoplasmic reticulum, and apoplast (Sahu et al. 2022). Various enzyme systems including reduced form of nicotinamide adenine dinucleotide phosphate (NADPH) oxidase, oxalate oxidases, peroxidases, lipoxygenases, and amine oxidases has been associated to the ROS production upon pathogen detection in the plant (Shetty et al. 2008). As mentioned earlier under such conditions dangerous by-products including superoxide radicals, hydrogen peroxide, hydroxyl radicals, and singlet oxygen are generated which are highly detrimental to plant health and productivity. These harmful products produced affect the proteins and lipids, consequently, resulting in cellular damage and death. The increased accumulation of ROS in plants affects cellular homeostasis. The ROS generated in the plant cell can be associated with electron leakage that occurred during photosynthesis and respiration (Nadarajah 2020).

In plants, chloroplasts are considered one of the main ROS generation sites. Photosynthetic electron transfer chains generate superoxide which consequently results in the electron transfer chains being overenergization. Superoxide (O_2) is produced due to electron leakage from Fe-S centers of photosystem I or decreased ferredoxin to O_2 (Mehler reaction), which is subsequently promptly metabolized by superoxide dismutase (SOD) to hydroperoxide. Although too much ROS generation can be hazardous, in this instance, the ability of oxygen to receive electrons avoids too much reduction of the electron transport chains, reducing the likelihood of Singlet oxygen production. In photosystem II, especially in conditions of intense light, energy is transferred from excited triplet state chlorophyll to triplet oxygen, resulting in the production of singlet oxygen. Tocopherols and carotenoids both have the ability to directly quench singlet oxygen or inhibit the generation of singlet oxygen via quenching chlorophyll in its excited triplet state (Gechev et al. 2006).

One of the main sources of cellular ROS is the mitochondrial respiratory chain. During typical oxygen metabolism, ATP production generates ROS. Hence, ROS are typically thought of as by-products of energy perfusion to cell processes. The production of ROS is the main function of NADPH oxidase (NOX) enzymes. NOX2 is a member of the NOX family and, when active, may encourage ROS generation by activating ryanodine receptors, which may result in Ca^{2+} sparks. The production of ROS in all types of cells is caused by the participation of NADPH and NADH in the respiratory chain and cellular metabolisms. Furthermore, Toll-like receptors (TLR) TLR1, TLR2, and TLR4 can promote the generation of ROS via recruiting mitochondria to macrophage phagosomes and translocating tumor necrosis factor receptor-associated factor 6 (*TRAF6*) to mitochondria to engage in evolutionarily conserved signaling intermediate in Toll pathways (ECSIT). Though, mitochondria are one of the major sites of ROS production, however, the detailed mechanisms of ROS generation still remain unclear (Yang and Lian 2020).

5.2.2 Effect of Pathogen Invasion and Pathogen-Induced Reactive Oxygen Species on Plant Agro-Physiological and Biochemical Traits

From the physiological standpoint, the word stress can be considered as a response to different situations. Generally, stress can be considered as a disruption or change in the normal physiological processes induced by a wide range of factors that tend to alter plant stability (Oladosu et al. 2019). Generally, the emergence of stress significantly affects plant functionality and induces biochemical, physiological, molecular, and morphological mechanisms to compensate for the induced stress. These bring about the accumulation of ROS including superoxide anion radicals, and hydrogen peroxide (H_2O_2) which consequently result in decreased photosynthesis activities, increase lipid peroxidation as well as increased programmed cell death and retarded growth in the plant cell (Rane et al. 2022). The accumulation of ROS in plants is considered an important regulator of plant stress responses. Several environmental problems encountered by the plant bring about oxidative stress as a result of enhanced production of ROS or disruption and degradation of ROS scavengers. Pathogens attacks (nematode and fungi) and other abiotic stress (drought, heat, heavy metal) are among the environmental factors that lead to ROS production in plants (Gechev et al. 2006) which consequently disrupts the biochemical, physiological, molecular, and morphological processes of the host or invaded species.

Morphologically, biotic stress generally can affect plant growth and development by disrupting cell enlargement and divisions process. Meristematic tissues in cells are essential for plant expansion via active cell division. However, any disruption in the cell water potential causes a reduction in the size of the cell and cell division processes, consequently affecting plant growth and development (Oyebamiji et al. 2022). For instance, the invasion of sweet corn by pathogens coupled with water stress caused a decrease in the ear weight, leaf area and plant height compared to the control treatments. Pathogen attack on Arabidopsis plants coupled with other stress results in a decrease in plant leaf number and weight (Ramegowda and Senthil-Kumar 2015). Furthermore, Nishio and Kusano (1975) reported that *Pyrenochaeta* sp. (fungal flora) alters the growth of upland rice seedlings by producing toxic substances. Pathogen-infected crops display various symptoms of nutrient deficiency which include retarded growth and development, leaves yellowing, wilting, and in severe cases death of the plant. Severe pathogen infection on mature crops leads to unexpected wilting and early death. Furthermore, nematode invasion of the root causes its cell expansion, swelling, or the development of galls on the roots of the inoculated plants (Ralmi et al. 2016). The size of galls may vary from slight thickenings to lumps 5–10 cm across. Galls are also observed in the leaves and stem region though this is quite uncommon in crops grown outdoors. The galls produced by nematode destroy the root's vascular tissues and also disrupts the transportation of nutrients and water within the crop. Furthermore, the root system of the infected crop is vulnerable to attack by other pathogens such as fungi and bacteria (Ralmi et al. 2016). When plants are subjected to stress, ROS are produced by the root tissues which may hinder normal plant processes. For instance, exogenous use of H_2O_2

prevents via closing aquaporin pores or accumulation of plasma membrane intrinsic proteins (PIPs) in internal structures called vesicles and vacuoles (Rane et al. 2022). Kesba and El-Beltagi (2012) reported that nematode infection caused a drastic rise in of both lipid peroxidation (MDA) and H_2O_2 in roots of all rootstocks (Kesba and El-Beltagi 2012). The invasion of lettuce farmland by pathogens has become prevalent due to increasing lettuce planting grown under plastic tunnels. This invasion causes poor growth of lettuce due to the gall formation in the roots which inhibits the movement of sufficient water and nutrients for normal lettuce growth.

These disruptions in plant development affect different biochemical and physiological processes such as ion uptake, respiration, photosynthesis, growth promoters, carbohydrate, source–sink relationship, and nutrient metabolism (Oladosu et al. 2019) and consequently the overall plant yield. Global agriculture experiences significant economic losses due to the invasion of pathogens including viruses, bacteria, nematodes, and fungi. Among the biotic stress, the nematode is believed to be the foremost candidate responsible for massive yield reduction on agricultural farmland. (Ralmi et al. 2016) reported that *Meloidogyne* species destroy around 5% of world crop production annually. In USA from 1950 to 2000 the yield loss of cotton lint increased from 1% to over 4% which can be attributed to an invasion by these pathogens (Starr et al. 2007). Elsewhere, approximately 8.8% yield loss was recorded in advanced countries while 14.6% was documented in tropical and subtropical regions which were mainly associated with nematodes invasion (Pulavarty et al. 2021). Pathogens attack rice at the tillering stage causing about 20–40% yield loss and while infection at the vegetative stage results in a more severe yield reduction of approximately 50% (Miah et al. 2013). In 1989, it was estimated that plant parasitic nematode species caused massive potato production losses of about 16.7% in South Africa alone costing $seven million annually while in Uganda its accounts for about 50% of banana losses (Pandey et al. 2017). Pathogen attacks also cause a reduction in produces quality and marketability. For example, severely infected lettuce plants by nematode produce small and loose consequently making them unmarketable (Ralmi et al. 2016). In the Philippines, fungal exposures such as mycotoxins contribute significantly to a reduction in the quantity and quality of farm produce. For example, exposure of rice to fungal contamination brings about stunted seedling and elongation known as bakanae diseases (Balendres et al. 2019).

The co-occurrence of more than one pathogen is considered more detrimental to plants compared to single pathogen invasion. In 1892, a study by Atkinson that focused on the influence of root-knot nematodes on the expression of *Fusarium* wilt of cotton advanced the study of nematode-inoculated crop susceptibility to fungal invasion. Since then, research has focused on the interactions between nematodes and several fungi on various host crops (Ragozzino and d'Errico 2012). In the USA, the inoculation of tomatoes with *Fusarium* and *M. incognita* causes complete wilting of the crops while only 60% wilt was observed with only nematode-infected plants. This could imply that the combined nematode-fungus effects are more disastrous to plants than individual pathogen effects. Furthermore, the study also indicated that physiological effects induced by nematodes on tomatoes depend on the types of nematodes (Ragozzino and d'Errico 2012). Nematodes and fungi combination caused plant disease as a result of continuous monocropping. Furthermore, the interaction between nematodes and fungi was suspected to exist because the presence of nematode alone may not be enough to cause a significant decrease in plant growth as a result of continuous cropping (Kyndt et al. 2014). With continuous cropping, it was observed that nematode invasion increases the pathogenicity of other microorganisms (Kyndt et al. 2014) which may possibly cause more damage to agricultural commodities. Moreover, the aforementioned nematode-fungus interaction promotes the development of nematodes due to damaged roots' release of CO_2 which is widely recognized to be a potent nematode attractant. For instance, the presence of *Verticillium* sp. in tomato and eggplant encourage *Pratylenchus* spp. to reproduce more frequently (Ragozzino and d'Errico 2012).

When a plant is subjected to stress, certain alteration happens at physiological levels including a reduction in photosynthesis, stomatal conductance, and transpiration rate (Oyebamiji et al. 2022). Pathogen attacks often result in the death of the plant, chlorotic and necrotic lesions development, and consequently a reduction in the photosynthesis rate. According to previous studies, using chlorophyll fluorescence imaging, alterations in photosynthesis caused by pathogen invasion are localized. For instance, in pathogen-infected Arabidopsis and tomato leaves, a ring of improved photosynthesis rate was detected surrounding the area while a reduction in photosynthesis rate was observed at the infected region (Selvaraj and Fofana 2012). Furthermore, the characterization of photosynthetic components infected grapes leaves indicates that foliar signs were attributed to stomatal closure and change of photosynthetic components. A reduction in the assimilation of CO_2 and transpiration, a significant rise in the concentration of intercellular CO_2, a significant reduction in the maximum fluorescence yield and the effective Photosystem II quantum yields, and a decrease of the total chlorophyll but a stable carotenoid content were documented (Selvaraj and Fofana 2012).

Several studies have reported the detrimental impact of ROS production in plants. One of the primary sources of ROS production is the chloroplast. The excessive accumulation of ROS in this region significantly affects the photosynthesis processes and associated apparatus and is indirectly due to altered stomatal mechanisms. ROS production caused the degradation of proteins, chlorophyll, and lipid peroxidation in the leaves which results in reduced photosynthetic electron transport chain, membrane integrity, and cell death (Rane et al. 2022). For instance, an increase in ROS production was observed in the leaves of wheat plant inoculated with the pathogen bringing about a decrease in catalase production (Debona et al. 2012) possibly affecting the overall plant growth. Another study reported a significant reduction in the chlorophyll contents of the plant due to the nematode invasion on the roots. As the stress level increases, the chlorophyll content reduces. Furthermore, both the level of ROS production and MDA content was enhanced significantly in nematode-infected plants (Sharma and Sharma 2017). Since the leaves are known as photosynthetic sites as well as ROS accumulation, plant survival during these adverse conditions depends on its ability to maintain homeostasis. Based on published literature, ROS a role in leaf senescence through programmed cell death and degradation of the cell membrane and hence disruption of physiological processes and functions in plants. The leaf senescence triggered by ROS can potentially affect the photosynthesis process though this depends on the leaf age (Rane et al. 2022). Apart from chlorophyll degradation experienced by the plant due to stress, loss of pollen viability of sorghum is documented due to increased membrane degradation and rise in ROS accumulation in the pollen grains as well as reduced antioxidant enzymes in pollen and pistils (Rane et al. 2022). Chhabra et al. (2020) reported that pathogen invasion of aromatic and non-aromatic rice seedlings caused a decrease in the total chlorophyll, chlorophyll a, chlorophyll b, and carotenoid content. Finally, pathogen infection results in the enhanced intensity of respiration processes in the infection regions. For instance, infected maize leaves with *Colletotrichum graminicola* showed enhanced expression of respiration-related genes at infection sites (Chhabra et al. 2020). In order to mitigate oxidative damage caused due to ROS accumulation, plants have developed ROS scavenging systems which include non-enzymatic and enzymatic mechanisms (Debona et al. 2012).

5.3 Techniques Used in ROS Detection

The detection of ROS accumulation in a plant cell can be achieved using several techniques. These techniques are based on histochemical staining, fluorescence, luminescence, electron paramagnetic resonance (EPR) spectroscopy or ROS sensors (Table 5.1). To identify ROS within the cells, fluorescent probes, $CeCl_3$, and ROS sensors are also employed (Lehmann et al. 2015). Singlet oxygen can be detected using dansyl-2,2,5,5-tetramethyl-2,5-dihydro1H-pyrrole (DanePy) (a fluorescence probe) and an electron spin resonance (ESR) probe. Moreover, singlet oxygen can also be determined via its extremely weak luminescence at 1270 nm (Schweitzer and Schmidt 2003) or by sensitive laser spectroscopy. Hydrogen peroxide can be identified by histochemical staining, using a 3,3′-diaminobenzidine (DAB) solution (Zulfugarov et al. 2011). However, due to their relatively short

Table 5.1 Summary of different techniques used to detect ROS in plants

Products	Reactive oxygen species	Detection
ABTS	H_2O_2	Spectrophotometrical
Luminol	ROS	Chemiluminescence
Amplex red	H_2O_2	Spectrophotometrical
3–30 diaminobenzidine (DAB)	H_2O_2	Histochemical
Cerium (III) chloride (CeCl3)	H_2O_2	Cytochemical
Dansyl-based fluorescence sensors	O_2	Fluorescence
Ferrous ion oxidation (FOX)	H_2O_2	Spectrophotometrical
3-Methylbenzothiazoline hydrazine	H_2O_2	Spectrophotometrical
Dihydroethidium (DHE)	O_2	Fluorescence
2-deoxyribose (DOR)	Hydroxyl radical	Spectrophotometrical
Nitroblue tetrazolium (NBT)	O_2	Histochemical
Dihydrorhodamine123	ROS	Fluorescence

life-times coupled with the capacity of plant cells to scavenge ROS, ROS detection is challenging. Moreover, the handling of tissue that results in tissue injury or degradation may bring about the production of ROS artifacts. Another challenge faced is the dearth of probes with high selectivity. This means no probe is guaranteed to function for a certain tissue under specific circumstances. Due to various drawbacks encountered, scientists are therefore encouraged to make the necessary controls and preliminary tests to establish and determine the reliability and validity of the probes they are employing (Lehmann et al. 2015; Table 5.1).

5.4 ROS Detoxification

5.4.1 Enzymatic Antioxidants

The enzymes involved in the antioxidant mechanisms and are distributed throughout many types of subcellular components including peroxisomes, mitochondria, cytosol, chloroplast, and endoplasmic reticulum. The various enzymes involved in the antioxidant mechanisms are comprises of superoxide dismutase (SOD), ascorbate proxidase (APX), monodehydroascorbate reductase (MDHAR), guaiacol peroxidase (GPX), peroxidase (POD), catalases (CATs), dehydroascorbate reductase (DHAR), glutathione reductase (GR), and glutathione-S-transferase (GST) (Rasool et al. 2013; Das and Roychoudhury 2014).

5.4.1.1 Superoxide Dismutase (SOD)

SOD is a member of the metalloenzyme family and is ubiquitous in all aerobic organisms. As the first line of defense to combat ROS under stress, SOD impels the elimination of O_2 by dismutating it into O_2 and H_2O_2 and prevents the formation of OH via the Haber-Weiss reaction. In plants, there are three SOD forms, which have been classified based on the metal ion that binds to them and can be found in specific cell components, including Fe-SOD and Cu/Zn-SOD in the cytosol (Kliebenstein et al. 1998; Dvořák et al. 2020), chloroplast stroma (Dvořák et al. 2020), chloroplast thylakoids (Kliebenstein et al. 1998; Myouga et al. 2008) peroxisome; and mitochondria. Meanwhile, Mn-SOD is found in mitochondria (Morgan et al. 2008) and peroxisomes only (Mittler 2002; Pilon et al. 2011). Overproduction of SOD generation has been associated with the tolerance mechanism of the plant under oxidative stress (Mittler 2017). There is research being done to increase tomato superoxide dismutase (SOD) production, which provides resistance to *Alternaria solani* (Gulzar et al. 2021). They discovered that the infected tomato plants had drastically elevated SOD activity. These results are also consistent with prior research that discovered an increase in SOD activity in cucumber, turfgrass, and peas when the environment was stressed (Zhu et al. 2004; He et al. 2010; Tripathi et al. 2015). Other than that, SOD helps roots development, germination, subcellular organelle development such as chloroplast; and in the flowering process (Rizhsky et al. 2003; Myouga et al. 2008; Dvořák et al. 2020).

5.4.1.2 Catalases (CAT)

The kinetic characteristics that enable detoxification of overabundance amount of H_2O_2 (produced by SOD) (Tuzet et al. 2019), CAT is an enzyme that have high affinity toward H_2O_2 and consists of tetrametic heme with iron protein, which is accountable for catalyzing the dismutated H_2O_2 into H_2O and O_2 with good energy efficiency at the rate of 6×10^6 per minute. CAT is positioned in all important sites where H_2O_2 had been produced in the subcellular compartments such as peroxisomes (Du et al. 2008), mitochondria, cytosol and chloroplast in higher plants. Shortage of CAT in plants could lead to chlorosis, sterility and sensitivity even to normal photorespiratory surrounding (Sharma and Ahmad 2014). CAT is one of the key players that are required under the adverse environment, such as photooxidative stress (Zhang et al. 2020a, b), in plants (Bhaduri and Fulekar 2012). CAT is present in three different types of isoforms, with CAT1 expressed in pollen and seeds, CAT2 in photosynthetic tissues present in peroxisomes and cytosol, and CAT3 in leaves and vascular tissues localized within mitochondria. Recent studies have shown that CAT activity goes up in barley leaves at certain stages of the plant's growth, making them resistant to the pathogen *Bipolaris sorokiniana.* Furthermore, it was also discovered that APX, GR, and polyphenols rose, leading the H_2O_2 level to fall at that specific development stage (Kaur et al. 2021). Additional research on *Sclerotium rolfsii* resistance in chickpea, particularly in the leaves and stems, is offered by boosting CAT activity as the first line of defense (Sahni and Prasad 2020). They discovered that even with POD highly activated during infection, ROS would still accumulate and ultimately cause damage to plant tissue if CAT levels of expression were low at that time. CATs also help in developing leaf and its senescence (Zhang et al. 2020a, b), improving growth of shoot, ovule, pollen, and the seeds of a plant (Sharma and Ahmad 2014; Su et al. 2018; Palma et al. 2020).

5.4.1.3 Ascorbate Proxidase (APX)

H_2O_2 in peroxisomes had been scavenged predominantly by CAT enzyme, meanwhile, APX performs the same function in cytosol and chloroplast (with higher specificity for H_2O_2) to protect cells from scarred by H_2O_2 and OH. In the article by Dvořák et al. (2020), other than APX, there are other enzymes implicated in an ascorbate-glutathione cycle, including MDHAR, DHAR, and GR, in balancing H_2O_2 levels in plant cells.

APX utilizes AsA as a reducing agent to reduce H_2O_2 to H_2O and MDHA via electron transfer. APX also presents throughout the subcellular compartments, same as CAT, but it has greater affinity toward H_2O_2, thus, APX has high ability to scavenge H_2O_2 efficiently under stresses (Granlund et al. 2009). Five isoforms of APX with different amino acids located at different locations such as cytosolic, mitochondrial, peroxisomal, and chloroplast (stroma and thylakoid) (Sharma and Dubey 2004; Maruta et al. 2012, 2016). Jogaiah et al. (2020) found that powdery mildew-resistant cucumbers had higher levels of POX activity. This meant that the cucumbers were less likely to get sick from powdery mildew. According to a recent study by Siddaiah et al. (2017), who discovered POX was expressed in accordance with phytohormone production in *Trichoderma hamatum,* peroxidase was stated to stimulate the salicylic acid pathway to boost the plants defensive mechanisms against pathogenic attack.

5.4.1.4 Monodehydroascorbate Reductase (MDHAR) and Dehydroascorbate Reductase (DHAR)

Sahu et al. (2022) stated MDHAR enzymes are generated in subcellular components such as chloroplasts, cytosols, peroxisomes, mitochondria, and glyoxysomes. Another enzyme called as DHAR, functions similar to MDHAR. It reduces the DHA into AsA, and uses GSH as reducing agent resulting in the production of GSSG. DHAR always be found in dry seeds, roots, and shoots (etiolated and green).

5.4.1.5 Glutathione Reductase (GR)

GR is a part of the flavoenzyme class and it assists plants under stress by eliminating ROS by producing GH from GSSG (produce by oxidation of MDHAs and DHAs) with the presence of NADPH as a reducing agent (Gill et al. 2013). GR is an essential and last enzyme in the cycles of ASC-GSH, where it has been found prevalently in chloroplasts, some in cytosol, mitochondria, nuclei, endoplasmic reticulum, vacuoles, and peroxisomes (Dvořák et al. 2021). GR escalates the production of the disulfide bond in glutathione disulfide, sustaining the big cellular ratio of GSH/GSSG (Dvořák et al. 2021). GSH plays a big role as a reducing agent in avoiding oxidized activity in thiol groups and reacting with the disastrous ROS. GR was reported to be found in a process implicating the development of embryos, germination of seeds, and development of root systems (Sumugat et al. 2010; Yu et al. 2013; Marty et al. 2019). Gulzar et al. (2021) reported that an increase in GR activity increases in resistance against tomato plants against *Alternaria solani*.

5.4.1.6 Guaiacol Peroxidase (GPX)

GPX is an enzyme consisting of hemes and comprised of 40–50 kDa monomers which could banish the overexpression of H_2O_2 under normal and stressed conditions, with the presence of GSH (Das and Roychoudhury 2014). GPX actively involves as electron donors/reducing agents in the intracellular area, including cytosol and vacuole; in the cell wall, since it is one of the vital part in lignin biosynthesis; and in extracellular (Asada 1992; Sharma and Dubey 2004; Caverzan et al. 2016). GPX needs molecules like guaiacol or pyragallol as electron donor as they have aromatic molecules properties (Asada 1992). As GPX enzyme implicates itself in peroxidative and hydroxylic reactions, logically it can be involved in biosynthetic processes for lignin and ethylene, auxin breakdown, cell wall expansion, IAA degradation, injury repair, and resistance to pathogen (Passardi et al. 2004).

5.4.2 Non-Enzymatic Antioxidants

All ROS could not showed the scavenge by enzymatic antioxidants because they are greatly reactive including O_2 and OH, thus, non-enzymatic antioxidants highly required to scavenge those type of ROS which those molecules usually smaller in size of the organic molecules with even lower molecule weight. Plants are provided with non-enzymatic defense systems and low molecular weight antioxidants such as reduced glutathione (GSH), ascorbic acid (AsA), flavonoids, alpha-tocopherol, alkaloids, thioredoxin, major cellular redox buffers, phenolic compounds, carotenoids, and osmolyte proline. Involvement of non-enzymatic defense systems in regular process in cellular process includes cell division, extension, aging and cell death for sustaining redox homeostasis of the cell (Das and Roychoudhury 2014).

5.4.2.1 Ascorbic Acid (AsA)

AsA is the most prevalent and being studied broadly among all antioxidants existed. Considering it has pivotal roles in plants development (Do et al. 2016), it had been known to be electrons donor not only to enzymatic reactions, but also non-enzymatic reactions. During stress in the adverse environment, redox buffer AsA could breakdown the ROS and eventually protect the cells from free oxygen radicals that can cause damages to it (Dell'Aglio and Mahmdi 2020). The majority of plant cell types, apoplasts, and organelles contain large amounts of AA, especially in tissues that are photosynthetic. Being the first lines in defense mechanism of immune system from disastrous ROS where it can provide membrane's safeguard by rapidly eliminate deleterious ROS and recycling other non-enzymatic oxidants including vitamin E and carotenoids (Shao et al. 2008). There are two enzymes implicated in balancing the level of AsA in plants under stress or not which are MDHAR and DHAR (Chen et al. 2003). MDHA, which is fundamental forms of oxidized AA, can bind to one and another producing AA or DHA (Do et al. 2016). NADPH acted as MDHAR's (reduced forms of AsA) electron donor producing AsA from MDHA (this is in reaction involving enzymatic components) (Potters et al. 2002). DHAR, with oxidation of GSH as it will donor its hydrogen, will reduce DHA into AsA.

5.4.2.2 Carotenoids

Carotenoids are a molecule that attracted to lipids, or being called "lipophilic" antioxidants to avoid the cells from oxidative damage by ROS. In plants, carotenoids will impede the reactions implicated with ROS, especially peroxyl radicals, to guard the membrane and lipid of the plant cells (Moussa et al. 2019).

5.4.2.3 Vitamin E

Other than carotenoids, α-tocopherol or commonly called "Vitamin E" is also lipophilic antioxidants which it has capability to protect lipids consisting membranes and PSII structure from oxidative adverse effect due to ROS presence, by interfering oxygen molecules and quenching the surfeit energy (Ivanov and Khorobrykh 2003). Generally, α-tocopherol can impede efficiently the ROS to avoid the impact of its toxicity, by confined ROS and put to a stop of the lipid autooxidation chain (Moussa et al. 2019).

5.4.2.4 Flavonoid

Flavonoid also a phenolic antioxidants type of metabolite that been produced in plant and other organisms with photosynthetic properties. There are four types of flavonoid including flavonols, flavones, isoflavones, and anthocyanins. To prevent injury due to oxidizing activity involving H_2O_2, OH, and O_2, flavonoids is being considered as second lines defense in scavenging ROS in injured photosynthetic tissue cells because of overabundance agitated energy. As singlet oxygen quencher, flavonoids will alleviate the damage membrane in chloroplast (Agati et al. 2012).

5.4.2.5 Tannins Lignin and Proline

Other than flavonoid, tannins and lignin also are antioxidants phenolic with scavenging ability which it capable to restrain peroxyl radicals in lipid peroxidation (Sharma et al. 2012). Another antioxidants, namely, proline, an osmolyte who have reported numerously could diminish the disastrous impact of ROS under stress (Das and Roychoudhury 2014).

5.5 ROS and Plant Pathogen Defense

Reactive oxygen species (ROS) are considered necessary for plant growth and development as well as for plant adaptation to a wide range of stress which can be achieved via interactions with phytohormone signaling. Generally, excessive production of ROS is witnessed when plants are subjected to a wide range of environmental stresses including biotic and abiotic. Though the accumulation of ROS under such conditions is always considered detrimental to the plant, however, this simple perspective fails to consider the crucial role of ROS in plant growth and development (Considine and Foyer 2021). Evidence has shown that ROS can also act as a signaling molecule in maintaining an array of functions in the plant such as acting as a secondary messenger in plants by maintaining a diverse role in plant growth and development (Choudhury et al. 2013). From the seedling stage to the senescence, ROS are produced and expelled from the plant tissue dynamically allowing the plant to maintain its development in order to adapt to different stress conditions (Huang et al. 2019).

Seed quality has a significant impact on crop overall output and productivity. ROS are strongly linked with different aspects of seed physiology by playing a key role to prevent pathogen invasion as well as regulating dormancy. The dual function of ROS in seed establishment and in cell death has been highlighted in previous literature (Considine et al. 2015). It has also been reported that a high accumulation of oxygen favor seed dormancy and that loss of viability and breaking dormancy can be attributed to protein oxidation located in the cytosol rather than the oxidation of membrane lipids (Considine et al. 2015). Growing evidence has shown that the development of plant vegetative apex is influenced by ROS homeostasis. For instance, the accumulation of H_2O_2 in the elongation zone of *A. thaliana* triggers cell differentiation while O_2 production in the meristematic zone of the root tip helps in cell division (Huang et al. 2019). Furthermore, the accumulation of O_2 by NADPH oxidase in the root hair tip induces Ca2þ peaks which are required for the growth of the root hair (Gechev et al. 2006). One of the important regulators preserving this balance is the transcription factor UPB1 (UPBEAT1). Previous research revealed that ROS homeostasis and root growth are both influenced by the way H_2O_2 itself affects *UPB1* expression (Huang et al. 2019). Moreover, the quiescent center (QC) and distal stem cell (DSC) are necessary for the maintenance of the root apical meristem (RAM) size. The root stem cell niche (SCN) identity is impacted by *APP1* (*A. thaliana* P-loop NTPase1) via its regulation of local ROS homeostasis. Reduction in ROS levels brings about disruption in AP1, increases the cell division rate in the QC, and induces root DSC differentiation, indicating that ROS levels are strongly related to RAM size in *Arabidopsis* (Huang et al. 2019). ROS accumulation in cell organelles such as chloroplast and mitochondria caused alteration in the nuclear transcriptome. However, signal transduction mechanisms are still not fully understood yet. Furthermore, gene expression is influenced by ROS via changing and modifying transcription factors (Choudhury et al. 2013). Apart from the role of ROS alone, the interactions between ROS and reactive nitrogen species (RNS) are crucial in maintaining plant growth and development, regulating plant physiological and biochemical responses to a wide range of environmental stress, from germination, root development, and gravitropism to stomatal closure (Considine et al. 2015).

ROS interactions with plant hormones can be another determinant of specificity. For instance, plant hormones such as auxin, abscisic acid (ABA), and jasmonic acid together coupled with ROS help regulate different processes such as stomatal closure, growth, and development as well as wounding responses. Similarly, salicylic acid and ethylene work together with ROS. During pathogen invasion, the production of ethylene and salicylic acid increases which consequently enhance ROS accumulation (Gechev et al. 2006). Recent research has shown that interactions of ethylene coupled with ABA and H_2O_2 maintain stomatal closure. Furthermore, ROS also interacts with classical hormones, a number of polypeptides, including systemin, and the recently identified polypeptide *AtPep1*. These hormones can help trigger the synthesis of H_2O_2 subsequently activating defense gene expression in *A. thaliana* (Gechev et al. 2006). Production of ROS under stress conditions helps to activate phytohormone signaling cascades, such as abscisic acid (ABA) and jasmonic acid (JA) that which helps to reduce cyclin-dependent protein kinases (CDKs) that regulate cyclins expression or control the posttranslational modification of CDKs which results to cell cycle arrest or exit from the cell cycle. For instance, the high osmolarity glycerol (HOG) MAP kinase (*MAPK*) cascade is triggered under stress conditions, which results in cell cycle arrest by suppressing cyclin gene expression (Considine and Foyer 2021). Moreover, ROS accumulation is involved in controlling cell differentiation and cell cycle progression (Considine and Foyer 2021).

Plant-invaded by pathogens can also become resistant to upcoming pathogen invasion if they successfully survive by earlier attack by phytopathogenic viruses, bacteria, or fungi. Even though the plant might not have a gene conferring cultivar-specific resistance, it seems that the initial disease to infect the plant immunizes it against subsequent infections by similar pathogens. The readiness of plants to suppress further pathogen invasion is spread in the entire plant system. This response is known as systemic acquired resistance (SAR). The emergence of SAR is often attributed to diverse cellular defense responses like phytoalexins and PR proteins production, active oxygen species accumulation and quick change in the plant cell wall, and increased function of several defense-related enzymes (Thakur and Sohal 2013).

5.6 Conclusions and Future Prospects

At present the most challenging tasks faced by contemporary researchers is the development of plant resistance to multiple biotic stresses while maintaining high yield. Biotic stresses are recognized as significant factors that impede plant productivity on a global scale. Consequently, scientists worldwide have directed increased attention toward elucidating the mechanisms of abiotic stresses through the use of physiological, biochemical, molecular, and cellular approaches, in order to identify potential solutions that can mitigate adverse effects and enhance crop productivity. During biotic stress, many responses and reactions are triggered in the plant which includes reduced growth as well as ROS accumulation. A substantial amount of research has revealed that excessive production of ROS which is triggered by abiotic stresses poses a constant threat to plant survival due to oxidative stress it causes within the plant cellular components. ROS react with organic molecules consequently causing inhibition of enzymes, membrane lipid peroxidation, RNA and DNA damage, and protein oxidation. In order for plants to safeguard themselves and avert oxidative degradation, they have developed an effective mechanism with enzymatic and non-enzymatic antioxidants. Several literatures have established the positive impacts of enzymatic and non-enzymatic in combating oxidative damage to the cell. Recent studies have shown that ROS can also act as a signaling molecule in maintaining an array of functions in the plant such as acting as a secondary messenger in plants by maintaining a diverse role in plant growth and development.

References

Agati G, Azzarello E, Pollastri S, Tattini M (2012) Flavonoids as antioxidants in plants: location and functional significance. Plant Sci 196:67–76

Asada K (1992) Ascorbate peroxidase-a hydrogen peroxide-scavenging enzyme in plants. Physiol Plant 85:235–241

Ashraf M, Ahmad MSA, Öztürk M, Aksoy A (2012) Crop improvement through different means: challenges and prospects. In: Ashraf M, Öztürk M, Ahmad M, Aksoy A (eds) Crop production for agricultural improvement. Springer, Dordrecht, pp 1–15

Atkinson GF (1892) Some diseases of cotton. Bulletin 41:19–29

Balendres MAO, Karlovsky P, Cumagun CJR (2019) Mycotoxigenic fungi and mycotoxins in agricultural crop commodities in The Philippines: a review. Food Secur 8:249

Bellincampi D, Cervone F, Lionetti V (2014) Plant cell wall dynamics and wall-related susceptibility in plant-pathogen interactions. Front Plant Sci 5:228

Bhaduri AM, Fulekar MH (2012) Antioxidant enzyme responses of plants to heavy metal stress. Rev Environ Sci Biotechnol 11:55–69

Bongers T, Bongers M (1998) Functional diversity of nematodes. Appl Soil Ecol 10:239–251

Bridge J, Starr JL (2007) Plant nematodes of agricultural importance: a colour handbook, 1st edn. CRC Press, Boca Raton

Burros L (1996) In: Brady NS, Weil RR (eds) The nature and properties of soils, 11th edn. Prentice Hall Inc., New York, p 740

Caverzan A, Casassola A, Brammer SP (2016) Reactive oxygen species and antioxidant enzymes involved in plant tolerance to stress. In: Shanker AK, Shanker C (eds) Abiotic and biotic stress in plants-recent advances and future perspectives. Intech Open, London

Chen Z, Young TE, Ling J, Chang SC, Gallie DR (2003) Increasing vitamin C content of plants through enhanced ascorbate recycling. Proc Natl Acad Sci USA 100:3525–3530

Chhabra R, Kaur S, Vij L, Gaur K (2020) Exploring physiological and biochemical factors governing plant pathogen interaction: a review. Int J Curr Microbiol App Sci 9:1650–1666

Choudhury S, Panda P, Sahoo L, Panda SK (2013) Reactive oxygen species signaling in plants under abiotic stress. Plant Signal Behav 8:e23681

Considine MJ, Foyer CH (2021) Stress effects on the reactive oxygen species-dependent regulation of plant growth and development. J Exp Bot 72:5795–5806

Considine MJ, María Sandalio L, Foyer CH (2015) Unravelling how plants benefit from ROS and NO reactions, while resisting oxidative stress. Ann Bot 116:469–473

Das K, Roychoudhury A (2014) Reactive oxygen species (ROS) and response of antioxidants as ROS-scavengers during environmental stress in plants. Front Environ Sci 2:53

Davis EL, Hussey RS, Mitchum MG, Baum TJ (2008) Parasitism proteins in nematode plant interactions. Curr Opin Plant Biol 11:360–366

Debona D, Rodrigues FÁ, Rios JA, Nascimento KJT (2012) Biochemical changes in the leaves of wheat plants infected by *Pyricularia oryzae*. Phytopathology 102:1121–1129

Delaunois B, Jeandet P, Clément C, Baillieul F, Dorey S, Cordelier S (2014) Uncovering plant-pathogen crosstalk through apoplastic proteomic studies. Front Plant Sci 5:249

Dell'Aglio E, Mahmdi A (2020) What are the roles for dehydroascorbate reductases and glutathione in sustaining ascorbate accumulation. Plant Physiol 183:11–12

Do H, Kim IS, Jeon BW, Lee CW, Park AK, Wi AR, Shin SC, Park H, Kim YS, Yoon HS (2016) Structural understanding of the recycling of oxidized ascorbate by dehydroascorbate reductase (OsDHAR) from *Oryza sativa* L. *japonica*. Sci Rep 6:19498

Du YY, Wang PC, Chen J, Song CP (2008) Comprehensive functional analysis of the catalase gene family in *Arabidopsis thaliana*. J Integr Plant Biol 50:1318–1326

Dvořák P, Krasylenko Y, Ovečka M, Basheer J, Zapletalová V, Šamaj J, Takáč T (2020) In vivo light-sheet microscopy resolves localisation patterns of FSD1, a superoxide dismutase with function in root development and osmoprotection. Plant Cell Environ 44:68–87

Dvořák P, Krasylenko Y, Zeiner A, Šamaj J, Takáč T (2021) Signaling toward reactive oxygen species-scavenging enzymes in plants. Front Plant Sci 11:618835

Elling AA (2013) Major emerging problems with minor *Meloidogyne* species. Phytopathology 103:1092–1102

Foyer CH, Noctor G (2005) Redox homeostasis and antioxidant signaling: a metabolic interface between stress perception and physiological responses. Plant Cell 17:1866–1875

Gechev TS, Van Breusegem F, Stone JM, Denev I, Laloi C (2006) Reactive oxygen species as signals that modulate plant stress responses and programmed cell death. BioEssays 28:1091–1101

Gill SS, Anjum NA, Hasanuzzaman M, Gill R, Trivedi DK, Ahmad I, Pereira E, Tuteja N (2013) Glutathione and glutathione reductase: a boon in disguise for plant abiotic stress defense operations. Plant Physiol Biochem 70:204–212

Gonzalez-Chang M, Tiwari S, Sharma S, Wratten SD (2019) Habitat management for pest management: limitations and prospects. Ann Entomol Soc Am 112:302–317

Goverse A, Smant G (2014) The activation and suppression of plant innate immunity by parasitic nematodes. Annu Rev Phytopathol 52:243–265

Granlund I, Storm P, Schubert M, García-Cerdán JG, Funk C, Schröder WP (2009) The TL29 protein is lumen located, associated with PSII and not an ascorbate peroxidase. Plant Cell Physiol 50:1898–1910

Gulzar N, Ali S, Shah MA, Kamili AN (2021) Silicon supplementation improves early blight resistance in *Lycopersicon esculentum* Mill by modulating the expression of defense-related genes and antioxidant enzymes. 3 Biotech 11:1–13

He Y, Xiao H, Wang H, Chen Y, Yu M (2010) Effect of silicon on chilling-induced changes of solutes, antioxidants, and membrane stability in seashore *Paspalum turfgrass*. Acta Physiol Plant 32:487–494

Huang H, Ullah F, Zhou DX, Yi M, Zhao Y (2019) Mechanisms of ROS regulation of plant development and stress responses. Front Plant Sci 10:800

Ivanov B, Khorobrykh S (2003) Participation of photosynthetic electron transport in production and scavenging of reactive oxygen species. Antioxid Redox Signal 5:43–53

Jajic I, Sarna T, Strzalka K (2015) Senescence, stress, and reactive oxygen species. Plants 4(3):393–411

Jogaiah S, Satapute P, De Britto S, Konappa N, Udayashankar AC (2020) Exogenous priming of chitosan induces upregulation of phytohormones and resistance against cucumber powdery mildew disease is correlated with localized biosynthesis of defense enzymes. Int J Biol Macromol 162:1825–1838

Kaur S, Bhardwaj RD, Kaur J, Kaur S (2021) Induction of defense-related enzymes and pathogenesis-related proteins imparts resistance to barley genotypes against spot blotch disease. J Plant Growth Regul 41:682–696

Kesba HH, El-Beltagi HS (2012) Biochemical changes in grape rootstocks resulted from humic acid treatments in relation to nematode infection. Asian Pacific J Trop Biomed 2:287–293

Kliebenstein DJ, Monde RA, Last RL (1998) Superoxide dismutase in *Arabidopsis*: an eclectic enzyme family with disparate regulation and protein localization. Plant Physiol 118:637–650

Kort J (1972) Nematode diseases of cereals of temperate climates. In: Webster JM (ed) Economic nematology. Academic Press, Boca Raton, pp 97–126

Kyndt T, Fernandez D, Gheysen G (2014) Plant-parasitic nematode infections in rice: molecular and cellular insights. Annu Rev Phytopathol 52:135–153

Lamers J, Van der Meer T, Testerink C (2020) How plants sense and respond to stressful environments. Plant Physiol 182:1624–1635

Lehmann S, Serrano M, L'Haridon F, Tjamos SE, Metraux JP (2015) Reactive oxygen species and plant resistance to fungal pathogens. Phytochemistry 112:54–62

Maisarah MS, Azmi AR (2015) Pest management. In: Saad MM, Man A, Musa Z (eds) Bukit Paddy Books Sariam Othman. MARDI, Serdang, pp 65–80

Maisarah MS, Badrulhadza A, Mohd FM, Allicia J, Norsuha SM, Kogeethavani R, Azmi M (2018) Pests, diseases and weeds in Malaysia, 4th edn. Malaysian Agricultural Research and Development Institution (MARDI): Ministry of Agriculture and Agro-based Industry, Malaysia, Serdang

Mardiana-Jansar K, Bajrai F, Ishak MS, Ismail B (2019) Effect of onduty™ herbicide on weed populations and rice yields in selected rice fields, Melaka, Malaysia. Sains Malays 48:2355–2365

Marty L, Bausewein D, Müller C, Bangash SAK, Moseler A, Schwarzländer M, Müller-Schüssele SJ, Zechmann B, Riondet C, Balk J, Wirtz M, Hell R, Reichheld JP, Meyer AJ (2019) *Arabidopsis* glutathione reductase 2 is indispensable in plastids, while mitochondrial glutathione is safeguarded by additional reduction and transport systems. New Phytol 224:1569–1584

Maruta T, Inoue T, Noshi M, Tamoi M, Yabuta Y, Yoshimura K, Ishikawa T, Shigeoka S (2012) Cytosolic ascorbate peroxidase 1 protects organelles against oxidative stress by wounding- and jasmonate-induced H_2O_2 in *Arabidopsis* plants. Biochim Biophys Acta Gen Subj 1820:1901–1907

Maruta T, Sawa Y, Shigeoka S, Ishikawa T (2016) Diversity and evolution of ascorbate peroxidase functions in chloroplasts: more than just a classical antioxidant enzyme. Plant Cell Physiol 57:203

Melillo MT, Leonetti P, Leone A, Veronico P, Bleve-Zacheo T (2011) ROS and NO production in compatible and incompatible tomato-*Meloidogyne incognita* interactions. Eur J Plant Pathol 130:489–502

Miah G, Rafii MY, Ismail MR, Puteh AB, Rahim HA, Asfaliza R, Latif MA (2013) Blast resistance in rice: a review of conventional breeding to molecular approaches. Mol Biol Rep 40:2369–2388

Mittler R (2002) Oxidative stress, antioxidants and stress tolerance. Trends Plant Sci 7:405–410

Mittler R (2017) ROS are good. Trends Plant Sci 22:11–19

Morgan MJ, Lehmann M, Schwarzlander M, Baxter CJ, Sienkiewicz-Porzucek A, Williams TC et al (2008) Decrease in manganese superoxide dismutase leads to reduced root growth and affects tricarboxylic acid cycle flux and mitochondrial redox homeostasis. Plant Physiol 147(1):101–114

Moussa Z, Judeh ZM, Ahmed SA (2019) Nonenzymatic exogenous and endogenous antioxidants. In: Das K, Das S, Biradar MS, Bobbarala V, Tata SS (eds) Free radical medicine and biology, vol 3. IntechOpen, London, pp 1–22

Myouga F, Hosoda C, Umezawa T, Iizumi H, Kuromori T, Motohashi R, Shono Y, Nagata N, Ikeuchi M, Shinozaki K (2008) Heterocomplex of iron superoxide dismutases defends chloroplast nucleoids against oxidative stress and is essential for chloroplast development in *Arabidopsis*. Plant Cell 20:3148–3162

Nadarajah KK (2020) ROS homeostasis in abiotic stress tolerance in plants. Int J Mol Sci 21:5208

Nishio M, Kusano S (1975) Effect of fungi associated with roots on the growth of continuously cropped upland rice. Soil Sci Plant Nutri 21:161–171

Okorocha EOA, Ogbuji RO, Ijeoma OF, Okorocha CG (2014) Research article relationship between root-knot nematode *Meloidogyne javanica* inoculum densities and ginger (*Zingiber officinale* Roscoe). Acad J Biosci 2(11):809–812

Oladosu Y, Rafii MY, Samuel C, Fatai A, Magaji U, Kareem I, Kolapo K (2019) Drought resistance in rice from conventional to molecular breeding: a review. Int J Mol Sci 20:3519

Oyebamiji YO, Shamsudin NAA, Asmuni MI, Adebola LA (2022) Drought impacts and the tolerance mechanisms in rice (*Oryza sativa* L.): a review. Egypt J Agron 44:97–115

Palma JM, Mateos RM, López-Jaramillo J, Rodríguez-Ruiz M, González-Gordo S, Lechuga-Sancho AM, Corpas FJ (2020) Plant catalases as NO and H_2S targets. Redox Biol 34:101525

Palomares-Rius JE, Hedley PE, Cock PJ, Morris JA, Jones JT (2012) Comparison of transcript profiles in different life stages of the nematode *Globodera pallida* under different host potato genotypes. Mol Plant Pathol 13:1120–1134

Pandey P, Irulappan V, Bagavathiannan MV, Senthil-Kumar M (2017) Impact of combined abiotic and biotic stresses on plant growth and avenues for crop improvement by exploiting physio-morphological traits. Front Plant Sci 8:537

Passardi F, Longet D, Penel C, Dunand C (2004) The class III peroxidase multigenic family in rice and its evolution in land plants. Phytochemistry 65:1879–1893

Pilon M, Ravet K, Tapken W (2011) The biogenesis and physiological function of chloroplast superoxide dismutases. Biochim Biophys Acta Bioenerg 1807:989–998

Pinheiro JB, Amaro GB, Pereira RB (2010) Occurrence and control of nematodes in leafy vegetables. Embrapa, Brasília

Pinheiro JB, Amaro GB, Pereira RB (2011) Nematoides in *Capsicum* chili peppers. Embrapa Hortaliças, Brasília

Potters G, De Gara L, Asard H, Horemans N (2002) Ascorbate and glutathione: guardians of the cell cycle, partners in crime. Plant Physiol Biochem 40:537–548

Pulavarty A, Egan A, Karpinska A, Horgan K, Kakouli-Duarte T (2021) Plant parasitic nematodes: a review on their behaviour, host interaction, management approaches and their occurrence in two sites in the Republic of Ireland. Plan Theory 10:2352

Ragozzino A, d'Errico G (2012) Interactions between nematodes and fungi: a concise review. REDIA 94:123–125

Ralmi NHAA, Khandaker MM, Mat N (2016) Occurrence and control of root-knot nematode in crops: a review. Austral J Crop Sci 10:1649–1654

Ramegowda V, Senthil-Kumar M (2015) The interactive effects of simultaneous biotic and abiotic stresses on plants: mechanistic understanding from drought and pathogen combination. J Plant Physiol 176:47–54

Rane J, Singh AK, Tiwari M, Prasad PV, Jagadish SV (2022) Effective use of water in crop plants in dryland agriculture: implications of reactive oxygen species and antioxidative system. Front Plant Sci 12:3098

Rasool S, Ahmad A, Siddiqi TO, Ahmad P (2013) Changes in growth, lipid peroxidation and some key antioxidant enzymes in chickpea genotypes under salt stress. Acta Physiol Plant 35:1039–1050

Rizhsky L, Liang H, Mittler R (2003) The water-water cycle is essential for chloroplast protection in the absence of stress. J Biol Chem 278:38921–38925

Sahni S, Prasad BD (2020) Management of collar rot disease using vermicompost and a PGPR strain *pseudomonas* sp. and their effect on defense-related enzymes in chickpea. Indian Phytopathol 73:301–311

Sahu PK, Jayalakshmi K, Tilgam J, Gupta A, Nagaraju Y, Kumar A, Hamid S, Singh HV, Minkina T, Rajput VD, Rajawat MVS (2022) ROS generated from biotic stress: effects on plants and alleviation by endophytic microbes. Front Plant Sci 13:1042936

Schweitzer C, Schmidt R (2003) Physical mechanisms of generation and deactivation of singlet oxygen. Chem Rev 103(5):1685–1758

Selvaraj K, Fofana B (2012) An overview of plant photosynthesis modulation by pathogen attacks. Adv Photosynt Fundam Aspects 22:466–484

Shao HB, Chu LY, Lu ZH, Kang CM (2008) Primary antioxidant free radical scavenging and redox signaling pathways in higher plant cells. Int J Biol Sci 4:8

Sharma I, Ahmad P (2014) Catalase: a versatile antioxidant in plants. In: Ahmad P (ed) Oxidative damage to plants, antioxidant networks and signaling. Elsevier, Amsterdam, pp 131–148

Sharma P, Dubey R (2004) Ascorbate peroxidase from rice seedlings: properties of enzyme isoforms, effects of stresses and protective roles of osmolytes. Plant Sci 167:541–550

Sharma IP, Sharma AK (2017) Physiological and biochemical changes in tomato cultivar PT3 with dual inoculation of mycorrhiza and PGPR against root-knot nematode. Symbiosis 71:175–183

Sharma SB, Swarup G (1984) Cyst forming nematodes of India. Cosmo Publication, New Delhi, India, p 150

Sharma P, Jha AB, Dubey RS, Pessarakli M (2012) Reactive oxygen species, oxidative damage, and antioxidative defense mechanism in plants under stressful conditions. J Bot 2012:217030

Shetty NP, Jørgensen HJL, Jensen JD, Collinge DB, Shetty HS (2008) Roles of reactive oxygen species in interactions between plants and pathogens. Eur J Plant Pathol 121:267–280

Siddaiah CN, Satyanarayana NR, Mudili V, Gupta VK, Gurunathan S, Rangappa S, Huntrike SS, Srivastava RK (2017) Elicitation of resistance and associated defense responses in *Trichoderma hamatum* induced protection against pearl millet downy mildew pathogen. Sci Rep 7:43991

Singla P, Bhardwaj RD, Kaur S, Kaur J (2019) Stripe rust induced defence mechanisms in the leaves of contrasting barley genotypes (*Hordeum vulgare* L.) at the seedling stage. Protoplasma 257:169–181

Starr JL, Koenning SR, Kirkpatrick TL, Robinson AF, Roberts PA, Nichols RL (2007) The future of nematode management in cotton. J Nematol 39:283

Su T, Wang P, Li H, Zhao Y, Lu Y, Dai P (2018) The Arabidopsis catalase triple mutant reveals important roles of catalases and peroxisome-derived signaling in plant development. J Integr Plant Biol 60:591–607

Sumugat MR, Donahue JL, Cortes DF, Stromberg VK, Grene R, Shulaev V, Welbaum GE (2010) Seed development and germination in an *Arabidopsis thaliana* line antisense to glutathione reductase 2. J New Seeds 11:104–126

Swarup G, Sosa-Moss C (1990) Nematode parasites of cereals. In: Luc M, Sikora RA, Bridge J (eds) Plant parasitic nematodes in subtropical and tropical agriculture. CAB International, Wallingford, pp 109–136

Thakur M, Sohal BS (2013) Role of elicitors in inducing resistance in plants against pathogen infection: a review. Int Scholar Res Not 2013:762412

Topalovic O, Heuer H (2019) Plant-nematode interactions assisted by microbes in the rhizosphere. Curr Iss Mol Biol 30:75–88

Torres MA, Jones JDG, Dangl JL (2006) Reactive oxygen species signaling in response to pathogens. Plant Physiol 141:373–378

Tripathi DK, Singh VP, Prasad SM, Chauhan DK, Dubey NK (2015) Silicon nanoparticles (SiNp) alleviate chromium (VI) phytotoxicity in *Pisum sativum* (L.) seedlings. Plant Physiol Biochem 96:189–198

Tuzet A, Rahantaniaina MS, Noctor G (2019) Analyzing the function of catalase and the ascorbate-glutathione pathway in H_2O_2 processing: insights from an experimentally constrained kinetic model. Antioxid Redox Signal 30:1238–1268

Vellosillo T, Vicente J, Kulasekaran S, Hamberg M, Castresana C (2010) Emerging complexity in reactive oxygen species production and signaling during the response of plants to pathogens. Plant Physiol 154(2):444–448

Waszczak C, Akter S, Jacques S, Huang J, Messens J, Van Breusegem F (2015) Oxidative post-translational modifications of cysteine residues in plant signal transduction. J Exp Bot 66:2923–2934

Wesemael WML, Moens M, Viaene N, Taning LM (2014) Life cycle and damage of the root-knot nematode *Meloidogyne* minor on potato, *Solanum tuberosum*. Nematology 16:185–192

Wilkinson C (2014) Nematodes. Agriculture and Food. Department of Primary Industries and Regional Development, Government of Western Australia, Perth. https://www.agric.wa.gov.au/nematodes/. Accessed 27 Sept 2023

Wojtaszek P (1997) Oxidative burst: an early plant response to pathogen infection. Biochem J 322:681–692

Yang S, Lian G (2020) ROS and diseases: role in metabolism and energy supply. Mol Cell Biochem 467:1–12

Yu X, Pasternak T, Eiblmeier M, Ditengou F, Kochersperger P, Sun J, Wang H, Rennenberg H, Teale W, Paponov I (2013) Plastid-localized glutathione reductase2-regulated glutathione redox status is essential for *Arabidopsis* root apical meristem maintenance. Plant Cell 25:4451–4468

Zhang Y, Li S, Li H, Wang R, Zhang KQ, Xu J (2020a) Fungi-nematode interactions: diversity, ecology, and biocontrol prospects in agriculture. J Fungi 6:206

Zhang S, Li C, Ren H, Zhao T, Li Q, Wang S (2020b) BAK1 mediates light intensity to phosphorylate and activate catalases to regulate plant growth and development. Int J Mol Sci 21:1437

Zhu Z, Wei G, Li J, Qian Q, Yu J (2004) Silicon alleviates salt stress and increases antioxidant enzymes activity in leaves of salt-stressed cucumber (*Cucumis sativus* L.). Plant Sci 167:527–533

Zulfugarov IS, Tovuu A, Kim JH, Lee CH (2011) Detection of reactive oxygen species in higher plants. J Plant Biol 54:351–357

Functions of Flavonoids in Plant, Pathogen, and Opportunistic Fungal Interactions

6

Abhibandana Das, Srishti Choudhury, Vaithiyalingam Gopinath, Waqar Majeed, Sanjukta Chakraborty, K. Sindhura Bhairavi, Sanhita Chowdhury, Vinod Kumar Dubey, and Mohd. Sayeed Akhtar

6.1 Introduction

Flavonoids belong to low molecular weight polyphenolic plant secondary metabolites, which have broad-spectrum functions in plants. In other words, "flavonoids" are a group of chemical compounds which contain a diphenyl propane backbone where two aromatic rings (namely, A and B) are connected through a central 3-carbon chain or ring (Wang et al. 2022). About 9000 and above flavonoid compounds are reportedly

A. Das · K. S. Bhairavi
Department of Entomology, College of Agriculture, Assam Agricultural University, Jorhat, Assam, India

S. Choudhury
Department of Entomology, College of Agriculture, Punjab Agricultural University, Ludhiana, India

V. Gopinath
Department of Plant Pathology and Nematology, Post Graduate College of Agriculture, Dr. Rajendra Prasad Central Agricultural University, Pusa, Samastipur, Bihar, India

W. Majeed
Department of Zoology, Wildlife and Fisheries, University of Agriculture, Faisalabad, Pakistan

S. Chakraborty
Department of Plant Pathology, College of Agriculture, Assam Agricultural University, Jorhat, Assam, India

S. Chowdhury · V. K. Dubey (✉)
Department of Entomology, Post Graduate College of Agriculture, Dr. Rajendra Prasad Central Agricultural University, Pusa, Samastipur, Bihar, India

M. S. Akhtar
Department of Botany, Gandhi Faiz-e-Aam College, Shahjahanpur, Uttar Pradesh, India

M. S. Akhtar (ed.), *Opportunistic Fungi, Nematode and Plant Interactions*, https://doi.org/10.1007/978-981-97-2045-3_6

ubiquitously present in the plant kingdom (Buer et al. 2010; Mierziak et al. 2014a). Chalcones, flavonols, flavones, proanthocyanidins, anthocyanins, and aurones are the major flavonoid sub-groups. Flavonoid synthesis depends on the plants organs, tissue, and environmental stimulus (Abdel-Lateif et al. 2012). Flavonoids are structurally diverse, contributing to their broad array of functions in the plant, mainly their role in protecting plants against various abiotic and biotic stresses. Plants are invaded by many pathogens have been the cause of the severe economic crisis and epidemics throughout the world. The innate defence machinery of the plants becomes an area of interest to recognize the underlying molecules and the mechanisms involved, which can further be explored as a tool for managing the disease.

Similarly, the roles and processes played in the interaction between plants and beneficial microorganisms are also quintessential because the beneficial microbiome associated with the plant also forms an important component of the defence system. Flavonoids secreted naturally by plants or induced production due to pathogen invasion forms an important group of molecules with a powerful role in plant defence. Hence, it becomes crucial to study and understand the role of flavonoids in the multifaceted interaction between plants, beneficial microbes, and pathogenic microorganisms. The anti-pathogenic properties are attributed to their ability to quench reactive oxygen species (ROS) produced by the plant and the pathogens due to infection (Mierziak et al. 2014b). Upon invasion by a pathogen, the flavonoids are synthesized in the plants and transported to the infection site, which induces hypersensitive responses in the plants, causing cell death and necrosis. Flavonoids can also act against the pathogens by callus or tylose formation, degradation of cell wall degrading enzymes produced by the pathogen or by chelation of ions required for the growth and multiplication of the pathogen (Treutter 2005).

The plant–fungal interaction is considered as old as the evolution of vascular plants dating back to 400–460 million years ago (Zeilinger et al. 2016). Fungi can act as a friend or foe of the plants. Mostly, they survive in the plant microbiome as beneficial growth promoters, abiotic stress releasers or biocontrol agents and maintain stable association with the plants as against their survival as opportunistic or aggressive pathogens wherein the association is destabilized. Few genera of fungi can also survive as endophytes which may act as opportunistic pathogens over beneficial symbionts. On the other hand, fungal species that could break the mutual benefit balance developed further to plant pathogens surviving as biotrophs, hemibiotroph or necrotrophs. Flavonoids assume an important role in establishing the associated microbiota of a plant, especially the soil microbiota of which fungi form a significant group. Flavonoids may act as signalling molecules or chemo-attractants for beneficial and detrimental fungi. Thus, flavonoids released as root exudates of plants may increase or decrease a plant's vulnerability to a pathogenic fungus. Flavonoids have inhibited spore germination and hyphal growth in many root pathogens. In contrast, flavonoids have shown stimulatory effects for some fungi, as in the case of *Fusarium solani* infecting peas (Zeilinger et al. 2016). Opportunistic fungi and mycorrhizae form an efficient group of root symbionts producing biologically active compounds. In this chapter, we discussed the functionality of flavonoids in the plant and fungal interaction, including both pathogenic and opportunistic fungi. The major function of flavonoids is depicted in (Fig. 6.1).

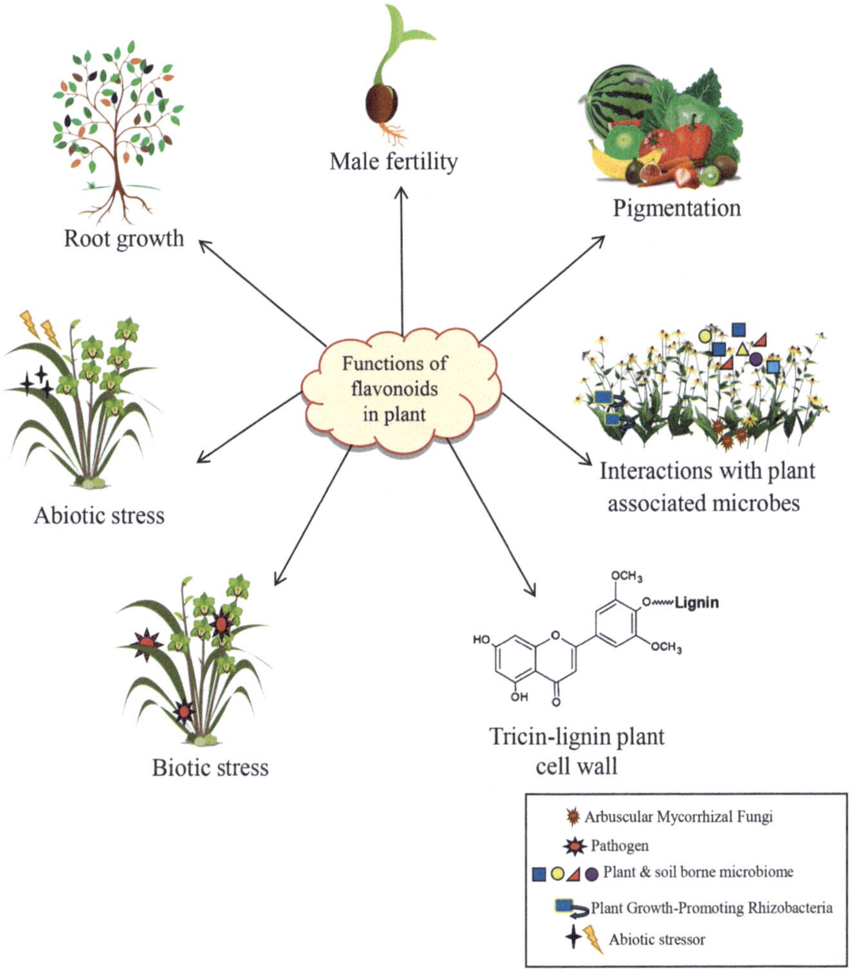

Fig. 6.1 Function of flavonoids in plant

6.2 Flavonoids

Flavonoids are polyphenolic compounds of low molecular weight, biologically active phytochemicals secreted as secondary metabolites. They are ubiquitously present in the whole plant kingdom (Donadio et al. 2021). They are naturally occurring compounds which gather mostly in plants' edible parts (Luthra et al. 2021; Kaur and Roy 2021). Flavonoids are hydroxylated phenolic substances frequently found in vacuoles, and plants usually produce these as a response to infections by microorganisms. Secondary metabolites, such as flavonoids, are produced through secondary pathways. These pathways involve synthesizing compounds the plant requires in trace amounts (Roy et al. 2021). For instance, hormones and coenzymes are primary pathways regulated by secondary metabolites. They can also function as antibiotics and toxins under stressful situations. Flavonoids are a crucial component of our everyday nutrition. It has been claimed that over 9000 of its derivatives, including those for auxin transport, signalling, flower colours, allelopathy, and ultraviolet (UV) protection, are crucial for plant life (Buer et al. 2010). Certain flavonoids are renowned as floral pigments that give flowers their colour and aroma and increase male fertility (Fig. 6.1). Moreover, they are also known to impart scent and flavour to fruits and seeds. An extensive study on flavonoids has highlighted how their different structures and activities can impact microbes, plants, and animals (D'Arcy 2022).

Due to their different chemical structure and variation from the connected substituent, flavonoids serve several significant roles in plants. The main function of flavonoids in plants is to protect against biotic stresses such as herbivores and pathogen damage and abiotic stresses like UV radiation, salinity, drought and heat. They also serve as colour descriptors and attractants for pollinators and seed dispersers, as attractants for insect feeding in host-species recognition, and as astringent or bitter taste characteristics to deter birds and other animals. Other functions include giving flowers and other components different hues, playing a role in the transfer of auxin and regulating reactive oxygen species (Buer et al. 2010; Agati et al. 2012; Davies et al. 2012). The scavenging ability of ROS is crucial to preserving the equilibrium of the redox state inside the plant cells. The structure of this molecule has conjugated double bonds and functional groups in the rings, which give them considerable antioxidant activity (Rice-Evans et al. 1996; Amić et al. 2003). Moreover, besides increasing plant biomass, flavonoids are also essential for developing *Rhizobium* spp. symbiotic interaction between plants and helpful microorganisms in leguminous plants (Zhang et al. 2009; Buer et al. 2010). Besides having a key function in plants, they are also crucial for human health due to various pharmacological effects (Ahmad et al. 2015).

6.2.1 Chemical Composition of Flavonoids

Diphenyl propane, which comprises 15C atoms, makes up the backbone of a flavonoid. Two 6-membered rings, A and B, are connected by three carbon units, which might or might not be a component of the third ring (Karak 2019). The fundamental flavonoid ring structure is also called C6-C3-C6 (Brodowska 2017). The flavonoid molecules have three rings as their fundamental structure, with several substitutions such as hydrogenation, glycosylation, hydroxylation, methylation, malonylation, and sulfation. In nature, biologically active flavonoids are essentially polyphenols. In 1936, to create a synergistic effect, Albert Szent-Gyorgyi worked with vitamin C and the lemon co-factor, which was then called citrine. This was the time when research on flavonoids originally began. The basic flavonoid structure comprises the flavane nucleus, or 2-phenyl-benzo[α]pyrane and two additional benzene rings connected by a heterocyclic pyrane ring (Kogawa et al. 2007). Enzymatic oxidation or fermentation can polymerize flavonoids, producing proanthocyanidins or condensed tannins of varying sizes (dimers to 17-unit flavonol polymers) (Prior et al. 2001; Gu et al. 2003). Plants often bind flavonols and flavones to sugar molecules to produce glycoside conjugates, bonding the sugar residue by carbon or an oxygen atom. Sugar moiety attachment to flavonoids makes the molecules more polar for storage in the vacuoles of plant cells. Glycosidic unit conjugates most frequently found include arabinose, glucorhamnose, galactose-rhamnose, and glucose (Yazaki et al. 2009). The degree of hydroxylation of the aglycone, the location of the hydroxyl group, and any associated derivative groups, like isoprenoids, alkyl, alcoholic or carboxylic groups, are the major causes of variation in flavonoids. Theoretically, there might be more than 2×10^6 flavonoids based on these variances.

6.2.2 Types of Flavonoids

While the occurrence of flavonoids in plants has been known since antiquity, little was known about their chemical composition. The major classification of flavonoids with their sources and suitable examples is given in Table 6.1. The chemical identification of flavonoids and their related compounds in different plants and their production in laboratories began in the early twentieth century. The basic flavonoid structure consists of three carbon rings. They are categorized into different classes

Table 6.1 Classification of flavonoids with their sources

Flavonoid classes	Source	Examples	Reference
Flavones	Fruit skins, red wine, buckwheat, red pepper, tomato skin, carrot, broccoli, parsley, and sweet bell peppers	Apigenin, luteolin, chrysin, isoorientin, diosmetin, orientin, and vitexin	Ahmad et al. (2015), Kwon et al. (2016)
Flavanols	Grape, black strawberry, green cloudberry, red raspberry	Quercetin, kaempferol, myricetin, and rutin	D'Andrea (2015)
Catechins	Apple, beans, apricot, black tea, blackberry, cherry, chocolate, cedar, grape, peach, red raspberry, red wine, strawberry, pecans, plums	Epicatechin, catechin, gallocatechin, epigallocatechin, gallate	Gadkari and Balaraman (2015), Scherr et al. (2016)
Flavanones	Orange juice, grapefruit juice, citrus juices, parsley, celery, and mint	Eriodictyol, hesperetin, naringenin, caryolifolin, kushenol, barachinin, kurarinone, sophoraflavanone, poncirin	Parihar et al. (2015)
Anthocyanidins	Berries, cherry, eggplant skin, black rice, black carrot, grapes, red cabbage, pomegranates, and violet flower petals	Delphinidin, pelargonidin, cyanidin, peonidin, and malvidin	Islam (2016)
Isoflavones	Soybean (*Glycine max* L.), red clover (*Trifolium pratense* L.) alfalfa (*Medicago sativa* L.), common beans (*Phaseolus vulgaris*), chickpeas, mature seeds, and pea	Daidzein and genistein, glycitein, medicarpin, formanantine	Eleazu et al. (2017)

based on their substitutes and oxidation level on the middle ring (Fig. 6.2). Anthocyanidins (found in red, blue, and purple berries), flavanols (found in red wines, red grapes, and tea), flavonols (often found in foods), flavanones (found in citrus), flavones (found in green leafy species), and isoflavones (found in soybean) are the major subclasses of flavonoids. Around 700 are flavones, and 3000 flavones may be found in plants (de Rijke et al. 2006).

6.2.2.1 Isoflavonoid

Their presence in plants is rare, mainly in leguminous plants, soybeans, and some microorganisms. Some studies have shown these compounds to be effective against cancer, cardiovascular problem, and menopausal health (Yeung and Yu 2003). Antioxidant properties of isoflavonoid affect blood vessels. These compounds show estrogenic activities in animal models. Thus, they can be used as phytoestrogens (D'Arcy 2022).

6.2.2.2 Neoflavonoid

Their composition is such that 4-phenylchromen is the backbone, and the substituted hydroxyl group is absent at the second position. In 1951, *Calophyllum inophyllum* was used as a neoflavonoid for the first time. It was found in the bark of the plant *Mesua thwaitesii* (D'Arcy 2022).

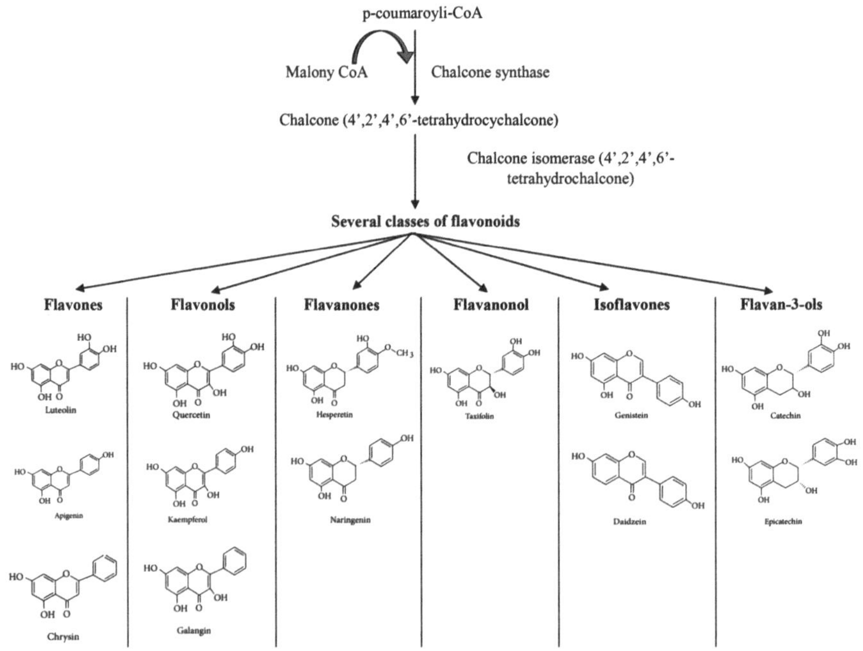

Fig. 6.2 Synthesis of flavonoids and their classification

6.2.2.3 Flavones

At 2C of the B ring in flavones, it is joined with the C ring. Leutein, luteolin, and rutin glycosides are the main members of this group. Some other flavones are apigenin and chrysin. The skins of fruits like tomato, red wine, red pepper, and buckwheat are some prominent dietary sources of flavones. It is seen that plants like *Aloe vera*, *Bacopa moneirra*, *Mentha longifolia*, and *Momordica charantia* also act as sources of flavones (Tajammal and Siddiqa 2022).

6.2.2.4 Flavonols

The B ring of flavonols is attached to the C ring at 2 positions, and a hydroxyl group in 3 positions of the C ring is attached. Quercetin, kaempferol, and galangin are some of the important instances of flavonols. Myricetin and tamarixetin are some additional ingredients. Onion, olive oil, red wine, grapefruit, and berries are some main natural sources of flavonols (Sahu et al. 2014). Their traces are also present in some plants, such as *Azadirachta indica*, *Acalypha indica*, *Cannabis sativa*, and *Betula pendula* (Gupta et al. 2013).

6.2.2.5 Flavanones

Flavanone is an important class of flavonoids. Some major natural sources of flavanones are citrus fruits like oranges, lemons, and grapes. Various types of flavanones, such as naringin, naringenin, and hesperidin, are present in this group. *Citrus medica* is an important source of hesperidin (Tajammal and Siddiqa 2022).

6.2.2.6 Flavanonols

Some other classes of flavonoids are flavanonols. The B ring is attached to the C ring at position 2, and the C ring has hydroxyl groups attached at the third and fourth positions. One of the important compounds of flavanonols is taxifolin (Miyake et al. 2000).

6.2.2.7 Flavanols

Due to the presence of an OH group at position third of the C ring flavanols can also be written as flavan-3-ols due. Double bonds between positions 2 and 3 are missing, the same as in many flavonoids. Apples and tea are good sources of flavanols. In the flavanols subgroup, proanthocyanidins, epicatechin, catechin, glau-san-3-epicatechin, and epigallocatechin are mainly present (Marsafari et al. 2020). Brysonimacrassa is an important plant source of flavanols.

6.2.2.8 Chalcones

These are also called open-chain flavonoids. Phloretin and chalconaringenin are present in the subgroup chalcones (D'Arcy 2022). C ring is absent in them, or it can be iterated that the C ring has an open structure in chalcones.

6.2.3 Production of Flavonoids

Flavonoids mostly appear in nature as polymerized, esterified, and glycosylated derivatives. They are all structural descendants of flavones. Flavones are found in the immature plant sap. Phenylalanine and malonyl coenzyme A play an important part in synthesizing flavonoids. In the biosynthesis of flavonoids, three molecules of malonyl-CoA are converted to p-coumaroyl-CoA to form naringenin chalcone (2,4,6,4-tetrahydroxychalcone) by using chalcone synthase (CHS). Chalcone is isomerized to flavanone by chalcone isomerase (CHI). The production of flavanone is then caused by chalcone isomerization utilizing chalcone isomerase (CHI) (Alzand and Mohamed 2012). Afterwards, the route diverges to produce several flavonoids, such as aurones, flavones, flavanonols (dihydro flavonols), dihydrochalcones, flavonols, isoflavones, anthocyanins, leucoanthocyanidins, and proanthocyanidins. Moreover, flavonoids are connected to several functional groups to influence their stability, reactivity, and solubility, including methylation by certain methyltransferases and glycosylation caused by glycosyltransferases (Martens and Mithöfer 2005). The process of flavonoid synthesis is described briefly in Fig. 6.2.

6.3 Interactions of Flavonoids with Opportunistic and Pathogenic Fungi

The physical and biochemical structural diversity of flavonoids, which also allows them to interact with a wide range of other molecules, is attributed to the multitude of roles flavonoids play in the dynamic interaction between plants and microorganisms, irrespective of the nature of the interaction, i.e., opportunistic or pathogenic. The flavonoids released by different plants and their role against phytopathogenic fungi (Table 6.2). The mechanisms underpinning such versatility in functions of flavonoids in the plant–microbe interaction is an area of extensive research and study. Flavonoids possess certain effective antifungal properties against a variety of pathogens. Flavonoids retard fungal growth using various mechanisms like the disruption of the plasma membrane, mitochondrial dysfunction, inhibition of cell division, and cell wall formation (Al Aboody and Mickymaray 2020). They can be credited with the endowment of resistance to plants. Invasive opportunistic fungi threaten plant life, and their management is paramount. The interaction of flavonoids with opportunistic fungi and its role in the functioning and defence mechanism of the plant provides hope for exploiting flavonoids to inhibit and control the growth and spread of these fungi. For this purpose, more extensive studies are required, which could give a proper understanding of all the different facets of flavonoid-pathogenic fungi interaction. The study revealed a significant association between flavonoid biosynthetic pathway and *Aspergillus flavus* infection, which suggests that flavonoids modulate toxin accumulation in maize kernels and thus assume a key role in plant resistance mechanisms (Castano-Duque et al. 2021). Flavonoids may act as signalling molecules at low concentrations, reducing fungal (*A. flavus*) proliferation and aflatoxin production in corn (Castano-Duque et al.

Table 6.2 Major flavonoids released by different plants and their role against phytopathogenic fungi

Host plant	Flavonoid compound	Pathogen	Reference
Rice	Naringenin, kaempferol, quercetin, hydroxyquercetin	*Pyricularia oryzae*	Padmavati et al. (1997)
Soybean	Isoflavone	*Phytophthora sojae*	Subramanian et al. (2005)
Barrel medic	Isoflavone	*Erysiphe pisi*	Foster-Hartnett et al. (2007)
Barley	Naringenin, kaempferol	*Gibberella zeae*	Bollina et al. (2010)
Sorghum	3-deoxyanthocyanidins, luteolin	*Colletotrichum sublineola*	Du et al. (2010)
Grapevine	Quercetin-3-O-galactoside	*Plasmopara viticola, Botrytis cinerea*	Ali et al. (2012)
Bean	Isoflavonoid phytoalexin: Genistein, daidzein, 2-hydroxygenistein, dalbergioidin, phaseollin, phaseollidin, phaseollin isoflavan, kievitone, coumestrol	*Colletotrichum lindemuthianum*	Durango et al. (2013)
Rice	Sakuranetin	*Fusarium fujikuroi, Magnaporthe oryzae, Rhizoctonia solani*	Hasegawa et al. (2014), Siciliano et al. (2015), Katsumata et al. (2018)
Flax	Isoorientin, isovitexin, vitexin	*Fusarium. Oxysporum, fusarium culmorum*	Mierziak et al. (2014b)
Cotton	Flavonoid	*Verticillium dahliae*	Long et al. (2019)

2022). Flavonoids like kaempferol and quercetin are well known for their efficiency against *A. fumigatus, A. niger, Candida,* and *S. cerevisiae* (Nguyen and Bhattacharya 2022). Some researchers have hypothesized that flavonoids like kaempferol and quercetin disrupt the plasma membrane and inhibit the mitochondrial function of pathogenic fungi (Al Aboody and Mickymaray 2020).

In the plant-pathogenic fungi interaction, the flavonoids can have a positive or a negative impact, albeit the negative impact, i.e., antagonistic or antifungal property of flavonoids, is widely studied. Besides being present within the plant, flavonoids are also found in plant root exudates (Cesco et al. 2010). They accumulate in the root cap cells and root tip. The antifungal activity of flavonoids, especially in the root exudates is known to negatively impact the distribution of pathogenic fungi in the rhizospheric niche of the plants. It is reported that the invasion by a fungal pathogen can induce de novo synthesis of flavonoid phytoalexins, which can further initiate hypersensitive response-mediated cell death, thus restricting the infection by the pathogen. The flavonoids can also be stored in the plant vacuoles in an inactive state as phytoanticipins and be released in case of any future attack by a pathogen (Lattanzio et al. 2006). The authors describe that during the programmed cell death, an oxidative burst is initiated due to the influx of calcium ions, followed by alkalization of apoplast and depolarization of the cell membrane, which causes the isoflavonoids to be oxidized and release free radicals, which in turn cause the cell death. Flavonoids can orchestrate the plant defence system against phytopathogen attack because of the ability to interfere with the mechanisms driving fungal virulence against their host. The flavonoids reduce the mycelial growth, spore germination, and activity of fungal enzymes such as cellulases, xylases and pectinases (Ghitti et al. 2022). The chelation of metal ions causes an increase in the availability of nutrients to the plants, starving the soil-borne phytopathogenic fungi causing their death. Quercetin, genistein, and kaempferol are examples of metal-chelating flavonoids which alter iron availability (Cesco et al. 2010). The defensive role of arbuscular mycorrhizae in a certain class of fungi can also result from the release of phytochemicals, including flavonoids. Pterocarpan flavonoids such as maackiain are reported to inhibit the growth of *Pythium graminicola,* while medicarpin in peas provides resistance against the fungus *Rhizoctonia solani* by affecting the elongation of germ tube and growth of fungal mycelium (Blount et al. 1992; Kapulnik et al. 2001; Jimenez-Gonzalez et al. 2008). However, there are also instances wherein the effect of the flavonoids is evaded by phytopathogenic fungi such as *R. solani,* thus making the plants susceptible to fungal infection (Pedras and Ahiahonu 2005). In certain cases, reduced accumul42ewt1ation of flavonoids can also make the plant susceptible to pathogen infection, e.g., the down regulation of flavonoids in barley increases the disease incidence of *F. graminearum* (Karre et al. 2019).

6.4 Strategies Employed by Flavonoids in the Plant Defence System

Plants are often at risk due to their interactions with pathogenic microbes and insect pests that may affect their growth and development. Due to the ever-present risk of harmful enemies, plants have developed, throughout evolution, a highly sophisticated immune system. This immune system comprises a combination of strategies such as restriction or termination of pathogen growth by either developing physical barriers (rigid cell wall) or with the help of secondary metabolites. Flavonoids play various roles in plants, ranging from signal molecules, detoxifying agents, pollinator attractants and allelochemical agents. Flavonoids mediate defence responses when exposed to abiotic stress. With respect to oxidative stresses, derivatives of isoflavan from *Glycyrrhiza glabra,* viz., glabridin, 4′-*O*-methylglabridin, 3′-hydroxy-4′-*O*-methylglabridin, hispaglabridin A and hispaglabridin B, successfully inhibited mitochondrial lipid peroxidation and dihydroxyfumarate-induced mitochondrial peroxidation (Haraguchi et al. 2010).

Compared to wild-type rice, transgenic rice line OxF3H was more tolerant to drought and UV-B radiation due to enhanced levels of kaempferol and quercetin. The over-accumulation of flavonoids increased the stress tolerance capacity of the plant by scavenging reactive oxygen species and capturing the UV-B radiation (Jan et al. 2022). During drought conditions, the Saharan plant, *Oudeneya africana* accumulated flavonoids and antioxidants, significantly improving its ability to cope with abiotic stress (Talbi et al. 2020). Similarly, drought was also found to promote the accumulation of flavonoids in two *Adonis* species, which provided drought resistance to plants by reducing the H_2O_2 content (Gao et al. 2020). Flavonoids in maize, *Zea mays,* improved drought resistance by reducing drought-induced oxidative damage and regulating stomatal movement (Li et al. 2021).

Apart from abiotic stresses, flavonoids also help the plant build resistance against different biotic stresses. Simple chemical compounds, namely, alkaloids, terpenoids, phenolic, stilbenoids, coumarins, and polyacetylenes make up a special group known as phytoalexins, which plants produce as a response mechanism to microbial attack (Iwashina 2003). In another study on insect-resistant compounds such as tannin, phenol, and flavonoids were studied before and after *Curculio nucum* chewing on *Corylus avellana.* The result showed a positive correlation between the flavonoids and hazelnut damage rate (Xingpeng et al. 2023).

6.4.1 Hypersensitive Response (HR) Mediated Cell Death

The major defence mechanism in plants involves triggered immunity against pathogens or microbes through pattern recognition receptors. Plants have evolved specialized genes that activate a second host immune response known as effector-triggered immunity through effector recognition. Interestingly, the effector-triggered immunity is comparatively faster and stronger than pattern recognition immunity. At the site of pathogen infection, it induces host-programmed cell death and produces ROS (Greenberg and Yao 2004). Hypersensitive response (HR) generation has been generally associated with activation by pathogen effectors or elicitors. HR is triggered due to interaction with pathogens, and the plant defends itself against the pathogen through regulated cell death. HR is mediated through multiple pathways and cell death features certain apoptosis-like traits (Salguero-Linares and Coll 2019). Cell death mediated by HR at the infection site initiates systemic signals that activate immunity in other plant parts. Systemic acquired resistance (SAR) (Shine et al. 2019). SAR is also due to the biosynthesis of salicylic acid, which is induced by HR synthesis by proteins (Malnoy et al. 2012).

Many studies have been conducted to study the resistance mechanism in tea plants for the role of HR in plant defence. Most upregulated genes related to HR and ROS production were found, mainly responsible for imparting disease resistance to *Camellia sinensis* against *Pestalotiopsis theae* (Senthilkumar et al. 2012). Additionally, studies conducted against Anthracnose in tea indicated that the HR mechanism is important in tea plant defence (Wang et al. 2018). In the *Marssonina* blotch of apples caused by the fungus, *Diplocarpon mali,* flavonoid, and antioxidant activity maintained free radicals and decreased the oxidative damage in plant cells. The flavonoids were mainly responsible for early response against *D. mali* through the HR mechanism (Park et al. 2023). A natural variant of *Xanthomonas citri,* the causal agent of Asian citrus canker triggered a host-specific hypersensitive response in *Citrus limon,* which was associated with the obtrusion of biofilm development and inhibition of bacterial growth (Roeschlin et al. 2017). Web blotch is a major disease occurring in peanuts caused by *Phoma arachidicola*. The fungus's appressorium formation and conidial germination rates were studied to study the plant's defence response. Analysis of the papillae formation as a response by the plant indicated HR followed by programmed cell death (Sun et al. 2022). The difference in the enzymatic responses of susceptible and resistant cultivars of pear indicates that HR contributes to the high resistance of the Dargazi cultivar to fire blight-causing bacterium, *Erwinia amylovora* (Erfani-Moghadam and Zarei 2022). In addition to defending the plant against various pathogens, HR-mediated cell death also arbitrates susceptibility to necrotrophic pathogens. Despite various works on HR-mediated cell death, it continues to be a complex phenomenon that seems influenced by various factors.

6.4.2 Reactive Oxygen Species (ROS)

The earliest accounts of plant response to the pathogenic attack have been reactive oxygen species (ROS). ROS are related to many important plant biochemical processes, and extensive studies indicate that ROS production results from different biotic and abiotic stresses (Faghıh et al. 2019). ROS are made up of a group of molecules derived from molecular oxygen and are present in the cells in various forms like singlet oxygen ($^1O_2$), superoxide (O_2^-), hydrogen peroxide (H_2O_2), and hydroxyl radical (HO^-). Flavonoids are canonically termed as scavengers of ROS as a part of their antioxidant action. The formation and removal of ROS are both equally important, as ROS is essential for basic biological functions, and excessive ROS may prove cytotoxic (Sies and Jones 2020; Sies 2020). Flavonoids interact with a wide range of ROS through redox-active phenolic moieties, and the transfer of electrons or hydrogen takes place as a consequence (Alov et al. 2015).

Flavonoids improve plant drought tolerance (Li et al. 2017) by alleviating drought-induced ROS damage (Li et al. 2021). A novel antioxidant glucurinated flavonoid (GF) isolated from spinach directly affected the scavenging of ROS and hence displayed the potential to combat oxidative damage (Bergman et al. 2003). Reactive oxygen species scavenging flavonoids, kaempferol and quercetin, influence the over-expression and induction of flavanone 3-hydroxylase in rice lines, which increases the resistance of the plant to bacterial leaf blight (Jan et al. 2021). Recent technological advancements have enabled the identification of ROS sensors and specific signalling pathways, and the hormones involved. Flavonoids can modulate the ROS-signalling cascade by buffering stress-induced alterations in ROS homeostasis. Hence, one of the basic functions of flavonoids is as development regulators in plants and animals (Brunetti et al. 2018).

6.4.3 Altering Ion Availability

Plant metabolites called flavonoids include flavanols, flavones, flavanones, and isoflavones. These compounds cause chemical changes in the soil, which influence the availability of nutrients for plants. A description of different flavonoids from root exudates, their release mechanisms, and their role in rhizosphere nutrient acquisition are provided (Cesco et al. 2012; Alseekh et al. 2020; Askari et al. 2021).

6.4.3.1 Nitrogen (N)

Plants require huge quantities of nitrogen (N), an essential macronutrient. Due to high values of immobilization/ mineralization ratios, it is often poorly available in soil (Tinker and Nye 2000; Scavo et al. 2019; Bouyahya et al. 2022). Presently, flavonoids are the best-studied compounds in root exudates due to their role as signal compounds for establishing N_2-fixing symbiosis with rhizobia. Seeds and roots of members of the Fabaceae family, including *Phaseolus vulgaris* (Linn.), *Glycine max* (Linn.), *Medicago sativa* (Linn.), and *Lotus japonicus* (Linn.) release flavonoids (Cooper 2007; Werner 2007). Exudation of flavonoids is affected by the form and quantity of the nitrogen supply from the outside, with particularly inhibiting effects coming from high nitrate fertilization levels (Wojtaszek et al. 1993). Furthermore, flavonoids modify bacterial surface polysaccharides, causing flavonoid exudation from the host plant and causing resistance, chemoattraction, and the induction of nod genes (Eckardt 2006).

6.4.3.2 Phosphorus (P)

It is fundamental for some molecules, including nucleic acids, phospholipids, and ATP. Due to their low solubility and diffusion rates, soils frequently lack P (Marschner 1995). Flavonoids have been implicated in establishing mycorrhizal interactions as a signal that induces fungal growth, spore germination and hyphae branching (Perry et al. 2007). Flavonols stimulate hyphal growth when elevated CO_2 concentrations are present in the rhizosphere, a characteristic root-induced modification (Bécard et al. 1992). Some studies suggest that flavonoids may also act as metal chelators, making it possible to chemically mobilize soluble soil-P forms. Based on the findings of metal-chelating carboxylates in root exudates of plants deficient in P, the mechanisms proposed include (I) exchange chelation at cationic P-binding sites (e.g., Al, Fe, Ca), (II) occupation of P-binding sites, and (III) iron-phosphate splitting caused by flavonoids' reducing properties (Pinton et al. 2007).

6.4.3.3 Micronutrients

Iron is an essential micronutrient important in respiration, photosynthesis and major metabolic processes like sulphur and nitrogen metabolism. Some microorganisms can use flavonoids as C sources since they reduce microbial respiration in the presence of flavonoids. Rhizosphere Fe levels could be increased by CO_2 produced in this case (Tomasi et al. 2008; Colombo et al. 2014). In response to low Fe availability, caffeic acid is released, triggering a redox reaction which reduces Fe (III) to Fe (II) (Olsen et al. 1981). Genistein and hydroxy genistein derivatives are released in huge quantities from root clusters of white lupin (Weisskopf et al. 2006). Changes in pH and redox levels can limit the solubility of minerals containing manganese (Mn) in soils (Lindsay 1991). Mn-efficient plants release more Mn-reducing compounds into the rhizosphere. Through chemical reduction, phenolic-containing exudates contribute to increased Mn availability (Marschner 1988). It is still not clear if flavonoids play a role in this phenomenon. Copper (Cu) is found in Fe-Cu oxides and cation-exchange surfaces of soil colloids. Like Fe, flavonoids can also make Cu relatively immobile (Hartley et al. 2007), although, the reduction of Cu by flavonoids and its chemical characterization has already been demonstrated (Hajji et al. 2006).

6.4.4 Biosynthetic Pathway

Plants share a flavonoid biosynthesis pathway that begins with the catalysation of chalcone isomerase (CHI) and chalcone synthase (CHS). Plants that produce flavonoids need CHS and CHI for flavonoid biosynthesis. CHS and CHI form the middle C ring of flavanones, such as naringenin. Several downstream flavonoid biosynthesis enzymes use flavanones as substrates, producing a variety of flavonoid metabolites. Different types of downstream enzymes contribute to the formation of different flavonoid molecules in different plants. In dicot plants, flavonol synthase (FLS) and flavanone 3-hydroxylase (F3H) are required for flavonol synthesis and accumulation. In the flower corolla and fruits, anthocyanidin synthase (ANS) and dihydroflavonol 4-reductase (DFR) are required for anthocyanidin accumulation. Conversely, monocots utilize flavone synthase (FNS) to synthesize and accumulate flavones. Additionally, legumes contain isoflavone synthase (IFS), which produces isoflavones. Diversity in flavonoids is achieved by the numerous enzymes in plants to catalyse hydroxylation, acylation, methylation and glycosylation of flavonoids (Nabavi et al. 2020).

A key feature of flavonoid biosynthesis enzymes is their ability to form complexes with each other. The cytochrome P450 proteins bound to the endoplasmic reticulum (ER) bind to these proteins (Jørgensen et al. 2005). As a result of these enzyme complexes, a metabolon is formed that facilitates the channelling of those intermediates, which aids in enhancing the production of flavonoids. Flavonoid biosynthesis occurs within the nucleus, ER, and tonoplast, as these cellular structures contain the enzymes required for flavonoid biosynthesis (Saslowsky and Winkel 2001; Saslowsky et al. 2005). Anthocyanin synthesis in tomatoes can be significantly increased by ectopic expression of MYB transcription factors (Butelli et al. 2008). Maize transcription factors also increased isoflavonoid synthesis in soybean by ectopically expressing and inhibiting the anthocyanin pathway (Yu et al. 2003). Environmental stress, such as UV-B radiation and drought, has little effect on the flavonoid biosynthetic pathway (Liu et al. 2013).

6.4.5 Other Biochemical Strategies

The endogenous signalling molecules are ethylene (ET) and abscisic acid (ABA), which are associated with plant defence signalling against biotic stress (Janda and Ruelland 2015; Naskar et al. 2021). The induction of defence against biotrophic or necrotrophic pathogens is dependent on two signalling pathways: salicylic acid and jasmonic acid, respectively (Glazebrook 2005). Plants respond to abiotic attacks by synthesizing SA, contributing to systemic and local resistance (Janda and Ruelland

2015). SAR is a plant immune response that develops after a local infection. It is generally related to both increased levels of SA and additional small metabolites that SAR begins to develop. When the plant is exposed to a pathogen, these metabolites rapidly activate SAR defences (Shah et al. 2014). Induced systemic resistance (ISR) is activated when a plant is under attack by necrotrophic bacteria. ISR has been seen in action when protecting *Arabidopsis* sp. against necrotrophic fungi, viz., *Plectosphaerella cucumerina* (Segarra et al. 2009), *Botrytis cinerea* (VanderEnt et al. 2008), and *Alternaria brassicicola* (Ton et al. 2002). Studies have shown the involvement of JA and ET in ISR regulation (Kazan and Lyons 2014). Also, methyl JA and SA are involved with ISR-related action in activating defence enzymes to protect plants cell walls and increase their antioxidant capacity (Derksen et al. 2013).

6.5 Phytomicrobiome Interactions

Like the human body, most plant parts like leaf, flower, stem, and root are inhabited by microbes interacting, affecting plant health (Fig. 6.3). Distinct phytomicrobiomes are present in different parts of a plant. For instance, there are differences in the community levels of the phyllo microbiome (shoot-associated microbes), rhizomicrobiome (root-associated microbes), and endo microbiome (inner plant tissue microbial community) (Brisson et al. 2019; Da Silva et al. 2014). Plants share their environment with numerous microorganisms, and these complex multilateral interactions influence the composition of the plant microbiota (Solanki et al. 2020). Plant microbiomes influence the response to abiotic and biotic stresses. Microbes are classified as mutualistic, commensal or pathogenic based on the outcome of their interaction with their host. During periods of abiotic and biotic stresses, the symbiotic microbiotas present in the plants are deeply involved in various biochemical processes, starting from nutrient uptake to improvised plant defence systems. To successfully implement microbiota-mediated crop protection, it is required to understand the mechanisms involved, like how the microorganisms interact with their hosts and other microbes present in the natural environment (Srivastava et al. 2012). Plant-associated microbiomes generally benefit their hosts; however, the microbial composition in the rhizosphere and their colonization efficiency are influenced by the environmental parameters and the genetic and physiological state of the host.

6.5.1 Flavonoids as Signalling Molecules

One of the most extensive groups of specialized metabolites, flavonoids is derived from phenylpropanoid pathways and includes over 9000 derivatives (Williams et al. 2004). These derivatives are ubiquitous in the plant kingdom. They are recognized for their wide-ranging and crucial roles in plant development, physiology, and the productiveness of associations between plants and rhizobia. The presence of flavonoids is crucial in the interaction between plants and microbes, particularly in interactions between plants and rhizobia, as well as in arbuscular mycorrhizal symbioses (Singla and Garg 2017). The

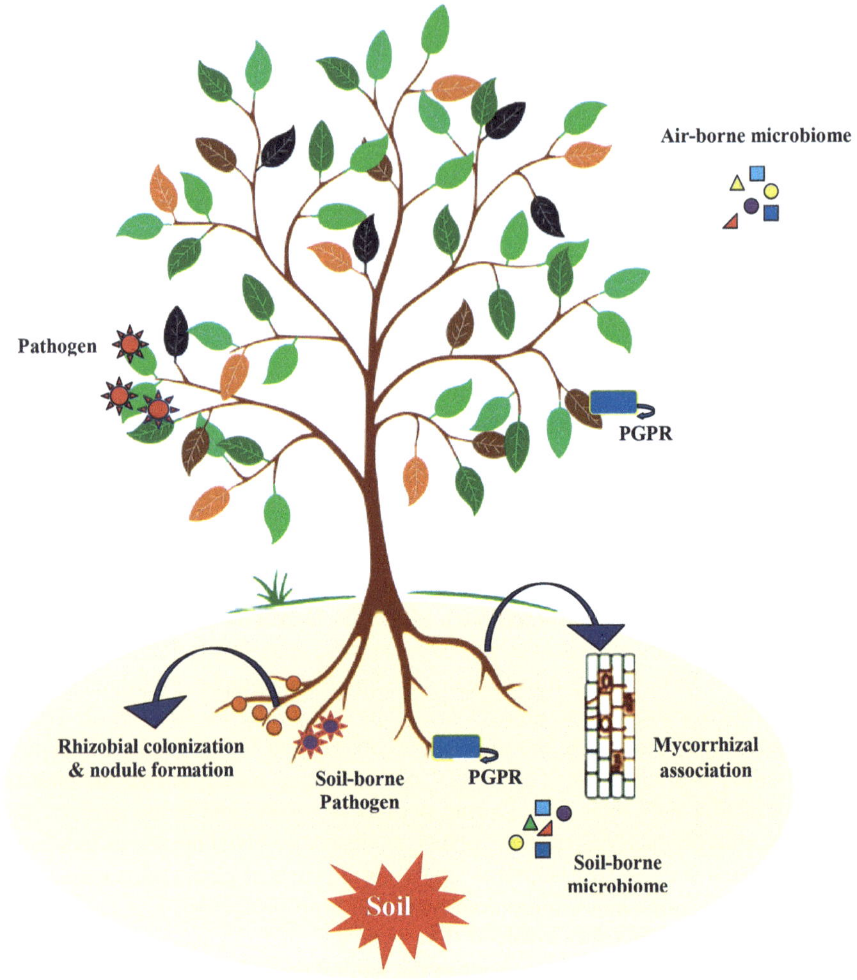

Fig. 6.3 Role of flavonoids mediating plants–microbe interaction

flavonoid family consists of flavonols, flavones, anthocyanins, proanthocyanidins, and chalcones, in addition to the standard C6-C3-C6 core structure of polyphenolics (Yonekura-Sakakibara et al. 2019). As signalling compounds that affect plants at the phytobiomes level, flavonoids were found to mediate between planta and its surrounding environment. Flavonoids like glycosides and aglycones are secreted with root exudates that serve as communication signals. They can govern allelopathic interference between species (Weston and Mathesius 2013). They are involved in mediating root association with soil microbiomes present below ground. Flavonoids also serve as signalling compounds for bacteria and arbuscular mycorrhizal fungi, affecting hyphal growth, spore germination, and root colonization efficiency (Larose et al. 2002). Flavonoids such as luteolin and chrysin, secreted by legumes such as *Medicago sativa*, are a specific signal for initiating symbiosis in *Rhizobium* bacteria (Hartwig et al. 1990).

Most of the release occurs at the root tips, emerging root hair zone, and at the target sites for Rhizobium infection (Abdel-Lateif et al. 2012). Most flavonoids act as signalling compounds for recruiting specific microbial taxa and as substrates for microbial growth (He et al. 2022). Flavonoids may signal soil microbes more broadly (Yu et al. 2021) by their massive impact on rhizosphere chemistry (e.g., 37% secondary metabolites in *Arabidopsis*) (Narasimhan et al. 2003). Plant defence involves exuded metabolites and chemical signals exchanged between plants and microorganisms, particularly those in the rhizosphere and endosphere, to maintain the holo biont's health and performance (Olanrewaju et al. 2019).

According to reports, flavanols induce an influx of reactive oxygen species (ROS), which triggers the Ca^{2+} signal cascade and root system death (Bais et al. 2018). In plant defence, flavonoids can act as quorum sensing inhibitors (QSI) through various mechanisms, such as biofilm formation, cell-to-cell signal inhibition, reduction of signal molecule biosynthesis, and competition for receptor binding sites by mimicking signal molecule structures (Truchado et al. 2015; Vikram et al. 2010). It can be noted that mycorrhiza helper bacteria (MHB), which are found to be closely related to specific arbuscular mycorrhizal fungi, can not only produce Nod factors but can also induce exudation of flavonoids which act as signalling compounds for attracting arbuscular mycorrhizal fungi, thus facilitating in root colonization (Vierheilig 2004; Frey-Klett et al. 2007).

6.5.2 Flavonoids as Chemo-Attractants

Most legumes can form root nodules containing rhizobia. These root nodules are developed via a signal exchange mechanism between the bacteria and plant roots. Flavonoids can induce root nodulation and other genes, act as a chemoattractant, regulate development, determine host specificity, and affect phytoalexin resistance in rhizobia during symbiosis (Cooper 2004). It is well documented that flavonoids are a chemoattractant to accumulate compatible Rhizobia at the plant root surface. They are released in huge quantities near the root tips (Hartwig et al. 1990; Graham 1991), and their optimal concentrations are found at the site of Rhizobium infection (near the emerging root hair zone). Studies with *Sinorhizobium meliloti* have demonstrated that some flavonoids are chemo-attractants for rhizobia, which promote the movement of the bacteria towards the roots, resulting in nodule formation (Zuanazzi et al. 1998).

6.5.3 Role of Flavonoids in the Build-Up and Regulation of Soil Microbiota and Symbiotic Relations

As explained in detail in the previous sections, flavonoids are found to affect nutrient availability by modifying the physicochemical properties of soil (Cesco et al. 2010). Leguminous roots produce flavonoids as their primary, species-specific chemical compound (Lira Jr et al. 2015). Since different legumes use different flavonoid components for nodulation, it is intriguing to consider the possibility of differences in transporter expression or selectivity between legumes. Moreover, it has been reported that plant–microbe associations are influenced by the transfer of phenylpropanoid metabolites. Root cell plasma membrane flavonoid transporters are thought to be necessary for releasing flavonoids into the rhizosphere and the subsequent transcriptional activation of rhizobial nodulation genes, which in turn initiates legume-rhizobium symbiosis (Banasiak et al. 2021).

The role of flavonoids in forming arbuscular mycorrhizal (AM) fungi-plant association is well-recognized, where phosphorus-deficient soils are especially crucial for most plant species (Nagata et al. 2016). The flavones (luteolin and chrysin) and flavonols (morin) increased the number of entry points and colonization in roots. In contrast, rutin and kaempferol did not change the patterns of root colonization or the growth of pre-symbiotic AMFs (Scervino et al. 2007). Hesperetin, naringenin, and apigenin are flavonoids that boost *Gigaspora margarita* spore germination percentage, hyphal growth, and root colonization patterns (Nakabayashi et al. 2014). It has also been demonstrated that root-secreted flavonoids stimulate mycorrhizal symbioses by promoting fungal spore germination, hyphal growth, and root colonization (Shah and Smith 2020). Flavonoids (quercetin) in root exudates AM fungal colonization, and plant productivity was all higher in plants from imported communities. Studies have shown that AM symbiosis positively affects the development, trichome quantity, and accumulation of flavonoids in *Stevia rebaudiana*. AM-colonized plant species often experience a significant increase in total leaf surfaces and leaf and root dry mass (Sarmiento-López et al. 2021). It has also been found that the flavonoid quercetin can promote mycorrhizal growth in plants regardless of genotype (Gomes et al. 2021). From all the successful examples of flavonoid and plant–microbe interaction, it is now well-established that through flavonoids plants can regulate the recruitment or selection of various microorganisms from their rhizosphere on a need or purpose basis.

The need to understand the mechanisms that draw microbes to plant roots has also increased. Additionally, we need to better understand how signalling pathways such as quorum sensing and others contribute to establishing and maintaining a complex root microbiome in the rhizosphere. Metabolic engineering can be used to change the rhizosphere microbiome by targeting the transport and synthesis of numerous enzymes and associated genes (Mishra et al. 2021).

Improved availability of plant hormones, and nutrients, alleviation of biotic and abiotic stress and soil health are just some of the positive effects of plant growth promoting rhizobacteria (PGPR) based inoculants that can be enhanced through the use of this effective and environmentally sustainable rhizosphere engineering technique. Plant gene expression can be modified to produce higher targeted flavonoid levels that colonize the rhizosphere and promote growth. By manipulating the rhizosphere, flavonoids can increase crop yield and soil quality by attracting more beneficial microbes.

6.6 Negative Impact of Flavonoids on Beneficial Rhizosphere Organisms

Till now, we have discussed the positive impact of flavonoids in managing disease-causing fungal pathogens. But, given the capable strength of flavonoids, it is not impossible to think they can also harm beneficial microorganisms. Many studies have been conducted to determine if flavonoids can cause a different response in other organisms. Significantly reduced diversity in bacterial communities was observed in daidzein-treated soils, an important flavonoid present in root exudates of *Medicago* spp., indicating a possible impact on specific soil microbial populations (Okutani et al. 2020). Despite the growth of beneficial taxa, quercetin reduced soil microbiome diversity (Schütz et al. 2021). An additional microbiome modulator and root exudates contain molecules like phenolic acid derived from degraded flavonoid compounds (Clocchiatti et al. 2021). Moreover, the antioxidant activity of the resulting molecules is often higher than that of the precursor molecules (Catalán et al. 2020), which may affect specialized taxa and the plant rhizosphere microenvironment. Both positive and negative effects can be exerted by flavonoids on the expression level in bacterial nod genes of nitrogen-fixing bacteria. In *B. japonicum*, daaidzein and genistein induce nod genes, while in *S. meliloti*, they inhibit them. Similarly, quercetin inhibits nod formation in *R. leguminosarum*, while naringenin stimulates production (Mierziak et al. 2014a). It is worth considering that some negative effects can result from the combination of antibiotics and flavonoids. For example, tests with *E. coli* 27 strain showed that isoquercetin antagonized aminoglycoside antibiotics such as amikacin, kanamycin, neomycin, and gentamicin (Górniak et al. 2019).

6.7 Flavonoids in Plant Stress and Disease Management

The yield output of crops is directly related to the productivity of the soil. The basic inputs required for the proper growth of crops are water, nutrient supply, and physiochemical properties of the soil. Besides these requirements, the phytomicrobiome is recognized as a major contributor to crop production. The microorganisms present in the phytomicrobiome play a big role in aiding plant growth during both stressed and unstressed conditions. Phytomicrobiome associations largely depend on signal transmission, achieved via flavonoids. Effective plant–microbe interaction in the rhizosphere is made possible by the activities of different types of flavonoids. The role of flavonoids in the rhizosphere and phytomicrobiome has been discussed in depth in the earlier subunits. Biotic and abiotic stresses often threaten sustainable agriculture. These plants produce flavonoids in response to these stresses to induce resistance against them.

6.7.1 Flavonoids Against Abiotic Stresses

6.7.1.1 Flavonoids against Salt and Drought Stress

ROS are produced as by-products of oxidative metabolism during abiotic stress conditions like salt and drought. It has been found that flavonoids play an important role in detoxifying and scavenging ROS. Studies suggest that increasing anthocyanin accumulation by 40% may be a phytochemical strategy for combating salt stress and the following toxic reactions (Kaliamoortiiy and Rao 1994). A study compared the defensive properties of anthocyanin in two rice varieties: salt-sensitive and salt-tolerant. Results showed that salt-tolerant varieties had higher total anthocyanin content than salt-sensitive varieties, with antioxidant activities ranging from 125–199% and 106–113%, respectively. It was concluded from the study that in rice plants, anthocyanins help protect cells by detoxifying accumulated salts (Sugiyama and Yazaki 2014). In another study of wheat leaves under drought stress, flavonoid biosynthesis and accumulation increased, suggesting wheat drought resistance is closely related to flavonoid accumulation (Ma et al. 2014).

6.7.1.2 Flavonoids as UV Scavengers

Based on the wavelength of light, UV radiation is divided into three segments: UV-A (315–400 nm), UV-B (280–315 nm), and UV-C (˂280 nm). Now, UV-B has the potential to cause severe damage to the plant system. It can negatively affect photosynthesis, transpiration, and starch concentration, promote cellular damage, and increase susceptibility to diseases by weakening plant defence mechanism, affecting cell division, and inhibiting plant growth. However, flavonoids like anthocyananins absorb UV-B radiation, allowing only little radiation to pass through the epidermal cells of the leaves. An experiment involving two apple varieties—Granny Smith and Braeburn showed that anthocyanin-free Granny Smith apples were significantly low in chlorophyll and carotenoid content. At the same time, sun-light exposed Braeburn had a higher content of chlorophyll and carotenoid (Solovchenko and Schmitz-Eiberger 2003).

6.7.1.3 Influence of Flavonoids on Auxin and Essential Elements

By controlling auxin transport, flavonoids can change how plants respond to stress. Auxins are important phytohormones that play a role in gene transcription, mitotic transition, and root development. Plant stress-induced morphogenic responses, such as flight strategies of sessile organisms when exposed to unfavourable environments, may be greatly enhanced by flavonoids via auxin movement. In sunny environments, the dwarf, bushy phenotype with few, small, and thick leaves protects deep-in-the-canopy leaves from light-induced disturbances of their cellular homeostasis. Additionally, flavonoids may increase the availability of nutrients. Flavonoids are transported to the soil by ABC transporters during times of low nutrition availability to allow them to bind with necessary metal ions for plant growth. Quercetin, genistein, and kaempferol, for instance, alter Fe availability by chelating iron oxides containing the inaccessible iron and reducing Fe^{3+} to Fe^{2+} (Kumar and Pandey 2013).

6.7.2 Flavonoids against Biotic Stresses

6.7.2.1 Flavonoids as Nematicides

Parasitic nematodes significantly reduce crop yield by forming cysts or galls on the roots. Nematode invasion frequently stimulates flavonol synthesis and accumulation in plant root systems. Plants can better deal with nematode infections due to the induced synthesis of these phytoalexins. When infected with *Pratylenchus penetrans*, lima beans produce coumestrol, an isoflavonoid that kills this nematode.

When infected by *Meloidogyne incognita*, soybean roots produce glyceollin, a protective isoflavonoid. O-methyl-apigenin-C-deoxyhexoside-O-hexoside (a flavone-C-glycoside) was identified as the primary phytoalexin in oat roots and shoots when infected with major cereal nematodes (Shah and Smith 2020).

6.7.2.2 Antibacterial Effects of Flavonoids

Several studies have been well-understood regarding the interaction of flavonoids with fungal microorganisms. However, phytoalexin synthesis in response to phytopathogenic bacterial invasion is poorly understood. Hence, attempts have been made by several scientists to identify antibacterial flavonoids. By inoculating bean plants with *Pseudomonas* spp., researchers investigated the production of antibacterial phytoalexins in the bean leaves. The phytoalexin isoflavonoid coumestrol inhibited the growth of two pathogenic bacterial species: *P. phaseolicola* and *P. marsprunorum,* by accumulating in infected bean leaves. Therefore, the inhibition of bacterial colonization in susceptible and hypersensitive bean leaf lesions was explained by the accumulation of coumestrol (Lyon and Wood 1975). New anti-disease products were developed due to recent research into natural remedies for biotic stress. The isoflavonoids isolated from *Erythrina poeppigiana* roots were tested against *Staphylococcus aureus* and compared with five other root isolates. The results showed the isoflavonoids exhibited strong inhibitory properties against the inoculated pathogens. Additionally, it was believed that novel compounds could be effective antibiotics against *S. aureus* infections (Tanaka et al. 2004). Another study showed that both pathogenic (*Pseudomonas glycinea*) and non-pathogenic (*P. lachrymans*) inoculations of soybean leaves resulted in the accumulation of phytoalexins-coumestrol and daidzein. Coumestrol inhibits colonization by pathogenic bacteria. It was also suggested that soybean leaves were resistant to *P. glycinea* because of the resistance provided by the induced accumulation of isoflavonoids (Keen and Kennedy 1974).

6.7.3 Flavonoids in Disease Management

In the previous units, we discussed the resistance strategies flavonoids employ against fungal pathogens. There has been a growing emphasis on using safer biopesticides in recent years, and flavonoids appear to be ideal candidates for safe application in the fields. The presence of flavonoids imparts resistance against a wide array of pathogenic fungi. Flavanes, flavanones, and isoflavones are considered efficient anti-microbial agents. More and more research has been conducted to identify the flavonoids responsible for imparting resistance against specific fungi species. Proanthocyanidins or dihydroquercetin, even in trace amounts, have been shown to protect barley mutants from *Fusarium* sp. To form a physical barrier against pathogen attack, mechanisms such as inhibition of pathogen cellulases, pectinases, and xylanases, cross-linking of microbial enzymes, chelation of metal ions necessary for enzyme activity, or tightening of cell walls are involved (Skadhauge et al. 1997).

Fungal disease cedar-apple rust, caused by *Gymnosporangium yamadai,* significantly damages apple trees, resulting in fruit of lower quality and size, ultimately lowering fruit value in the marketplace. A study showed that in symptomatic rust-infected tissue (RIT), the content of flavonoid compounds, particularly anthocyanin and catechin, significantly increased. The expression levels of MYB transcription factors and structural genes connected with flavonoid biosynthesis were one to seven-fold higher in the RIT. The genes CHS, ANS, FLS, DFR, and MYB10 demonstrated more than a tenfold increase, indicating that their expression was significantly higher in the RIT (Lu et al. 2017). Another study demonstrated insecticidal properties against *Drosophila melanogaster* and *Spodoptera frugiperda* were demonstrated by Calceolaria extracts containing several secondary metabolites, including flavonoids. These extracts also had antifungal and antibacterial properties, according to the researchers. Cultivation of flax suffers the greatest losses from fusarium infection. Genetic engineering with traditional cross-breeding can produce new plant varieties with increased flavonoid concentrations to show resistance against these species (Mierziak et al. 2014b).

Du et al. (2010) studied resistance activity in sorghum varieties when inoculated with anthracnose pathogen *Colletotrichum sublineolum.* Seedlings of the sorghum-resistant cultivar SC748–5 accumulated luteolin more rapidly and more extensively than those of the susceptible cultivar BTx623, where apigenin was present as the major flavone in the infected seedlings (Bag et al. 2022a). The experiment showed that luteolin inhibited *C. sublineolum* spore germination more effectively than apigenin in a dose-dependent manner. At 20 μM, spore germination of the fungal pathogen was reduced by about 60%; at 50 μM, it was reduced by more than 80%. Both flavone aglycones were regarded as sorghum phytoalexins due to their antifungal and pathogen-inducible properties. The study also identified a pathogen-induced gene encoding cytochrome P450 protein (CYP93G3).

Charcoal rot disease affects many plant species and is brought on by the soil-borne necrotrophic fungi *Macrophomina phaseolina.* This disease cannot be effectively controlled, and till now, no *M. phaseolina*-resistant host cultivar has been found. By using real-time quantitative PCR technique (RT-qPCR), the expression of 14 *M. truncatula* genes in response to *M. phaseolina* infection was examined to study the host–pathogen interaction on a molecular level. Flavonoid/isoflavonoid biosynthesis was one of these genes' various disease response pathways. According to their findings, the phenylpropanoid pathway in legumes contains several essential enzymes required for the plant biosynthesis of flavonoids and isoflavonoids. Many genes involved in the biosynthesis of isoflavonoids and dihydro flavonols were also analysed, including flavone synthase (FS), isoflavone synthase (IFS), isoflavone reductase (IFR), and vestitone reductase (VR). At 1 dpi, most of these genes showed more expressed, and at 2 dpi, all were induced more than two-fold. These genes were activated more strongly in the shoot than in the root. In two dpi shoot samples, the expression of several genes had increased by more than 50-fold, including CHI, CHR, CHS, IFR, FS, and VR. *M. truncatula* plants were also found to have partial resistance when treated with ET or methyl jasmonate (MJ) (Gaige et al. 2010).

Flavonoids can also indirectly improve plant resistance against fungi by increasing essential ion availability and aiding in the overall development of plants. In a pot study comparing faba bean as mono-crop to faba bean intercropped with wheat, it was found that the intercropping increased the number of nodules and the dry weight of the beans (Bag et al. 2022b). Flavonoids, isoflavones, chalcones, and hesperetin were likely responsible for the increase in faba bean nodulation. In addition, it helped in producing five different kinds of flavonoids in wheat than that found in monocropping. Their also suggested that faba bean nodulation and nitrogen fixation were positively affected by flavonoids in root exudates, present only in the intercropping system (Liu et al. 2017).

6.8 Challenges in the Use of Flavonoids

Flavonoids play a crucial role in root–rhizosphere interaction. But there are many challenges before they can be used in managing fungal pathogens and other microorganisms (Hassan and Mathesius 2012).

- Flavonoids have typically been studied under laboratory conditions. Still, records of their functionality in natural rhizosphere conditions need to be collected to determine how effective they remain under natural soil conditions of degradation, adsorption, digestion, metabolism and altered solubility under natural states of breakdown, adsorption, digestion, metabolism, and altered solubility in the soil.
- Flavonoids may have unanticipated effects on non-target organisms. Some flavonoids may improve host plant nodulation or mycorrhization. However, they can also indirectly affect bacterial quorum sensing, plant–soil interactions, and soil biochemistry. Catechin, for instance, is a potent allelopathic signal that may suppress plant growth and prevent quorum sensing in host-related soil bacteria. Similarly, quorum sensing regulation in non-target bacteria may be affected by naringenin, which is also responsible for causing the expression of nod genes in some rhizobia. Soybean isoflavonoids can attract the destructive pathogen *P. sojae* and its symbiont *B. japonicum* to the plant (Xiao et al. 2022).
- Any change in the form or concentration of the flavonoids found in plant tissue has altered effects on the plant. These alterations might be beneficial or detrimental to the whole plant-microbe interaction system. It also altered the cell morphology, root growth, gravitropism, and muted responses by influencing auxin transport.
- Flavonoids affect the structure of the soil microbial community. It may serve as energy source for the existing microbial populations or newly introduced species and also inhibited the production of phytoalexins. Mutation also affect exudates, it was demonstrated that the *Arabidopsis* abcg30 mutant's increased the exudation of phenolics had a wide range of effects on the structure of bacterial and fungal communities.
- Altering fluxes through other branches of flavonoid or related pathways may be a disadvantage of manipulating particular branches of the flavonoid biosynthesis pathway. This has been cited as one factor in the failure to achieve sufficient flavonoid accumulation required for production at a large scale in genetically modified plants.

- To change the biology of the rhizosphere, manipulating the flavonoid pathway will require more in-depth knowledge of how flavonoid transport and exudation are controlled. For specific flavonoids to be exuded into the rhizosphere, the altered expression must be coupled with rhizosphere signals that specifically induce the desired pathways. Knowing which promoters and transcription factors are explicit for triggering flavonoid induction would be necessary.
- The current extraction and chemical synthesis methods used in flavonoid production are inadequate to meet the commercial demand. Flavonoid production through microbe may provide a promising alternative way. However, production titers are restricted to mgL^{-1}, making industrial use of these strains difficult and expensive. In addition to engineering the production strain to maximize the availability of precursors, extensive engineering is required for the pathway itself. Also, the strategies currently used result in various microbial strains; efficient and thorough screening techniques are required to identify the suitable strains with the best performance results (Delmulle et al. 2018).

6.9 Conclusions and Future Prospects

Flavonoids are biologically active molecules that play various roles in eukaryotic and prokaryotic organisms, especially regarding plant physiology, growth, and development. Flavonoids can interact with other plants and microorganisms thanks to their physical and biochemical properties, which also protect from abiotic factors. Very low concentrations are required for the activity, indeed much lower than is necessary for effective radical quenching, suggesting a regulatory function for flavonoids, which influence plant development via interaction with an auxin transport network. Their nuclear localization suggests that they function as transcriptional regulators.

Mycorrhizal interactions already serve this purpose through the regulation of microbial genes. To efficiently engineer plant-beneficial microbiomes, modulating specific microbial populations through flavonoid metabolism to include loss-of-function or gain-of-function in the system may emerge as a promising approach. Research into the manipulation of flavonoids is expected to expand, allowing farmers to take advantage of these bioactive chemicals and further their agricultural progress.

References

Abdel-Lateif K, Bogusz D, Hocher V (2012) The role of flavonoids in the establishment of plant roots endosymbioses with arbuscular mycorrhiza fungi, rhizobia and *Frankia* bacteria. Plant Signal Behav 7:636–641

Agati G, Azzarello E, Pollastri S, Tattini M (2012) Flavonoids as antioxidants in plants: location and functional significance. Plant Sci 196:67–76

Ahmad A, Kaleem M, Ahmed Z, Shafiq H (2015) Therapeutic potential of flavonoids and their mechanism of action against microbial and viral infections-a review. Food Res Int 77:221–235

Al Aboody MS, Mickymaray S (2020) Anti-fungal efficacy and mechanisms of flavonoids. Antibiotics 9:45

Ali K, Maltese F, Figueiredo A, Rex M, Fortes AM, Zyprian E, Pais MS, Verpoorte R, Choi YH (2012) Alterations in grapevine leaf metabolism upon inoculation with *Plasmopara viticola* in different time-points. Plant Sci 191:100–107

Alov P, Tsakovska I, Pajeva I (2015) Computational studies of free radical-scavenging properties of phenolic compounds. Curr Top Med Chem 15:85–104

Alseekh S, de Souza LP, Benina M, Fernie AR (2020) The style and substance of plant flavonoid decoration; towards defining both structure and function. Phytochemistry 174:112347

Alzand KI, Mohamed MA (2012) Flavonoids: chemistry, biochemistry, and antioxidant activity. J Pharm Res 5:4013–4020

Amić D, Davidović-Amić D, Bešlo D, Trinajstić N (2003) Structure-radical scavenging activity relationships of flavonoids. Croat Chem Acta 76:55–61

Askari SH, Ashraf MA, Ali S, Rizwan M, Rasheed R (2021) Menadione sodium bisulfite alleviated chromium effects on wheat by regulating oxidative defense, chromium speciation, and ion homeostasis. Environ Sci Pollut Res 28:36205–36225

Bag S, Mondal A, Majumder A, Mondal SK, Banik A (2022a) Flavonoid mediated selective cross-talk between plants and beneficial soil microbiome. Phytochem Rev 3:1739–1760

Bag S, Mondal A, Majumder A, Mondal SK, Banik A (2022b) Flavonoid mediated selective cross-talk between plants and beneficial soil microbiome. Phytochem Rev Proc Phytochem Soc Eur 21:1739–1760

Bais AF, Lucas RM, Bornman JF, Williamson CE, Sulzberger B, Austin AT, Wilson SR, Andrady AL, Bernhard G, McKenzie RL, Aucamp PJ (2018) Environmental effects of ozone depletion, UV radiation and interactions with climate change: UNEP environmental effects assessment panel, update 2017. Photochem Photobiol Sci 17:127–179

Banasiak J, Jamruszka T, Murray JD, Jasiński M (2021) A road map of plant membrane transporters in arbuscular mycorrhizal and legume-rhizobium symbioses. Plant Physiol 187:2071–2091

Bécard G, Douds DD, Pfeffer PE (1992) Extensive in vitro hyphal growth of vesicular-arbuscular mycorrhizal fungi in the presence of CO_2 and flavonols. Appl Environ Microbiol 58:821–825

Bergman M, Perelman A, Dubinsky Z, Grossman S (2003) Scavenging of reactive oxygen species by a novel glucurinated flavonoid antioxidant isolated and purified from spinach. Phytochemistry 62:753–762

Blount JW, Dixon RA, Paiva NL (1992) Stress responses in alfalfa (*Medicago sativa* L). Antifungal activity of medicarpin and its biosynthetic precursors implications for the genetic manipulation of stress metabolites. Physiol Mol Plant Pathol 41:333–349

Bollina V, Kumaraswamy GK, Kushalappa AC, Choo TM, Dion Y, Rioux S, Faubert D, Hamzehzarghani H (2010) Mass spectrometry-based metabolomics application to identify quantitative resistance-related metabolites in barley against *Fusarium* head blight. Mol Plant Pathol 11:769–782

Bouyahya AE, Omari N, Hakkour M, El Menyiy N, Benali T, Kulikov D, Chamkhi I (2022) A review on transcriptomic and metabolomic responses of plants to nanopollution. Environ Sci Pollut Res 29:22913–22929

Brisson VL, Schmidt JE, Northen TR, Vogel JP, Gaudin ACM (2019) Impacts of maize domestication and breeding on rhizosphere microbial community recruitment from a nutrient depleted agricultural soil. Sci Rep 9:15611

Brodowska KM (2017) Natural flavonoids: classification, potential role, and application of flavonoid analogues. Eur J Biol Res 7:108–123

Brunetti C, Fini A, Sebastiani F, Gori A, Tattini M (2018) Modulation of phytohormone signaling: a primary function of flavonoids in plant-environment interactions front. Plant Sci 9:1042

Buer CS, Imin N, Djordjevic MA (2010) Flavonoids: new roles for old molecules. J Integr Plant Biol 52:98–111

Butelli E, Titta L, Giorgio M, Mock HP, Matros A, Peterek S, Martin C (2008) Enrichment of tomato fruit with health-promoting anthocyanins by expression of select transcription factors. Nat Biotechnol 26:1301–1308

Castano-Duque L, Gilbert MK, Mack BM, Lebar MD, Carter-Wientjes CH, Sickler CM, Cary JW, Rajasekaran K (2021) Flavonoids modulate the accumulation of toxins from *Aspergillus flavus* in maize kernels. Front Plant Sci 12:761446

Castano-Duque L, Lebar MD, Carter-Wientjes C, Ambrogio D, Rajasekaran K (2022) Flavonoids modulate *Aspergillus flavus* proliferation and aflatoxin production. J Fungi 8:1211

Catalán M, Ferreira J, Carrasco-Pozo C (2020) The microbiota-derived metabolite of quercetin, 3,4-dihydroxyphenylacetic acid prevents malignant transformation and mitochondrial dysfunction induced by hemin in colon cancer and normal colon epithelia cell lines. Molecules 25:18

Cesco S, Neumann G, Tomasi N, Pinton R, Weisskopf L (2010) Release of plant-borne flavonoids into the rhizosphere and their role in plant nutrition. Plant Soil 329:1–25

Cesco S, Mimmo T, Tonon G, Tomasi N, Pinton R, Terzano R, Nannipieri P (2012) Plant-borne flavonoids released into the rhizosphere: impact on soil bio-activities related to plant nutrition. Biol Fertil Soils 48:123–149

Clocchiatti A, Hannula SE, van den Berg M, Hundscheid MPJ, de Boer W (2021) Evaluation of phenolic root exudates as stimulants of saptrophic fungi in the rhizosphere. Front Microbiol 12:644046

Colombo C, Palumbo G, He JZ, Pinton R, Cesco S (2014) Review on iron availability in soil: interaction of Fe minerals, plants, and microbes. J Soils Sediments 14:538–548

Cooper JE (2004) Multiple responses of rhizobia to flavonoids during legume root infection. Adv Bot Res 41:1–62

Cooper JE (2007) Early interactions between legumes and rhizobia: disclosing complexity in a molecular dialogue. J Appl Microbiol 103:1355–1365

D'Andrea G (2015) Quercetin: a flavonol with multifaceted therapeutic applications. Fitoterapia 106:256–271

da Silva DAF, Cotta SR, Vollú RE, Jurelevicius D, Marques JM, Marriel IE, Seldin L (2014) Endophytic microbial community in two transgenic maize genotypes and in their near-isogenic non-transgenic maize genotype. BMC Microbiol 14:332

D'Arcy MS (2022) A review of biologically active flavonoids as inducers of autophagy and apoptosis in neoplastic cells and as cytoprotective agents in non-neoplastic cells. Cell Biol Int 46:1179–1195

Davies KM, Albert NW, Schwinn KE (2012) From landing lights to mimicry: the molecular regulation of flower colouration and mechanisms for pigmentation patterning. Funct Plant Biol 39:619–638

de Rijke E, Out P, Niessen WMA, Ariese F, Gooijer C, Brinkman UAT (2006) Analytical separation and detection methods for flavonoids. J Chromatogr A 1112:31–63

Delmulle T, De Maeseneire SL, De Mey M (2018) Challenges in the microbial production of flavonoids. Phytochem Rev 17:229–247

Derksen H, Rampitsch C, Daayf F (2013) Signaling cross-talk in plant disease resistance. Plant Sci 207:79–87

Donadio G, Mensitieri F, Santoro V (2021) Interactions with microbial proteins driving the antibacterial activity of flavonoids. Pharmaceutics 13:660–2021

Du Y, Chu H, Wang M, Chu IK, Lo C (2010) Identification of flavone phytoalexins and a pathogen-inducible flavone synthase II gene (SbFNSII) in sorghum. J Exp Bot 61:983–994

Durango D, Pulgarin N, Echeverri F, Escobar G, Quinones W (2013) Effect of salicylic acid and structurally related compounds in the accumulation of phytoalexins in cotyledons of common bean (*Phaseolus vulgaris* L.) cultivars. Molecules 18:10609–10628

Eckardt NA (2006) The role of flavonoids in root nodule development and auxin transport in *Medicago truncatula*. Plants Cell 18:1539–1540

Eleazu C, Eleazu K, Kalu W (2017) Management of benign prostatic hyperplasia: could dietary polyphenols be an alternative to existing therapies. Front Pharmacol 8:234

Erfani-Moghadam J, Zarei A (2022) Differential responses of pear cultivars to *Erwinia amylovora* infection; evidences of involvement the hypersensitivity response in pear resistance to fire blight. Eur J Plant Pathol 162:927–943

Faghih S, Zarei A, Ghobadi C (2019) Positive effects of plant growth regulators on physiology responses of *Fragaria × ananassa* cv. 'Camarosa' under salt stress. Int J Fruit Sci 19:104–114
Foster-Hartnett D, Danesh D, Penuela S, Sharopova N, Endre G, Vandenbosch KA, Young ND, Samac DA (2007) Molecular and cytological responses of *Medicago truncatula* to *Erysiphe pisi*. Mol Plant Pathol 8:307–319
Frey-Klett P, Garbaye J, Tarkka M (2007) The mycorrhiza helper bacteria revisited. New Phytol 176:22–36
Gadkari PV, Balaraman M (2015) Catechins: sources, extraction and encapsulation: a review. Food Bioprod Process 93:122–138
Gaige AR, Ayella A, Shuai B (2010) Methyl jasmonate and ethylene induce partial resistance in *Medicago truncatula* against the charcoal rot pathogen *Macrophomina phaseolina*. Physiol Mol Plant Pathol 74:412–418
Gao SS, Wang YL, Yua S, Huang YQ, Liu H, Chen W, He XY (2020) Effects of drought stress on growth, physiology and secondary metabolites of two *Adonis* species in Northeast China. Sci Horti 259:108795
Ghitti E, Rolli E, Crotti E, Borin S (2022) Flavonoids are intra-and inter-kingdom modulator signals. Microorganisms 10:2479
Glazebrook J (2005) Contrasting mechanisms of defense against biotrophic and necrotrophic pathogens. Annu Rev Phytopathol 43:205–227
Gomes B, Castro F, Santos R, Figueiredo P, Silva M, Vidal M, Ferreira I, Nunes J, Machado H, Gomes F (2021) Effect of quercetin on mycorrhizal synthesis between *tuber borchii* and *Arbutusu nedo* L. *in vitro* plants. Microbiol Res 12:69–81
Górniak I, Bartoszewski R, Króliczewski J (2019) Comprehensive review of antimicrobial activities of plant flavonoids. Phytochem Rev 18:241–272
Graham TL (1991) Flavonoid and isoflavonoid distribution in developing soybean seedling tissues and in seed and root exudates. Plant Physiol 95:594–603
Greenberg JT, Yao N (2004) The role and regulation of programmed cell death in plant-pathogen interactions. Cell Microbiol 6:201–212
Gu L, Kelm MA, Hammerstone JF, Beecher G, Holden J, Haytowitz D, Prior RL (2003) Screening of foods containing proanthocyanidins and their structural characterization using LC-MS/MS and thiolytic degradation. J Agric Food Chem 51:7513–7521
Gupta AD, Bansal VK, Babu V, Maithil N (2013) Chemistry, antioxidant and antimicrobial potential of nutmeg (*Myristica fragrans* Houtt). J Genet Eng Biotechnol 11:25–31
Hajji HE, Nkhili E, Tomao V, Dangles O (2006) Interactions of quercetin with iron and copper ions: complexation and autoxidation. Free Radic Res 40:303–320
Haraguchi H, Yoshida N, Ishikawa H, Tamura Y, Mizutani K, Kinoshita T (2010) Protection of mitochondrial functions against oxidative stresses by isoflavans from *Glycyrrhiza glabra*. J Pharm Pharmacol 52:219–223
Hartley A, Barger N, Belnap J, Okin GS (2007) Dryland ecosystems. In: Marschner P, Rengel Z (eds) Nutrient cycling in terrestrial ecosystems. Springer-Verlag, Berlin, pp 271–308
Hartwig UA, Maxwell CA, Joseph CM, Phillips DA (1990) Chrysoeriol and luteolin released from alfalfa seeds induce nod genes in *Rhizobium meliloti*. Plant Physiol 92:116–122
Hasegawa M, Mitsuhara I, Seo S, Okada K, Yamane H, Iwai T, Ohashi Y (2014) Analysis on blast fungus-responsive characters of a flavonoid phytoalexin sakuranetin; accumulation in infected rice leaves, antifungal activity and detoxification by fungus. Molecules 19:11404–11418
Hassan S, Mathesius U (2012) The role of flavonoids in root–rhizosphere signalling: opportunities and challenges for improving plant–microbe interactions. J Exp Bot 63:3429–3444
He D, Singh SK, Peng L, Kaushal R, Vílchez JI, Shao C, Wu X, Zheng S, Morcillo RJL, Pare PW, Zhang H (2022) Flavonoid attracted *Aeromonas* sp. from the *Arabidopsis* root microbiome enhances plant dehydration resistance. ISME J 16:2622–2632
Islam S (2016) Anthocyanin compositions in different colored *gladiolus* species: a source of natural food colorants. Am J Food Sci Technol 4:109–114
Iwashina T (2003) Flavonoid function and activity to plants and other organisms. Biol Sci Space 17:24–44

Jan R, Khan AM, Asaf S, Park JR, Lee IJ, Kim KM (2021) Flavonone 3-hydroxylase relieves bacterial leaf blight stress in rice via overaccumulation of antioxidant flavonoids and induction of defense genes and hormones. Int J Mol Sci 22:6152

Jan R, Khan MA, Asaf S, Waqas M, Park JR, Asif S, Kim N, Lee IJ, Kim KM (2022) Drought and UV radiation stress tolerance in rice is improved by over accumulation of non-enzymatic antioxidant flavonoids. Antioxidants 11:917

Janda M, Ruelland E (2015) Magical mystery tour: salicylic acid signalling. Environ Exp Bot 114:117–128

Jimenez-Gonzalez L, Alvarez-Corral M, Munoz-Dorado M, Rodriguez-Garcia I (2008) Pterocarpans: interesting natural products with antifungal activity and other biological properties. Phytochem Rev 7:125–154

Jørgensen K, Rasmussen AV, Morant M, Nielsen AH, Bjarnholt N, Zagrobelny M, Møller BL (2005) Metabolon formation and metabolic channeling in the biosynthesis of plant natural products. Curr Open Plant Biol 8:280–291

Kaliamoortiiy S, Rao A (1994) Effect of salinity on anthocyanin accumulation in the root of maize. Science 248:1637–1638

Kapulnik Y, Guenoune D, Galili S, Phillips DA, Volpin H, Chet I, Okon Y (2001) The defense response elicited by the pathogen *Rhizoctonia solani* is suppressed by colonization of the AM-fungus *Glomus intraradices*. Plant Sci 160:925–932

Karak P (2019) Biological activities of flavonoids: an overview. Int J Pharm Sci Res 10:1567–1574

Karre S, Kumar A, Yogendra K, Kage U, Kushalappa A, Charron JB (2019) HvWRKY23 regulates flavonoid glycoside and hydroxycinnamic acid amide biosynthetic genes in barley to combat *Fusarium* head blight. Plant Mol Biol 100:591–605

Katsumata S, Toshima H, Hasegawa M (2018) Xylosylated detoxification of the rice flavonoid phytoalexin sakuranetin by the rice sheath blight fungus *Rhizoctonia solani*. Molecules 23:276

Kaur S, Roy A (2021) A review on the nutritional aspects of wild edible plants. Curr Trad Med 7:552–563

Kazan K, Lyons R (2014) Intervention of phytohormone pathways by pathogen effectors. Plant Cell 26:2285–2309

Keen N, Kennedy B (1974) Hydroxyphaseollin and related isoflavanoids in the hypersensitive resistance reaction of soybeans to *Pseudomonas glycinea*. Physiol Plant Pathol 4:173–185

Kogawa K, Kazuma K, Kato N, Noda N, Suzuki M (2007) Biosynthesis of malonylated flavonoid glycosides on the basis of malonyltransferase activity in the petals of *Clitoria ternatea*. J Plant Physiol 164:886–894

Kumar S, Pandey AK (2013) Chemistry and biological activities of flavonoids: an overview. Sci World J 2013:162750

Kwon SM, Kim S, Song NJ, Chang SH, Hwang YJ, Yang DK, Hong JW, Park WJ, Park KW (2016) Antiadipogenic and proosteogenic effects of luteolin, a major dietary flavone, are mediated by the induction of DnaJ (Hsp40) homolog, subfamily B, member 1. J Nutri Biochem 30:24–32

Larose G, Chênevert R, Moutoglis P, Gagné S, Piché Y, Vierheilig H (2002) Flavonoid levels in roots of Medicago sativa are modulated by the developmental stage of the symbiosis and the root colonizing arbuscular mycorrhizal fungus. J Plant Physiol 159:1329–1339

Lattanzio V, Lattanzio VMT, Cardinali A (2006) Role of phenolics in the resistance mechanism of plants against fungal pathogens and insects. In: Imperato F (ed) Phytochemistry: advances in research. Research Signpost, Trivandrum, pp 23–67

Li B, Fan R, Huang S, Lei P, Guo JX, Zhan QD, Zhao X, Song CP (2017) Far infrared imaging, an effective way to screen maize seedling mutants for drought stress response. Biology 72:1010–1016

Li B, Fan R, Sun G, Sun T, Fan Y, Bai S, Guo S, Huang S, Liu J, Zhang H, Wang P (2021) Flavonoids improve drought tolerance of maize seedlings by regulating the homeostasis of reactive oxygen species. Plant Soil 461:389–405

Lindsay WL (1991) Inorganic equilibria affecting micronutrients in soils. Micronutri Agric 4:89–112

Lira MA Jr, Nascimento LR, Fracetto GG (2015) Legume-rhizobia signal exchange: promiscuity and environmental effects. Front Microbiol 6:945

Liu M, Li X, Liu Y, Cao B (2013) Regulation of flavanone 3-hydroxylase gene involved in the flavonoid biosynthesis pathway in response to UV-B radiation and drought stress in the desert plant, *Reaumuria soongorica*. Plant Physiol Biochem 73:161–167

Liu YC, Xiao MQ, Jing XX, Li T, Chi ZW, Jin JW, Yi Z (2017) Intercropping influences component and content change of flavonoids in root exudates and nodulation of Faba bean. J Plant Interact 12:187–192

Long L, Liu J, Gao Y, Xu FC, Zhao JR, Li B, Gao W (2019) Flavonoid accumulation in spontaneous cotton mutant results in red coloration and enhanced disease resistance. Plant Physiol Biochem 143:40–49

Lu Y, Chen Q, Bu Y, Luo R, Hao S, Zhang J, Tian J, Yao Y (2017) Flavonoid accumulation plays an important role in the rust resistance of malus plant leaves. Front Plant Sci 8:1286

Luthra R, Roy A, Pandit S, Prasad R (2021) Biotechnological methods for the production of ginsenosides. South Afr J Bot 141:25–36

Lyon FM, Wood R (1975) Production of phaseollin, coumestrol and related compounds in bean leaves inoculated with *Pseudomonas* spp. Physiol Plant Pathol 6:117–124

Ma D, Sun D, Wang C, Li Y, Guo T (2014) Expression of flavonoid biosynthesis genes and accumulation of flavonoid in wheat leaves in response to drought stress. Plant Physiol Biochem 80:60–66

Malnoy M, Martens S, Norelli JL, Barny MA, Sun-din GW, Smits THM, Dufy B (2012) Fire blight: applied genomic insights of the pathogen and host. Annu Rev Phytopathol 50:475–494

Marsafari M, Samizadeh H, Rabiei B, Mehrabi A, Koffas M, Xu P (2020) Biotechnological production of flavonoids: an update on plant metabolic engineering, microbial host selection, and genetically encoded biosensors. Biotechnol J 15:1900432

Marschner H (1988) Mechanisms of manganese acquisition by roots from soils. In: Manganese in soils and plants: proceedings of the international symposium on 'manganese in soils and plants' held at the Waite agricultural research institute, the University of Adelaide, Glen Osmond, South Australia, August 22–26th, 1988 as an Australian bicentennial event. Springer, Amsterdam, pp 191–204

Marschner H (1995) Mineral nutrition of higher plants, 2nd edn. Institute of Plant Nutrition University of Hohenheim, Stuttgart

Martens S, Mithöfer A (2005) Flavones and flavone synthases. Phytochemistry 66:2399–2407

Mierziak J, Kostyn K, Kulma A (2014a) Flavonoids as important molecules of plant interactions with the environment. Molecules 19:16240–16265

Mierziak J, Wojtasik W, Kostyn K, Czuj T, Szopa J, Kulma A (2014b) Cross breeding of transgenic flax plants overproducing flavonoids and glucosyltransferase results in progeny of improved antifungal and antioxidative properties. Mol Breed 34:1917–1932

Mishra M, Singh SK, Kumar A (2021) Microbial consortia: approaches in crop production and yield enhancement. In: Microbiome stimulants for crops. Woodhead Publishing, Cambridge, London, pp 293–303

Miyake Y, Shimoi K, Kumazawa S, Yamamoto K, Kinae N, Osawa T (2000) Identification and antioxidant activity of flavonoid metabolites in plasma and urine of eriocitrin-treated rats. J Agric Food Chem 48:3217–3224

Nabavi SM, Šamec D, Tomczyk M, Milella L, Russo D, Habtemariam S, Shirooie S (2020) Flavonoid biosynthetic pathways in plants: versatile targets for metabolic engineering. Biotechnol Adv 38:107316

Nagata M, Yamamoto N, Miyamoto T, Shimomura A, Arima S, Hirsch AM, Suzuki A (2016) Enhanced hyphal growth of arbuscular mycorrhizae by root exudates derived from high R/FR treated *Lotus japonicus*. Plant Sig Behav 11:e1187356

Nakabayashi R, Mori T, Saito K (2014) Alternation of flavonoid accumulation under drought stress in *Arabidopsis thaliana*. Plant Sig Behav 9:e29518

Narasimhan K, Basheer C, Bajic VB, Swarup S (2003) Enhancement of plant-microbe interactions using a rhizosphere metabolomics driven approach and its application in the removal of polychlorinated biphenyls. Plant Physiol 132:146–153

Naskar S, Roy C, Ghosh S, Mukhopadhyay A, Hazarika LK, Chaudhuri RK, Chakraborti D (2021) Elicitation of biomolecules as host defense arsenals during insect attacks on tea plants (*Camellia sinensis* (L.) Kuntze). Appl Microbiol Biotechnol 105:7187–7199

Nguyen TLA, Bhattacharya D (2022) Antimicrobial activity of quercetin: an approach to its mechanistic principle. Molecules 27:2494

Okutani F, Hamamoto S, Aoki Y, Nakayasu M, Nihei N, Nishimura T, Yazaki K, Sugiyama A (2020) Rhizosphere modelling reveals spatiotemporal distribution of daidzein shaping soybean rhizosphere bacterial community. Plant Cell Environ 43:1036–1046

Olanrewaju OS, Ayangbenro AS, Glick BR, Babalola OO (2019) Plant health: feedback effect of root exudates-rhizobiome interactions. Appl Microbiol Biotechnol 103:1155–1166

Olsen RA, Bennett JH, Blume D, Brown JC (1981) Chemical aspects of the Fe stress response mechanism in tomatoes. J Plant Nutri 3:905–921

Padmavati M, Sakthivel N, Thara KV, Reddy AR (1997) Differential sensitivity of rice pathogens to growth inhibition by flavonoids. Phytochemistry 46:499–502

Parihar A, Grotewold E, Doseff AI (2015) Flavonoid dietetics: mechanisms and emerging roles of plant nutraceuticals. In: Parihar A, Grotewold E, Dodeff AI (eds) Pigments in fruits and vegetables. Springer, New York, NY, pp 93–126

Park SE, Ban S, Han AR, Lee SY, Choi C (2023) Identifying early response genes against apple Marssonina blotch via comparative gene expression analysis between resistant and susceptible cultivars. Physiol Mol Plant Pathol 125:101975

Pedras MSC, Ahiahonu PWK (2005) Metabolism and detoxification of phytoalexins and analogs by phytopathogenic fungi. Phytochemistry 66:391–411

Perry LG, Alford ER, Horiuchi J, Paschke MW, Vivanco JM (2007) Chemical signals in the rhizosphere: root-root and root--microbe communication. In: Pinton R, Varanini Z, Nannipieri P (eds) The rhizosphere: biochemistry and organic substances at the soil-plant Interface. CRC Press, Boca Raton, p 297

Pinton R, Varanini Z, Nannipieri P (2007) The rhizosphere: biochemistry and organic substances at the soil-plant Interface, 2nd edn. CRC Press, Boca Raton

Prior RL, Lazarus SA, Cao G, Muccitelli H, Hammerstone JF (2001) Identification of procyanidins and anthocyanins in blueberries and cranberries (*Vaccinium* spp.) using high performance liquid chromatography/mass spectrometry. J Agric Food Chem 49:1270–1276

Rice-Evans CA, Miller NJ, Paganga G (1996) Structure antioxidant activity relationships of flavonoids and phenolic acids. Free Radic Biol Med 20:933–956

Roeschlin RA, Favaro MA, Chiesa MA, Alemano S, Vojnov AA, Castagnaro AP, Filippone MP, Gmitter FG Jr, Gadea J, Marano MR (2017) Resistance to citrus canker induced by a variant of *Xanthomonas citri* sp. *citri* is associated with a hypersensitive cell death response involving autophagy associated vacuolar processes. Mol Plant Pathol 18:1267–1281

Roy A, Datta S, Bhatia KS, Bhumika JP, Prasad R (2021) Role of plant derived bioactive compounds against cancer. South Afr J Bot 149:1017–1028

Sahu BD, Kalvala AK, Koneru M, Mahesh Kumar J, Kuncha M, Rachamalla SS, Sistla R (2014) Ameliorative effect of fisetin on cisplatin-induced nephrotoxicity in rats via modulation of NF-κB activation and antioxidant defence. PLoS One 9:e105070

Salguero-Linares J, Coll NS (2019) Plant proteases in the control of the hypersensitive response. J Exp Bot 70:2087–2095

Sarmiento-López LG, López-Meyer M, Sepúlveda-Jiménez G, Cárdenas L, Rodríguez-Monroy M (2021) Arbuscular mycorrhizal symbiosis in *Stevia rebaudiana* increases trichome development, flavonoid and phenolic compound accumulation. Biocat Agric Biotechnol 31:101889

Saslowsky D, Winkel SB (2001) Localization of flavonoid enzymes in *Arabidopsis* roots. The Plant J 27:37–48

Saslowsky DE, Warek U, Winkel BS (2005) Nuclear localization of flavonoid enzymes in Arabidopsis. J Biol Chem 280:23735–23740

Scavo A, Abbate C, Mauromicale G (2019) Plant allelochemicals: agronomic, nutritional and ecological relevance in the soil system. Plant Soil 442:23–48

Scervino JM, Ponce MA, Erra-Bassells R, Bompadre J, Vierheilig H, Ocampo JA, Godeas A (2007) The effect of flavones and flavonols on colonization of tomato plants by arbuscular mycorrhizal fungi of the genera *Gigaspora* and *glomus*. Can J Microbiol 53:702–709

Scherr RE, Keen CL, Zidenberg-Cherr S (2016) Nutrition and health info sheet: Catechins and epicatechins. https://cns.ucdavis.edu/sites/g/files/dgvnsk416/files/inline-files/fact-pro-catechins.pdf. Accessed 10 Sept 2023

Schütz V, Frindte K, Cui J, Zhang P, Hacquard S, Schulze-Lefert P, Knief C, Schulz M, Dörmann P (2021) Deferential impact of plant secondary metabolites on the soil microbiota. Front Microbiol 12:666010

Segarra G, Van der Ent S, Trillas I, Pieterse CMJ (2009) MYB72, a node of convergence in induced systemic resistance triggered by a fungal and a bacterial beneficial microbe. Plant Biol 11:90–96

Senthilkumar P, Thirugnanasambantham K, Mandal AK (2012) Suppressive subtractive hybridization approach revealed differential expression of hypersensitive response and reactive oxygen species production genes in tea (*Camellia sinensis* (L.) O. Kuntze) leaves during *Pestalotiopsis thea* infection. Appl Biochem Biotechnol 168:1917–1927

Shah A, Smith DL (2020) Flavonoids in agriculture: chemistry and roles in, biotic and abiotic stress responses, and microbial associations. Agronomy 10:1209

Shah J, Chaturvedi R, Chowdhury Z, Venables B, Petros RA (2014) Signaling by small metabolites in systemic acquired resistance. Plant J 79:645–658

Shine MB, Xiao X, Kachroo P, Kachroo A (2019) Signaling mechanisms underlying systemic acquired resistance to microbial pathogens. Plant Sci 279:81–86

Siciliano I, Amaral Carneiro G, Spadaro D, Garibaldi A, Gullino ML (2015) Jasmonic acid, abscisic acid, and salicylic acid are involved in the phytoalexin responses of rice to *Fusarium fujikuroi*, a high gibberellin producer pathogen. J Agric Food Chem 63:8134–8142

Sies H (2020) Oxidative stress: concept and some practical aspects. Antioxidants 9:852

Sies H, Jones DP (2020) Reactive oxygen species (ROS) as pleiotropic physiological signalling agents. Nat Rev Mol Cell Biol 21:363–383

Singla P, Garg N (2017) Plant flavonoids: key players in signaling, establishment, and regulation of rhizobial and mycorrhizal endosymbiosis. In: Varma A, Prasad R, Tuteja N (eds) Mycorrhiza-function, diversity, state of the art. Springer, Cham, pp 133–176

Skadhauge B, Thomsen K, Von Wettstein D (1997) The role of barley testa layer and its flavonoid content in resistance to Fusarium infections. Hereditas 126:147–160

Solanki S, Ameen G, Sanyal D, Jain S, Elakhdar A, Lall S, Chittem K, Brueggeman L, Kumar A, Brueggeman R (2020) Friends and foes phyto-microbial interactions. In: Kumar M, Kumar V, Prasad R (eds) Phyto-microbiome in stress regulation, environmental and microbial biotechnology. Springer, Singapore, pp 81–98

Solovchenko A, Schmitz-Eiberger M (2003) Significance of skin flavonoids for UV-B-protection in apple fruits. J Exp Bot 54:1977–1984

Srivastava S, Chaudhry V, Mishra A, Chauhan PS, Rehman A, Yadav A, Tuteja N, Nautiyal CS (2012) Gene expression profiling through microarray analysis in *Arabidopsis thaliana* colonized by *pseudomonas putida* MTCC5279, a plant growth promoting rhizobacterium. Plant Sig Behav 7:235–245

Subramanian S, Graham MY, Yu O, Graham TL (2005) RNA interference of soybean isoflavone synthase genes lead to silencing in tissues distal to the transformation site and to enhanced susceptibility to *Phytophthora sojae*. Plant Physiol 137:1345–1353

Sugiyama A, Yazaki K (2014) Flavonoids in plant rhizospheres: secretion, fate and their effects on biological communication. Plant Biotechnol 31:431–443

Sun Z, Cheng Y, Qi F, Zhang M, Tian M, Wang J, Wu X (2022) Resistance of peanut to web blotch caused by *Phoma arachidicola* is related to papillae formation and the hypersensitive response. Plant Pathol 71:1921–1931

Tajammal A, Siddiqa A (2022) Antioxidant, molecular docking and computational investigation of new flavonoids. J Mol Struct 1254:132189
Talbi S, Rojas J, Sahrawy M, Rodríguez-Serrano M, Cárdenas K, Debouba M, Sandalio L (2020) Effect of drought on growth, photosynthesis and total antioxidant capacity of the Saharan plant *Oudeneya africana*. Environ Exp Bot 176:104099
Tanaka H, Sato M, Oh-Uchi T, Yamaguchi R, Etoh H, Shimizu H, Sako M, Takeuchi H (2004) Antibacterial properties of a new isoflavonoid from *Erythrina poeppigiana* against methicillin-resistant *Staphylococcus aureus*. Phytomed 11:331–337
Tinker PB, Nye PH (2000) Solute movement in the rhizosphere. Oxford University Press, USA, p 678
Tomasi N, Weisskopf L, Renella G, Landi L, Pinton R, Varanini Z, Cesco S (2008) Flavonoids of white lupin roots participate in phosphorus mobilization from soil. Soil Biol Biochem 40:1971–1974
Ton J, Van Pelt JA, Van Loon LC, Pieterse CM (2002) Differential effectiveness of salicylate dependent and jasmonate/ethylene-dependent induced resistance in *Arabidopsis*. Mol Plant-Microbe Interact 15:27–34
Treutter D (2005) Significance of flavonoids in plant resistance and enhancement of their biosynthesis. Plant Biol 7:581–591
Truchado P, Larrosa M, Castro-Ibáñez I, Allende A (2015) Plant food extracts and phytochemicals: their role as quorum sensing inhibitors. Trends Food Sci Technol 43:189–204
VanderEnt S, Verhagen BWM, VanDoorn R, Bakker D, Verlaan MG, Pel MJ, Pieterse CM (2008) MYB72 is required in early signaling steps of rhizobacteria-induced systemic resistance in *Arabidopsis*. Plant Physiol 146:1293–1304
Vierheilig H (2004) Regulatory mechanisms during the plant-arbuscular mycorrhizal fungus interaction. Can J Bot 82:1166–1176
Vikram A, Jayaprakasha GK, Jesudhasan PR, Pillai SD, Patil BS (2010) Suppression of bacterial cell–cell signalling, biofilm formation and type III secretion system by citrus flavonoids. J Appl Microbiol 109:515–527
Wang Y, Hao X, Lu Q, Wang L, Qian W, Li N, Ding C, Wang X, Yang Y (2018) Transcriptional analysis and histochemistry reveal that hypersensitive cell death and H_2O_2 have crucial roles in the resistance of tea plant (*Camellia sinensis* (L.) O. Kuntze) to anthracnose. Hortic Res 5:18
Wang L, Chen M, Lam PY, Dini-Andreote F, Dai L, Wei Z (2022) Multifaceted roles of flavonoids mediating plant-microbe interactions. Microbiome 10:233
Weisskopf L, Tomasi N, Santelia D, Martinoia E, Langlade NB, Tabacchi R, Abou-Mansour E (2006) Isoflavonoid exudation from white lupin roots is influenced by phosphate supply, root type and cluster-root stage. New Phytol 171:657–668
Werner D (2007) Molecular biology and ecology of the rhizobia-legume symbiosis. The rhizosphere. Biochemistry and organic substances at the soil-plant Interface, 2nd edn. CRC Press, Boca Raton, pp 237–266
Weston LA, Mathesius U (2013) Flavonoids: their structure, biosynthesis and role in the rhizosphere, including allelopathy. J Chem Ecol 39:283–297
Williams RJ, Spencer JPE, Rice-Evans C (2004) Flavonoids: antioxidants or signalling molecules. Free Rad Biol Med 36:838–849
Wojtaszek P, Stobiecki M, Gulewicz K (1993) Role of nitrogen and plant growth regulators in the exudation and accumulation of isoflavonoids by roots of intact white lupin plants (*Lupinus albus* L.). J Plant Physiol 142:689–694
Xiao L, Sun Y, Tsao R (2022) Paradigm shift in phytochemicals research: evolution from antioxidant capacity to anti-inflammatory effect and to roles in gut health and metabolic syndrome. J Agric Food Chem 70:8551–8568
Xingpeng L, Xiu D, Huang J, Yu B, Jia S, Song L (2023) Nutshell physicochemical characteristics of different hazel cultivars and their defensive activity toward *curculio nucum* (Coleoptera: Curculionidae). Forests 14:319
Yazaki K, Sasaki K, Tsurumaru Y (2009) Prenylation of aromatic compounds, a key diversification of plant secondary metabolites. Phytochemistry 70:1739–1745

Yeung J, Yu TF (2003) Effects of isoflavones (soy phyto-estrolgens) on serum lipids: a meta–analysis of randomized controlled trials. Nutri J 2:1–8
Yonekura-Sakakibara K, Higashi Y, Nakabayashi R (2019) The origin and evolution of plant flavonoid metabolism. Front Plant Sci 10:943
Yu O, Shi J, Hession AO, Maxwell CA, McGonigle B, Odell JT (2003) Metabolic engineering to increase isoflavone biosynthesis in soybean seed. Phytochemistry 63:753–763
Yu P, He X, Baer M, Beirinckx S, Tian T, Moya YAT, Zhang X, Deichmann M, Frey FP, Bresgen V, Li C, Razavi BS, Schaaf G, von Wirén N, Su Z, Bucher M, Tsuda K, Goormachtig S, Chen X, Hochholdinger F (2021) Plant flavones enrich rhizosphere oxalobacteraceae to improve maize performance under nitrogen deprivation. Nat Plants 7:481–499
Zeilinger S, Gupta VK, Dahms TE, Silva RN, Singh HB, Upadhyay RS, Gomes EV, Tsui CK, Nayak SC (2016) Friends or foes? Emerging insights from fungal interactions with plants. FEMS Microbiol Rev 40:182–207
Zhang J, Subramanian S, Stacey G, Yu O (2009) Flavones and flavonols play distinct critical roles during nodulation of *Medicago truncatula* by *Sinorhizobium meliloti*. The Plant J 57:171–183
Zuanazzi JAS, Clergeot PH, Quirion JC, Husson HP, Kondorosi A, Ratet P (1998) Production of *Sinorhizobium meliloti* nod gene activator and repressor flavonoids from *Medicago sativa* roots. Mol Plant-Microbe Interact 11:784–794

7 Occupancy of Flavonoids in Plants, Pathogen, and Opportunistic Fungi Interactions

Maneesh Kumar, Priyanka Kumari, Arti Kumari, and Mohd. Sayeed Akhtar

7.1 Introduction

Flavonoids are a type of secondary metabolite that is exclusive to plants and is responsible for a wide range of biochemical processes. They are a class of polyphenolic chemicals and phytonutrients, both of which have antioxidant capabilities. They exist in plants as pigments and are responsible for the vibrant colors of many fruits and vegetables (Karak 2019; Swallah et al. 2020). Perhaps they contribute to better cardiovascular health and have anticancer, anti-viral, anti-inflammatory, and anti-inflammatory effects. They are effective in fending off infectious diseases and protecting against harmful ultraviolet rays. They are abundant in foods like tea, red wine, citrus fruits, onions, apples, and blueberries, to name a few of the places where you can find them. Some of the benefits associated with them include better heart health, protection against cancer and other viruses, and reduced inflammation. They are crucial in the fight against plant diseases and the relationship between plants and fungi (Guven et al. 2019; Kumari et al. 2019).

The word flavonoid comes from the Latin word flavus, which means yellow. They are a type of secondary metabolite found in plants, and their red, blue, and purple hues are the result of this compound. Notwithstanding their physiological roles in plants, they are also significant elements in the human diet, but they have

M. Kumar (✉)
Department of Biotechnology, Magadh University, Bodh Gaya, Bihar, India

P. Kumari
Department of Biotechnology, Gautam Buddha University, Greater Noida, Uttar Pradesh, India

A. Kumari
Department of Biotechnology, Patna Women's College, Patna, Bihar, India

M. S. Akhtar
Department of Botany, Gandhi Faiz-e-Aam College, Shahjahanpur, Uttar Pradesh, India

M. S. Akhtar (ed.), *Opportunistic Fungi, Nematode and Plant Interactions*, https://doi.org/10.1007/978-981-97-2045-3_7

not been the subject of a nutrient review. Despite having a common biosynthetic origin, they can be broken down into several different subclasses of structurally diverse moieties. More than 8000 flavonoids and derivatives have been identified, and they can be found in a variety of plant tissues. Among these, 4000 flavonoids have been classified as isoflavones, catechins, anthocyanidins, flavones, and flavonols. Flavonoids are a class of phenolic chemicals that act as antioxidants and free radical scavengers. Vegetables and fruits include a wide variety of plant flavonoids, which have been identified as numerous. Because of their low redox potential ($0.23 < E^{0'} < 0.75$ V), flavonoids are commonly used as chemical antioxidants. Because of their capacity to generate stable radicals, they show promising potential as free radical scavengers. Flavonoids have been shown to possess antioxidant properties, which can be attributed to their ability to scavenge free radicals. They are effective at scavenging peroxynitrite, hydroxyl radicals, superoxide anions, and lipid peroxides, making them effective cellular antioxidants. Furthermore, flavonoids are known to possess anti-inflammatory and anticancer properties, making them promising additives to pharmaceuticals and nutraceuticals (Amin et al. 2020; Arafah et al. 2020).

Flavonoids can also mediate the interactions between plants and beneficial microbes such as mycorrhizal fungi. They are involved in plant–microbe communication, helping to strengthen the symbiotic relationship between plants and beneficial microbes. Flavonoids also help protect plants from pathogens by signaling to beneficial microbes to activate defensive mechanisms such as the production of antimicrobial compounds (Bag et al. 2022). Additionally, flavonoids can act as attractants for beneficial insects, such as pollinators, which are essential for the production of fruits and nuts. They can also act as a deterrent for pest insects, as some of them have a bitter taste that can be used to protect plants from being eaten. Furthermore, flavonoids have been shown to have antioxidant and anti-inflammatory properties that can help protect plants from environmental stressors and disease (Khan et al. 2020; Shen et al. 2022). They are generally produced in response to environmental stimuli, such as pathogen attacks. When a plant detects a disease, it increases flavonoid synthesis as part of its defense reaction.

Flavonoids can prevent the pathogen from accessing the plant cell and causing harm by acting as a barrier. In addition to their antibacterial and antioxidant effects, they can also scavenge pathogen-generated reactive oxygen species (ROS). They can act as signals to activate defense responses in the plant, such as those mediated by Jasmonic acid (Shen et al. 2022; Tarfeen et al. 2022). They can also stimulate the production of defense compounds such as enzymes, phenolics, and terpenoids, which can inhibit the growth of the pathogen. Such compounds can induce systemic acquired resistance (SAR), which helps protect the entire plant from infection by pathogens. Flavonoid compounds have been shown to activate SAR by activating genes associated with the production of defense-related proteins (Jayaswall et al. 2016; Bukhat et al. 2020). This activation results in the production of defense-related proteins such as pathogenesis-related proteins, chitinases, and peroxidases. These proteins then help to protect the plant from infection by pathogens. Additionally, flavonoids can also act as antifungal agents, helping to prevent the spread of fungal diseases. SAR is triggered when the plant detects the presence of a pathogen and responds by activating defense mechanisms throughout the plant. This includes the production of defensive compounds such as phytoalexins, phytoanticipins, and other defensive proteins (Jayaswall et al. 2016; Tiku 2020). Also, the production of signaling molecules such as salicylic acid and Jasmonic acid helps activate and coordinate the plant's defense response. These molecules cause the production of proteins and enzymes that help the plant fight against pathogens. Plants produce pathogenesis-related proteins (PR proteins) in response to pathogenic threats. Specific genes encode these proteins and can be induced by infection (Jain and Khurana 2018; Zribi et al. 2021). PR proteins are known to act as an immune response to pathogens by triggering a defense mechanism that inhibits the growth of the pathogen, thus helping to protect the plant. PR proteins can also be used to detect the presence of pathogens in plants. For example, antibody-based detection systems can detect the presence of PR proteins in a sample, thereby indicating the presence of a pathogen. Additionally, PR proteins can be used in vaccine production to induce immunity against a specific pathogen. These proteins help to protect the plant from damage caused by pathogens. They can do this by either inhibiting the growth and reproduction of pathogens or by directly killing them (Jain and Khurana 2018). PR proteins also help activate other defense responses in the plant, such as producing reactive oxygen species and antimicrobial compounds. These molecules help to activate defense pathways and induce the production of defensive molecules that can help protect the plant from infection. Thus, flavonoids are significant in the relationships between pathogens and plants because they help plants protect themselves from possible threats. They stop or slow the growth of pathogenic microbes, which is a kind of built-in defense. Also, flavonoids can help good microbes work better, which helps them stay alive and grow in the plant's environment (Moosa et al. 2018). The aim of this chapter is to describe the role of flavonoids between the opportunistic fungi, plants, and pathogens and also highlights the role of flavonoids in plant–environment interaction and pathogen defense.

7.2 Biosynthesis of Flavonoids

Flavonoids are a class of secondary metabolites found in plants and serve a variety of metabolic roles. Under the effect of pathogen attack, plants produce various secondary metabolites, such as flavonoids, which play a role in defense against pathogens (Jan et al. 2021). They are synthesized in the various metabolic pathways, which are triggered by the release of pathogen-associated molecular patterns (PAMPs) and the activation of defense-related genes. This pathway involves the synthesis of chorismate, which is then converted to various flavonoid precursors, such as naringenin, and finally to various flavonoids. Flavonoids play a role in defense against pathogens by acting as signaling molecules, scavenging reactive oxygen species (ROS), promoting defense-related gene expression, and inhibiting the growth of pathogens (Jain et al. 2019; Thakur et al. 2019). It biosynthesis process is complicated that is accomplished with various enzymes. The precursor compounds are later involved in many changes including methylation by methyltransferases and glycosylation by certain glycosyltransferases (Sajid et al. 2021; Liu et al. 2022). As a result, their solubility, reactivity, and stability frequently change after being modified. In their native state, flavonoids are overwhelmingly found as glycosides. They serve a range of crucial purposes in plants due to their complex chemical make-up and the resultant diversity of their connected substituents. Due to their antioxidative capabilities, they also play a role in maintaining a redox state in cells while protecting plants from biotic (herbivores, pathogens) and abiotic (UV radiation, heat) challenges (Santisree et al. 2020). Flavonoids' antioxidative effect is tied to their molecular make-up, specifically their conjugated double bonds and the presence of organic compounds in the rings (Isah 2019; McAlpine et al. 2019; Khare et al. 2020).

In grasses such as rice (*Oryza sativa*), sorghum (*Sorghum bicolor*), and barley (*Hordeum vulgare*), the synthesis and accumulation of flavonoid phytoalexins are induced by a fungal infection. However, researchers have placed more an emphasis on benzoxazinoid and terpenoid phytoalexins in maize (*Zea mays*), and there is not much information about flavonoid induction in infection-related responses (Förster et al. 2022). Flavonoid phytoalexins are secondary metabolites produced by plants in response to pathogen attack. These compounds are believed to be involved in plant defense against pathogens by inhibiting the growth and metabolism of invading fungi (Cho and Lee 2015; Tiku 2020). They also provide a physical barrier to slow down the spread of infectious fungal agents. In general, flavonoid phytoalexins are derived from phenylpropanoid and flavonoid pathways. In maize, the main flavonoid phytoalexin is luteolin, which is derived from the flavonoid biosynthesis pathway (Nabavi et al. 2020). Luteolin is accumulated in the leaf, stem, and root of maize plants after infection and has been shown to inhibit the growth of several fungal species, such as *Fusarium graminearum* and *Fusarium culmorum* (Bouajila et al. 2020; Jaafari and Lekchiri 2022; Zhou et al. 2023). In addition, luteolin and other flavonoid phytoalexins have been shown to activate defense-related genes, such as peroxidase and chitinase, which help fight against pathogen attack. Other flavonoid phytoalexins, such as kaempferol, quercetin, and naringenin, have been reported in maize, but their levels are usually much lower than luteolin. These compounds are produced from the same flavonoid biosynthesis pathway as luteolin, but their accumulation is regulated differently in response to fungal infections (Abd El-Rahman et al. 2012; Haile et al. 2017).

With the manipulation of many end products with useful uses, metabolic engineering of plants has become possible thanks to the discovery of biosynthetic pathways and their control by MYB, basic helix-loop-helix (bHLH), and WD40-type transcription factors (Belwal et al. 2020). MYB transcription factors are responsible for the activation of structural genes involved in flavonoid biosynthesis, including the structural genes for chalcone synthase and dihydroflavonol 4-reductase. The bHLH transcription factors are responsible for activating genes involved in the regulation of flavonoid biosynthesis, including the structural genes for flavonoid 3′-hydroxylase, flavonoid 3′-monooxygenase, and flavonoid 3′,5′-hydroxylase. WD40-type transcription factors are involved in the control of flavonoid biosynthesis, including the structural genes for anthocyanidin synthase, flavanone 3-hydroxylase, and chalcone isomerase (Zhu et al. 2015; Kumar et al. 2018a; Belwal et al. 2020). Flavonoid compounds are used in the food industry to provide color and flavor to products, and have been shown to have antioxidant and anti-inflammatory activities. They are found in fruits, vegetables, nuts, and grains, and are believed to be beneficial to human health (Fig. 7.1). Examples of flavonoid compounds include apigenin, catechin, epicatechin, isorhamnetin, kaempferol, myricetin, and quercetin (Alagawany et al. 2017; Atala et al. 2017). Flavonoid compounds also have anticancer, anti-viral, and anti-allergenic properties, as well as being useful for the treatment of cardiovascular and neurological diseases and for reducing inflammation. They also have antioxidant properties, scavenging free radicals and protecting cells from damage (Barman et al. 2018). Flavonoids can also act as natural antibiotics and can reduce the risk of certain types of cancers. Additionally, they have been used in traditional Chinese medicine to treat colds, bronchitis, and

Fig. 7.1 Flavonoids containing fruits, vegetables, nuts, and grains

stomach problems (Alagawany et al. 2017; Saini et al. 2019). Higher plants make flavonoids through the shikimate pathway, the phenylpropanoid pathway, and the polyketide pathway.

The shikimate pathway is the metabolic pathway used by plants to synthesize the aromatic amino acids, which are the precursors of flavonoids. This pathway involves the sequential conversion of phosphoenolpyruvate and erythrose-4-phosphate to chorismate (Maeda and Dudareva 2012; Santos-Sánchez et al. 2019). This reaction is catalyzed by the enzyme chorismate mutase, which yields chorismate. This chorismate is then processed into the aromatic amino acids phenylalanine, tyrosine, and tryptophan, which serve as precursors for the synthesis of flavonoids (Qian et al. 2019). Phenylalanine is converted into cinnamic acid and its derivatives, which are used in the biosynthesis of flavonoids (Parmeggiani et al. 2015). Tyrosine is converted into coumaric acid, which is then converted into ferulic acid and its derivatives. These derivatives are then used to form flavonoid aglycones, which are the basic building blocks for many flavonoids (Pandey et al. 2016; Metsämuuronen and Sirén 2019). Tryptophan is converted into indole-3-acetic acid, which is then converted into chalcones and flavones. These are then converted into other flavonoids, such as flavonols and flavanones. Flavonoids are then further modified to form other classes of flavonoids, such as anthocyanins, flavanols, and isoflavones. These are then used to form the final flavonoid product (Lou et al. 2021; Shi et al. 2022).

The phenylpropanoid pathway is the pathway used by plants to synthesize aromatic compounds from the amino acid phenylalanine. The pathway in plants works as a rich supply of metabolites and provides these plants with a source of metabolites since it is required for the formation of flavonoids, phenolic acids, tannins, and other related compounds. It is one of the most studied and understood pathways in plants and has been used to engineer plants to produce various metabolites with beneficial properties. This pathway begins with the condensation of two molecules of malonyl-CoA to form the diketide malonate semialdehyde, followed by the formation of phenylalanine through a series of enzymatic steps. This pathway involves the transformation of phenylalanine into cinnamic acid, which is then further converted into various phenylpropanoid compounds, such as coumarins, flavones, flavanols, and anthocyanins. They can be further metabolized into other compounds, such as tannins and lignans. Organic compounds like cinnamic acid, coumaric acid, ferulic acid, and lignin are the important components of the cell wall. These molecules play a variety of roles in the plant, from providing structural support and acting as antioxidants to providing UV protection and even attracting pollinators and symbiotic microbes. These compounds are then used as precursors for the synthesis of flavonoids and other secondary metabolites.

The polyketide pathway is the metabolic route followed by plants for the production of polyketide molecules. The formation of these compounds begins with the condensation of acetyl-CoA molecules, which is followed by their further transformation into a variety of flavonoid compounds. These chemicals play an important role in the manufacture of flavonoids because they serve as precursors. Certain aromatic polyketides, like dietary flavonoids, have developed a reputation as miracle chemicals due to their extraordinary curative benefits for human health (Kumar et al. 2018b). These molecules are produced via the polyketide synthase (PKS) pathway. In the PKS pathway, the enzyme acetyl-CoA carboxylase (ACCase) catalyzes the conversion of acetyl-CoA to malonyl-CoA. Malonyl-CoA then acts as an extender unit to produce the aromatic polyketides. Other enzymes such as ketoacyl synthase (KS) and ketoacyl reductase (KR) are also involved in the PKS pathway. Certain aromatic polyketides with a narrow distribution have demonstrated more intriguing therapeutic characteristics, and further research should be conducted on them. These polyketides include the antifungal polyketide amphotericin B, the antitumor polyketide epothilone, and the anti-inflammatory polyketide microginin. These polyketides have shown promise as treatments for a variety of diseases, including cancer, fungal infections, and inflammation. Additional research should be focused on understanding the mechanism of action of these polyketides, as well as exploring potential new therapeutic applications. In addition, further investigation should be done to determine the safety and efficacy of these polyketides in clinical studies. Additionally, improved methods for the synthesis and purification of these polyketides should be developed to facilitate the manufacture and distribution of these compounds for therapeutic use.

7.3 Intermediates of Flavonoids

Two key intermediates on the way to producing various types of flavonoids are chalcones and flavanones. Chalcones are molecules that form the starting point in the synthesis of various flavonoids. They are usually yellow in color and are formed by the condensation of an aldehyde and a ketone. Flavanones are a class of compounds that are derived from chalcones. They are formed by the dehydration of chalcones, typically through the use of an acid catalyst. Flavanones are key intermediates in the synthesis of flavonoids such as flavones, flavonols, and anthocyanidins.

7.3.1 Chalcones

Chalcones are a type of naturally-occurring polyphenolic (flavonoid) compounds produced by the phenylpropanoid pathway, which is a metabolic pathway that produces various plant compounds from phenylalanine, an amino acid. They are often used as a precursor to other flavonoids such as flavonols, flavones, and anthocyanidins (Lou et al. 2021). They are found in a variety of plants including marigolds, chamomile, and hops. In addition to their use in flavonoid production, chalcones are also known for their antioxidant activities and are used in pharmaceuticals, cosmetics, and food additives. They are characterized by their ability to interact with free radicals and other reactive oxygen species, which makes them effective antioxidants. Studies have shown that some chalcones have a wide range of chemopreventive and modulatory effects that could be used to treat several diseases (Das et al. 2021; Liu et al. 2021). Their physicochemical qualities appear to determine how active they are biological. Chalcones are polyphenolic compounds found in a variety of plants such as cinnamon, ginger, and turmeric. Research has shown that these compounds have strong anti-inflammatory, anticancer, antioxidant, and antimicrobial properties (Asif 2015; Kumar et al. 2019). They have also been shown to affect the expression of genes associated with inflammation, cell growth, and apoptosis. In addition, chalcones have been shown to modulate the activity of enzymes involved in metabolic pathways and cellular signaling cascades, which can lead to the regulation of cell cycle, apoptosis, and cell proliferation (Jantan et al. 2021). These properties make them potentially useful in the treatment of a variety of diseases, including cancer, diabetes, and neurodegenerative diseases. In addition to their antioxidant properties, chalcones have been found to exhibit anti-inflammatory, antibacterial, and anti-viral activities. As such, they have been used in the formulation of pharmaceuticals, cosmetics, and food additives to improve the efficacy of these products.

7.3.2 Flavanones

One of the largest groups of flavonoids is flavones, which are a large group of compounds found in plants. They are found in fruits, vegetables, herbs, and other plant sources. Flavones are known for their antioxidant and anti-inflammatory properties. They are also thought to have a beneficial effect on cardiovascular health and help protect against certain types of cancer. They are also a type of flavonoid precursor produced by the polyketide pathway, a metabolic pathway that produces various plant compounds from acetyl-CoA, a molecule that is derived from carbohydrates (Wallis and Galarneau 2020). Flavones play multiple roles in plant–microbe and plant–insect interactions. They are a type of phenolic compound found in many plants and are known for their wide range of biological activities. They are involved in a variety of plant–microbe and plant–insect interactions, such as resistance to pathogens and pests, allelopathic effects, and attraction or repellence of pollinators (Wallis and Galarneau 2020; Raj et al. 2022). Flavones can act as antimicrobial agents by blocking the growth of bacteria and fungi, or by inhibiting the production of toxins by these organisms. They can also act as insecticides, repelling or killing certain pest species. In addition, flavones can act as allelochemicals, impacting the growth and development of other plants. Finally, flavones can act as pollinator attractants, luring beneficial insects to a plant in order to promote pollination. In microbes, flavones can be used as a source of carbon and nitrogen, and they also play a role in defense against microbial attack. For example, flavones can inhibit the growth of certain bacteria, such as *Escherichia coli*, and can act as an antibiotic. Insects can also be affected by flavones. They can be used as insecticides to repel or kill certain insect species, such as aphids, mites, and whiteflies. Flavones can also be used by other plants for defense against herbivores. They can act as an allelochemical, repelling or inhibiting certain insect species from feeding on the plant. They can also act as a deterrent to nematodes, which can be damaging to plant roots (Böttger et al. 2018; Mhlongo et al. 2018; Divekar et al. 2022; Prasad 2022).

Flavones play a wide variety of roles in plant biochemistry and physiology, and they are also crucial for human nutrition and health. They are important in plant biochemistry and physiology because they are plant secondary metabolites. These molecules are involved in a variety of processes such as defense against pests and pathogens, UV protection, pollinator attraction, and regulation of flowering and seed formation (Böttger et al. 2018; Raj et al. 2022). In addition, flavones play a role in human nutrition and health by providing essential antioxidants, anti-inflammatory effects, antiviral, and other beneficial properties. For example, flavones are found in abundance in foods such as onions, apples, and celery and are associated with a variety of health benefits. Antioxidants help to neutralize free radicals, which are substances in the body that can cause damage to your cells and lead to disease. Antioxidants also help to reduce inflammation, which is associated with many diseases (Kumar et al. 2018b; Akbari et al. 2022). Antivirals work by inhibiting the replication of viruses and can be used to treat infections caused by viruses (Ferraz et al. 2021). Finally, anti-inflammatory substances can help to reduce inflammation, which is associated with many diseases. By incorporating foods and supplements that contain these substances into your diet, you can help to protect your body from disease and maintain your overall health (Mhlongo et al. 2018; Kumar et al. 2020; Ferraz et al. 2021).

7.4 Plant–Pathogen Interactions

Diseases affecting plants are a major roadblock in the process of developing new income crops. Effectors are released as invasion weapons by diseases like oomycetes, fungus, and bacteria in order to enter and spread within host plants. They can cause a variety of damaging effects on the plants and reduce their yields significantly. Common diseases affecting plants include blight, mildew, rust, and smut. Other diseases include the potato late blight, the tomato late blight, powdery mildew, and downy mildew (Couto and Zipfel 2016). Pathogenic microorganisms, on the one hand, have been playing a silent but vicious game with plants ever since they first came into contact, stealing nutrients from host cells for survival and reproduction, while on the other, plants have been employing a wide range of defense mechanisms to prevent pathogen growth. Plants were able to limit the spread of viruses thanks to pattern-triggered immunity (PTI) defense responses that were initially activated by the recognition of pathogen-associated molecular patterns (PAMPs) by pattern recognition receptors (PRRs) (Couto and Zipfel 2016; Silva et al. 2018). According to the paradigm, the two primary components of plant innate immunity are the pathogen-associated molecular pattern (PAMP)-primed immune response (PTI) and the effector-primed immune response (ETI). The PTI response is the first line of defense, and it is triggered by pattern recognition receptors (PRRs) on the cell surface. These pattern recognition receptors are largely receptor-like kinases (RLKs) and receptor-like proteins (RLPs), which recognize pathogen-derived chemicals (PAMPs) (Pieterse et al. 2013; Pradhan et al. 2021; Zhang et al. 2022). The primary PAMPs that have received a lot of attention in recent years include flagellants, the elongation factor EF-Tu, a few lipopolysaccharides and oligosaccharides in bacteria, chitin and cell wall derivative glycans in fungi, and glutamine aminotransferase and -glycan in oomycetes. This detection of a pattern is often accompanied by a calcium influx, a deposition of callose, a burst of reactive oxygen species (ROS), activation of the miRNA pathway, activation of the MAPK cascade, and induction of the expression of a large number of defense-related genes, including disease-process-related proteins (PR) (Li et al. 2016). To prevent PTI, pathogens inject effectors into host plant cells, where they work to evade identification and dampen the plant's defenses. NLRs expressed by R genes in plants initiate ETI, a second line of plant immunological defense, when they recognize effectors released by pathogens (Couto and Zipfel 2016; Pradhan et al. 2021).

In recent years, the discovery of the microRNA (miRNA) family has revolutionized our understanding of the plant defense mechanisms. Mainly miRNAs are small non-coding RNAs that function as post-transcriptional gene regulators by targeting mRNA molecules for degradation or translational repression (Catalanotto et al. 2016; Waititu et al. 2020). The plant miRNA family is known to play a role in the regulation of various biological processes, including defense against pathogens. The miRNAs are known to regulate defense responses by modulating the expression of defense-related genes, resulting in a stronger immune response. For example, miR845 and miR393 have been shown to be involved in the regulation. They target the expression of the defense-related gene lipase-1 and induce the accumulation of the antimicrobial compound squalene (Boke et al. 2015). miR393, on the other hand, targets the expression of the defense-related gene adenosyltransferase-3 and induces the accumulation of the antimicrobial compound ethylene. In addition to miRNAs, plants also employ other defense mechanisms, such as phytoalexins, which are flavonoids produced by plants in response to pathogen attacks (Jiang et al. 2022). Phytoalexins are toxic to specific pathogens and can act as an additional layer of protection for the plant. Other plant defense mechanisms include the production of proteins that act as antifungal agents, or the production of enzymes that break down toxins produced by pathogens.

Plants also employ physical defenses, such as producing a waxy cuticle layer on the surface of their leaves, which can prevent the attachment of fungal spores. Plants also have mechanical defenses, such as thorns, spines, and prickles, which can deter animals from eating them. These structures can also make it difficult for herbivores to access and eat the plant's tissues. Additionally, some plants have the ability to produce toxins that can be poisonous or distasteful to animals, deterring them from feeding on the plant (Kumar et al. 2017a; Lou et al. 2021). In order to protect crops from diseases, it is important to understand the life cycle and transmission of the disease, and to use cultural, biological, and chemical control methods. Cultural control methods include crop rotation, early planting, and resistant varieties. Biological control methods involve the use of beneficial organisms such as predatory mites, parasitic wasps, and beneficial bacteria or fungi (Couto and Zipfel 2016; Kumar and Akhtar 2019). Chemical control methods involve the use of fungicides and insecticides. In addition to traditional methods of disease control, there are also new methods that are being developed. These include the use of biotechnology to produce resistant varieties, the use of RNA interference to regulate gene expression in plants, and the use of biopesticides to control insect pests. Overall, diseases affecting plants are a major challenge for farmers and agricultural scientists, but with the proper methods of disease control using flavonoid-based treatments, many of these diseases can be managed effectively (Kumar and Akhtar 2019; Kumar 2021). These properties make them especially effective at combating various diseases that can affect plants, such as fungi, viruses, and bacteria. Through the use of flavonoid-based treatments, farmers can protect their crops from disease and ultimately improve their yields:

- **Planting flavonoid-rich crops:** Planting crops like broccoli, cabbage, and spinach, which are rich in flavonoids, can help protect plants from disease. By planting these crops, you can help reduce the spread of diseases and pests that can damage your crops. These crops can provide important nutrients to your diet and help you stay healthy.
- **Adding flavonoids to soil**: Adding flavonoid-rich fertilizers and soil amendments to the soil can help protect plants from disease and improve crop yields. Flavonoid-rich fertilizers and soil amendments can also improve soil health by increasing beneficial bacteria and fungi in the soil, and by helping to reduce soil acidity. Moreover, flavonoid-rich fertilizers and soil amendments can increase the availability of micronutrients and minerals for plant growth, which can result in improved yields.
- **Spraying flavonoid solutions:** Spraying solutions containing flavonoids on the plants can help protect them from disease. Flavonoids are natural compounds found in many plants and vegetables, and they can act as an effective defense against fungal and bacterial diseases. Spraying solutions that contain flavonoids on plants can help protect them from disease by providing a physical barrier that can prevent pathogens from entering the plant, as well as by providing a natural antifungal and antibacterial action.
- **Introducing beneficial insects:** Introducing beneficial insects like ladybugs and parasitic wasps to the farm can help control pests and reduce the spread of disease. These insects are natural predators of disease-carrying pests like aphids and whiteflies, and they can help to reduce the need for chemical pesticides. Additionally, these insects can help to boost the farm's overall biodiversity, which can have positive impacts on the farm's overall health and productivity.
- **Utilizing crop rotation:** Crop rotation can help reduce the risk of disease by preventing the same disease-causing organisms from building up in the soil. It can also help prevent soil erosion and improve soil fertility by rotating crops with different nutrient needs. Crop rotation can also reduce the need for chemical fertilizers and pesticides, as well as help control weeds by planting certain crops in specific areas. Additionally, crop rotation can help reduce pests by changing the pest population.
- **Maintaining proper irrigation:** Proper irrigation can help reduce the spread of disease by preventing moisture from accumulating in the soil. This can help reduce the amount of standing water, which is the ideal breeding ground for

many types of disease-causing organisms. It can also help to reduce the presence of certain pests, such as mosquitoes, which can spread certain diseases. Additionally, proper irrigation can help to ensure that the soil is aerated, as well as to reduce the presence of weeds which can also spread disease.

- **Applying organic fungicides:** Applying organic fungicides that contain flavonoids can help protect plants from fungal diseases. These fungicides work by suppressing the growth of fungi and preventing them from reproducing. They also help to prevent the spread of the disease by creating a physical barrier between the infected and healthy parts of the plant. Additionally, they can also help to boost the plant's immune system, allowing it to fight off infection more effectively.
- **Introducing biological control agents:** Introducing biological control agents like certain bacteria and fungi can help reduce the spread of disease. Biological control agents are organisms that naturally occur in an ecosystem, and they can be used to reduce the population of a pest or disease-causing organism. These agents can be used to control the spread of diseases, such as viruses, bacteria, fungi, and parasites, by competing with the disease-causing organisms for resources. By introducing these agents into an ecosystem, the number of disease-causing organisms can be reduced, thus limiting the spread of the disease. Additionally, biological control agents can also be used to introduce beneficial organisms into an ecosystem, such as pollinators, which can help promote the growth of crops.

7.5 Flavonoids in Plant–Environment Interactions and Pathogen Defense

Flavonoid-rich plants have been used to treat a wide range of human health issues because of their ability to fight free radicals, bacteria, viruses, fungi, inflammation, and even cancer. Its mechanism of action in the cancer cell cycle has just recently been identified (Ahmad et al. 2015). They are bactericidal and bacteriostatic because they disrupt the synthesis of bacterial cell membranes, hinder the synthesis of nucleic acids, and interfere with bacterial metabolism. They are involved in a wide range of biological processes, including photosynthesis, pollination, and interactions with the environment (Kumar et al. 2018c). They are thought to play a role in plant–environment interactions by providing protection from environmental stressors. Furthermore, flavonoids may help plants to attract beneficial organisms such as pollinators and to repel harmful organisms such as pathogens. In addition, flavonoids are thought to be involved in the regulation of plant growth and development, and the regulation of plant–plant interactions. For example, flavonoids have been shown to play a role in allelopathy, a phenomenon in which plants produce chemicals that inhibit the growth of other plants. Thus, flavonoids are important for understanding and managing plant-environment interactions. Allelopathic compounds produced by plants include phenolic compounds, terpenoids, and flavonoids. These compounds are thought to act as a defense mechanism for the plant, deterring herbivores and competing plants from encroaching on the plant's space. Flavonoids are thought to be especially important allelochemicals, as they are relatively easy for plants to produce and are highly active in small concentrations. Studies have shown that flavonoids can be released from the roots, leaves, and other organs of a plant and can have a range of effects on other plants, including inhibiting seed germination, altering root growth and development, and reducing photosynthetic efficiency (Ohsugi et al. 1985). This suggests that flavonoids play a key role in plant allelopathy and can be used to manipulate the growth and development of neighboring plants. They are known to influence the growth and development of neighboring plants, typically through the production of hormones or other molecules that can alter the behavior or physiology of the other plants. For example, some flavonoids can act as allelochemicals, which are chemicals released by one plant that can modify the behavior of other plants, such as growth, reproduction, and germination. Moreover, flavonoids have been shown to affect the structure of the root systems of neighboring plants, as well as the growth of shoots and leaves. Overall, flavonoids appear to have a strong influence on the growth and development of neighboring plants.

7.6 Flavonoids in Plant and Fungi Interactions

Flavonoids are natural compounds found in plants that have been linked to inhibiting the growth of specific fungi. Studies have shown that flavonoids can act as a fungistatic agent, meaning they can stop the growth of fungi (Table 7.1). This is beneficial for plants because it can help protect them from fungal infections, which can cause plant diseases and reduce crop yields (Kumar et al. 2017b; Kumar 2021). They have been shown to have a variety of medicinal and antibacterial properties, including the inhibition of *Phytophthora infestans* and *Trichophyton mentagrophytes*. Studies have demonstrated that some flavonoids, such as quercetin, have strong inhibitory effects against both fungal species. Quercetin has been shown to reduce fungal growth, sporulation, and germination of *P. infestans*, as well as reduce the hyphal growth of *T. mentagrophytes*. Other flavonoids, such as kaempferol, have also been shown to inhibit the growth of *P. infestans* and *T. mentagrophytes*. Kaempferol has been demonstrated to reduce the spore germination of *P. infestans* and reduce the hyphal growth of *T. mentagrophytes*. In addition, certain flavonoids have been demonstrated to have synergistic effects when combined with other compounds, such as fungicides, further enhancing their inhibitory effects against the two fungal species. Thus, flavonoids are promising compounds for the inhibition of *P. infestans* and *T. mentagrophytes*. Studies have shown that flavonoids can inhibit the growth of *Botrytis cinerea*, *Colletotrichum lindemuthianum*, *F. oxysporum*, and *R. solani* in the plant growth system. The flavonoids can act as antifungal agents and can help to control the spread of the fungi in the plant system. These flavonoids can also be used as a preventative measure to reduce the occurrence of fungal infections. Flavonoids have also been found to reduce the production of toxin components by the fungi, thus reducing the potential for damage to the plant (Raji et al. 2022).

Table 7.1 Flavonoid control plant pathogens and opportunistic fungi

Structure of flavonoids	Properties	Role against plant pathogens	References
Luteolin 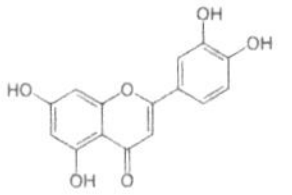	Antibacterial, anti-fungal, and antiviral	Inhibitory effect on *Fusarium, Alternaria*, *Pseudomonas*, and *Rhizoctonia*; reduce the growth of several species of Phytophthora; effective against other fungal plant pathogens, including *Botrytis*, *Sclerotinia*, and *Fusarium*; control bacterial plant diseases, such as *Pseudomonas* and *Xanthomonas*; also inhibit the spore formation in *Colletotrichum sublineola*	Du et al. (2010)
Sakuranetin 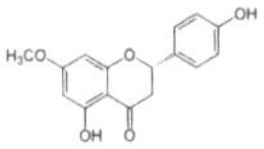	Antifungal behavior	Inhibit the growth of *F. fujikuroi*, *M. oryzae*, and *R. solani*	Hasegawa et al. (2014), Siciliano et al. (2015), Katsumata et al. (2018)
Rutin 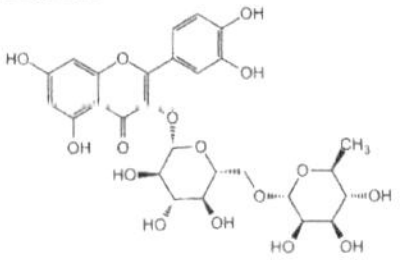	Anti-fungal and anti-bacterial	Strengthen the plant's immune system, which helps it to better resist infection from disease-causing organisms; reduce the symptoms of existing infections, such as discoloration and wilting, and can help to speed up the healing process	Mierziak et al. (2014), Radhakrishnan (2022)
Apigenin 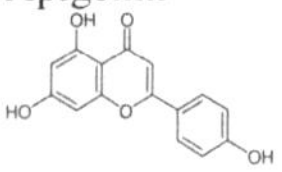	Antibacterial, antiviral, and antifungal agent	Inhibitory effect on *A. fumigatus*, *C. albicans*, *P. infestans* and *T. mentagrophytes*; reduces growth of *E. coli*, *V. cholerae*, and *S. aureus*; research into the effects of apigenin on plant pathogens is still ongoing	Nayaka et al. (2014)
Quercetin 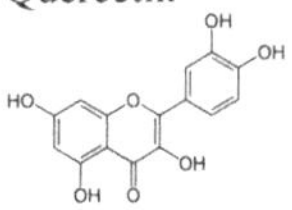	Antimicrobial and antifungal	Inhibitory effect on *F. oxysporum*, *A. solani*, *P. syringae*, and *Cercospora* spp. it is also used to control powdery mildew and other fungal diseases; foliar spray or incorporated into the soil; antifungal against *C. albicans*	Singh et al. (2015), Rocha et al. (2019)
Chlorogenic acid 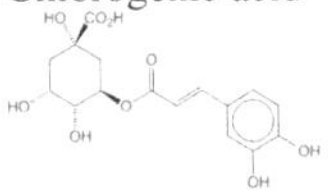	Antibacterial, anti-fungal, and antiviral; antioxidant	Inhibit fungal growth and reduce viral entry in plant; protect plants from oxidative damage and protection against plant diseases	Liu et al. (2017)

(continued)

Table 7.1 (continued)

Structure of flavonoids	Properties	Role against plant pathogens	References
Kaempferol 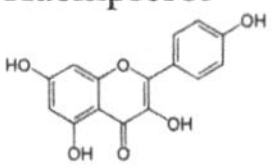	Antifungal, antiviral, insecticidal and antibacterial	Bacterial wilt caused by *Erwinia carotovora*; reduce the symptoms of several fungal and bacterial diseases, such as powdery mildew and anthracnose; inhibit viral entry in plant cell; inhibit growth plant-associated fungi, viruses, and bacteria, including phytopathogenic fungi, plant viruses, and plant-pathogenic bacteria; control of fungal (*C. parapsilosis*) biofilms complex	Rocha et al. (2019)
Iorhamnetin-3-glucoside 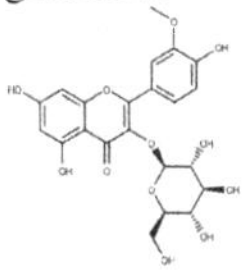	Anti-inflammatory activity and antifungal	Inhibition of *Bifidobacterium animalis* subsp. lactis AD011	Ahn et al. (2020)
Naringenin 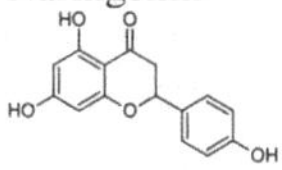	Antibacterial and antifungal	Control the growth of *A. solani, P. infestans,* and insect (*Papilio xuthus*); antifungal properties against *A. niger*	Amin et al. (2020)
Hesperidin 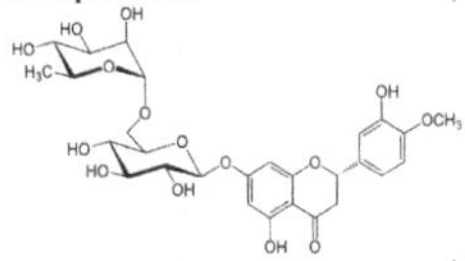	Antibacterial, anti-fungal, and antiviral	Control several diseases including powdery mildew, leaf blight, and black spot; enhance the effect of certain fungicides and herbicides and to increase the resistance of plants to certain pathogens	Arikan et al. (2022)
Chrysin 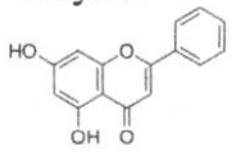	Antifungal and antibiofilm	Inhibitory effect the growth of *S. aureus*	Siddhardha et al. (2020)
Epicatechin 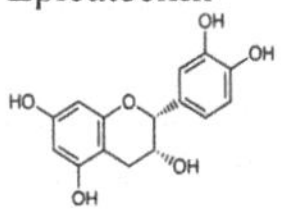	Antifungal synergy	Inhibitory effect the growth of *C. albicans*	Wang et al. (2021)
Catechin 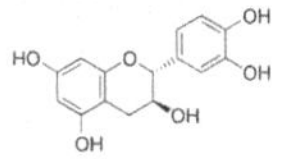	Antibacterial, anti-fungal, and antiviral; antioxidant	Inhibit the growth of *Botrytis cinerea, C. lindemuthianum, F. oxysporum*, and *R. solani*; inhibit the growth of tobacco mosaic virus; protect plants from oxidative damage	Li et al. (2022)
Myricetin 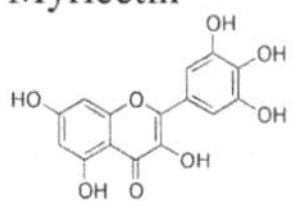	Antibacterial, anti-fungal, and antiviral	Inhibit the growth of certain fungi and bacteria, as well as viruses that infect plants; protect plants from environmental stressors, such as drought, high temperatures, and UV radiation; induce plant defense responses, such as increased production of phytoalexins and other antimicrobial compounds	Wang et al. (2022a, b)

Flavonoids can provide protection against oxidative damage caused by the fungi, which can result in improved plant growth and productivity. Hesperidin has been found to inhibit the proliferation of fungal pathogens in plants. Studies have shown that hesperidin can reduce the growth of various plant pathogenic fungi, including *F. graminearum*, *B. cinerea*, *A. solani*, *R. solani*, and *P. infestans*. It has been found to inhibit the production of fungal toxins and also to reduce the severity of fungal-induced diseases in plants. It can provide defense against oxidative harm caused by the fungi in agricultural crops, by scavenging free radicals (Mishra et al. 2019). Flavonoids are powerful antioxidants that can help to reduce oxidative stress, thereby protecting the crops from damage caused by the fungi. In addition, flavonoids can help reduce the growth of the fungi, as well as the amount of toxins they produce. Finally, flavonoids can help to regulate the metabolism of the crop and increase its resistance to disease.

7.7 Conclusions and Future Prospects

Plant secondary metabolites called flavonoids have many metabolic functions. They are synthesized in metabolic pathways activated by PAMPs and defense-related genes. They signal, scavenge ROS, promote defense-related gene expression, and decrease pathogen development. Methylation and glycosylation complicate their biogenesis. Due to their complicated chemical make-up, flavonoids are mostly glycosides and serve several important functions in plants. They eventfully mediate the interactions between plants and beneficial microbes, protect plants from pathogens, act as attractants for beneficial insects, and have antioxidant and anti-inflammatory properties. They are produced in response to environmental stimuli, such as pathogen attacks. Flavonoids are also involved in plant defense by acting as natural pesticides. They can reduce the ability of pathogens to colonize and penetrate the plant, reduce the growth of pathogens, and act as signaling molecules to alert other parts of the plant to the presence of a pathogen. They can help to reduce oxidative stress, protect crops from damage caused by fungi, reduce the growth of fungi, and increase resistance to disease. They also have anti-inflammatory, immune-modulating, and anti-allergic properties. Additionally, flavonoids may also have beneficial effects on the cardiovascular system, helping to reduce blood pressure and cholesterol levels. Flavonoids are found in a variety of fruits, vegetables, and other plant-based foods.

References

Abd El-Rahman SS, Mazen MM, Mohamed HI, Mahmoud NM (2012) Induction of defence related enzymes and phenolic compounds in lupin (*Lupinus albus* L.) and their effects on host resistance against *Fusarium* wilt. Eur J Plant Pathol 134:105–116

Ahmad A, Kaleem M, Ahmed Z, Shafiq H (2015) Therapeutic potential of flavonoids and their mechanism of action against microbial and viral infections—a review. Food Res Int 77:221–235

Ahn HJ, You HJ, Park MS, Li Z, Choe D, Johnston TV, Ku S, Ji GE (2020) Microbial biocatalysis of quercetin-3-glucoside and isorhamnetin-3-glucoside in *Salicornia herbacea* and their contribution to improved anti-inflammatory activity. RSC Adv 10:5339–5350

Akbari B, Baghaei-Yazdi N, Bahmaie M, Mahdavi Abhari F (2022) The role of plant-derived natural antioxidants in reduction of oxidative stress. Biofactors 48:611–633

Alagawany M, Abd El-Hack ME, Farag MR, Gopi M, Karthik K, Malik YS, Dhama K (2017) Rosmarinic acid: modes of action, medicinal values and health benefits. Anim Health Res Rev 18:167–176

Amin I, Majid S, Farooq A, Wani HA, Noor F, Khan R, Shakeel S, Bhat SA, Ahmad A, Madkhali H, Ahmad M (2020) Naringenin (4,5,7-trihydroxyflavanone) as a potent neuroprotective agent: from chemistry to medicine. Stud Nat Prod Chem 65:271–300

Arafah A, Rehman MU, Mir TM, Wali AF, Ali R, Qamar W, Khan R, Ahmad A, Aga SS, Alqahtani S, Almatroudi NM (2020) Multi-therapeutic potential of naringenin (4,5,7-trihydroxyflavonone): experimental evidence and mechanisms. Plants 9:1784

Arikan B, Ozfidan-Konakci C, Yildiztugay E, Zengin G, Alp FN, Elbasan F (2022) Exogenous hesperidin and chlorogenic acid alleviate oxidative damage induced by arsenic toxicity in *Zea mays* through regulating the water status, antioxidant capacity, redox balance and fatty acid composition. Environ Pollut 292:118389

Asif M (2015) Chemistry and antioxidant activity of plants containing some phenolic compounds. Chem Int 1:35–52

Atala E, Fuentes J, Wehrhahn MJ, Speisky H (2017) Quercetin and related flavonoids conserve their antioxidant properties despite undergoing chemical or enzymatic oxidation. Food Chem 234:479–485

Bag S, Mondal A, Majumder A, Mondal SK, Banik A (2022) Flavonoid mediated selective cross-talk between plants and beneficial soil microbiome. Phytochem Rev 21:1739–1760

Barman A, Marak CM, Barman RM, Sangma CS (2018) Nutraceutical properties of legume seeds and their impact on human health. In: Jimenez-Lopez JC, Clemente A (eds) Legume seed nutraceutical research. IntechOpen, London. https://doi.org/10.5772/intechopen.78799

Belwal T, Singh G, Jeandet P, Pandey A, Giri L, Ramola S, Bhatt ID, Venskutonis PR, Georgiev MI, Clément C, Luo Z (2020) Anthocyanins, multi-functional natural products of industrial relevance: recent biotechnological advances. Biotechnol Adv 43:107600

Boke H, Ozhuner E, Turktas M, Parmaksiz I, Ozcan S, Unver T (2015) Regulation of the alkaloid biosynthesis by miRNA in opium poppy. Plant Biotechnol J 13:409–420

Böttger A, Vothknecht U, Bolle C, Wolf A, Böttger A, Vothknecht U, Bolle C, Wolf A (2018) Plant secondary metabolites and their general function in plants. In: Lessons on caffeine, cannabis and co: learning materials in biosciences. Springer, Cham, pp 3–17

Bouajila A, Lamine M, Rahali F, Melki I, Prakash G, Ghorbel A (2020) Pearl millet populations characterized by *Fusarium* prevalence, morphological traits, phenolic content, and antioxidant potential. J Sci Food Agric 100:4172–4181

Bukhat S, Imran A, Javaid S, Shahid M, Majeed A, Naqqash T (2020) Communication of plants with microbial world: exploring the regulatory networks for PGPR mediated defense signaling. Microbiol Res 238:126486

Catalanotto C, Cogoni C, Zardo G (2016) MicroRNA in control of gene expression: an overview of nuclear functions. Int J Mol Sci 17:1712

Cho MH, Lee SW (2015) Phenolic phytoalexins in rice: biological functions and biosynthesis. Int J Mol Sci 16:29120–29133

Couto D, Zipfel C (2016) Regulation of pattern recognition receptor signalling in plants. Nat Rev Immunol 16:537–552

Das P, Srivastav AK, Muthukannan MR (2021) A study on molecular targeted approaches to cancer therapy and the role of chalcones in chemoprevention. Eur J Mol Clin Med 8:3254–3267

Divekar PA, Narayana S, Divekar BA, Kumar R, Gadratagi BG, Ray A, Singh AK, Rani V, Singh V, Singh AK, Kumar A (2022) Plant secondary metabolites as defense tools against herbivores for sustainable crop protection. Int J Mol Sci 23:2690

Du Y, Chu H, Wang M, Chu IK, Lo C (2010) Identification of flavone phytoalexins and a pathogen-inducible flavone synthase II gene (SbFNSII) in sorghum. J Exp Bot 61:983–994
Ferraz AC, Almeida LT, da Silva Caetano CC, da Silva Menegatto MB, Lima RL, de Senna JP, de Oliveira Cardoso JM, Perucci LO, Talvani A, de Lima WG, de Mello SB (2021) Hepatoprotective, antioxidant, anti-inflammatory, and antiviral activities of silymarin against Mayaro virus infection. Antivir Res 194:105168
Förster C, Handrick V, Ding Y, Nakamura Y, Paetz C, Schneider B, Castro-Falcón G, Hughes CC, Luck K, Poosapati S, Kunert G (2022) Biosynthesis and antifungal activity of fungus-induced O-methylated flavonoids in maize. Plant Physiol 188:167–190
Guven H, Arici A, Simsek O (2019) Flavonoids in our foods: a short review. J Basic Clin Health Sci 3:96–106
Haile ZM, Pilati S, Sonego P, Malacarne G, Vrhovsek U, Engelen K, Tudzynski P, Zottini M, Baraldi E, Moser C (2017) Molecular analysis of the early interaction between the grapevine flower and Botrytis cinerea reveals that prompt activation of specific host pathways leads to fungus quiescence. Plant Cell Environ 40:1409–1428
Hasegawa M, Mitsuhara I, Seo S, Okada K, Yamane H, Iwai T, Ohashi Y (2014) Analysis on blast fungus-responsive characters of a flavonoid phytoalexin sakuranetin; accumulation in infected rice leaves, antifungal activity and detoxification by fungus. Molecules 19:11404–11418
Isah T (2019) Stress and defense responses in plant secondary metabolites production. Biol Res 52:39
Jaafari A, Lekchiri S (2022) More insights into approaches with potential effect against pathogenic *Fusarium* species. Mycopathologia 19:51–63
Jain D, Khurana JP (2018) Role of pathogenesis-related (PR) proteins in plant defense mechanism. In: Singh A, Singh I (eds) Molecular aspects of plant-pathogen interaction. Springer, Singapore, pp 265–281
Jain C, Khatana S, Vijayvergia R (2019) Bioactivity of secondary metabolites of various plants: a review. Int J Pharm Sci Res 10:494–504
Jan R, Asaf S, Numan M, Kim KM (2021) Plant secondary metabolite biosynthesis and transcriptional regulation in response to biotic and abiotic stress conditions. Agronomy 11:968
Jantan I, Haque MA, Arshad L, Harikrishnan H, Septama AW, Mohamed-Hussein ZA (2021) Dietary polyphenols suppress chronic inflammation by modulation of multiple inflammation-associated cell signaling pathways. J Nutr Biochem 93:108634
Jayaswall K, Mahajan P, Singh G, Parmar R, Seth R, Raina A, Swarnkar MK, Singh AK, Shankar R, Sharma RK (2016) Transcriptome analysis reveals candidate genes involved in blister blight defense in tea (*Camellia sinensis* L. Kuntze). Sci Rep 6:30412
Jiang J, Zhu H, Li N, Batley J, Wang Y (2022) The miR393-target module regulates plant development and responses to biotic and abiotic stresses. Int J Mol Sci 23:9477
Karak P (2019) Biological activities of flavonoids: an overview. Int J Pharm Sci Res 10:1567–1574
Katsumata S, Toshima H, Hasegawa M (2018) Xylosylated detoxification of the rice flavonoid phytoalexin sakuranetin by the rice sheath blight fungus *Rhizoctonia solani*. Molecules 23:276
Khan A, Ikram M, Hahm JR, Kim MO (2020) Antioxidant and anti-inflammatory effects of citrus flavonoid hesperetin: special focus on neurological disorders. Antioxidants 9:609
Khare S, Singh NB, Singh A, Hussain I, Niharika KM, Yadav V, Bano C, Yadav RK, Amist N (2020) Plant secondary metabolites synthesis and their regulations under biotic and abiotic constraints. J Plant Biol 63:203–216
Kumar M (2021) *In silico* efficacy of [S]-8-gingerol (a derivative) with 6-gingerol against PT-domain of Polyketide synthase A (PksA). IP Int J Med Microbiol Trop Dis 7:62–64
Kumar M, Akhtar MS (2019) Application of microbial biotechnology in improving salt stress and crop productivity. In: Akhtar MS (ed) Salt stress, microbes, and plant interactions: mechanisms and molecular approaches, vol 2. Springer, Singapore, pp 133–159
Kumar M, Rana S, Kumar H, Kumar J, Mansuri R, Chandra Sahoo G (2017a) Inhibition of product template (PT) domain of aflatoxin producing polyketide synthase enzyme of *Aspergillus parasiticus*. Lett Drug Des Discov 14:811–818

Kumar H, Kumar M, Kumar B, Bharti S, Kumar P, Kumar M, Kumar J (2017b) Effect of acetone extract of *Zea mays* L. leaves on the growth of *Aspergillus flavus* and aflatoxin production. Int J Adv Res 5:1667–1671

Kumar A, Mosa KA, Ji L, Kage U, Dhokane D, Karre S, Madalageri D, Pathania N (2018a) Metabolomics-assisted biotechnological interventions for developing plant-based functional foods and nutraceuticals. Crit Rev Food Sci Nutr 58:1791–1807

Kumar M, Topno RK, Kumar H, Kumar J (2018b) Biological conversion of carcinogenic aflatoxin B1 to AFB1-8,9-exo-epoxide and its potential prevention. Int J Res Anal Rev 5:935–946

Kumar M, Sahoo G, Topno R, Pandey K, Ravidas V, Dikhit M, Kumar H, Das P (2018c) Molecular interaction of antifungal compounds with fungal type I Polyketide synthase and preventing hepatocarcinogenesis. Int J Infect Dis 73:285

Kumar M, Kumar H, Topno RK, Kumar J (2019) Analysis of impact of anaerobic condition on the aflatoxin production in *Aspergillus parasiticus* Speare. Agric Sci Digest 39:75–78

Kumar S, Abedin MM, Singh AK, Das S (2020) Role of phenolic compounds in plant-defensive mechanisms. In: Lone R, Shuab R, Kamili A (eds) Plant phenolics in sustainable agriculture. Springer, Singapore, pp 517–532

Kumari A, Kumar M, Sahoo GC, Kumar J (2019) *In silico* prediction of Glabridin potency against human tyrosinase in hyperpigmentation condition. Int J Res Anal Rev 6:332–341

Li L, Kim P, Yu L, Cai G, Chen S, Alfano JR, Zhou JM (2016) Activation-dependent destruction of a co-receptor by a *Pseudomonas syringae* effector dampens plant immunity. Cell Host Microbe 20:504–514

Li X, Zhang J, Lin S, Xing Y, Zhang X, Ye M, Chang Y, Guo H, Sun X (2022) (+)-Catechin, epicatechin and epigallocatechin gallate are important inducible defensive compounds against *Ectropis grisescens* in tea plants. Plant Cell Environ 45:496–511

Liu X, Vrieling K, Klinkhamer PG (2017) Interactions between plant metabolites affect herbivores: a study with pyrrolizidine alkaloids and chlorogenic acid. Front Plant Sci 8:903

Liu W, Feng Y, Yu S, Fan Z, Li X, Li J, Yin H (2021) The flavonoid biosynthesis network in plants. Int J Mol Sci 22:12824

Liu Y, Fernie AR, Tohge T (2022) Diversification of chemical structures of methoxylated flavonoids and genes encoding flavonoid-O-methyltransferases. Plants 11:564

Lou H, Hu L, Lu H, Wei T, Chen Q (2021) Metabolic engineering of microbial cell factories for biosynthesis of flavonoids: a review. Molecules 26:4522

Maeda H, Dudareva N (2012) The shikimate pathway and aromatic amino acid biosynthesis in plants. Annu Rev Plant Biol 63:73–105

McAlpine JB, Chen SN, Kutateladze A, MacMillan JB, Appendino G, Barison A, Beniddir MA, Biavatti MW, Bluml S, Boufridi A, Butler MS (2019) The value of universally available raw NMR data for transparency, reproducibility, and integrity in natural product research. Nat Prod Rep 36:35–107

Metsämuuronen S, Sirén H (2019) Bioactive phenolic compounds, metabolism and properties: a review on valuable chemical compounds in Scots pine and Norway spruce. Phytochem Rev 18:623–664

Mhlongo MI, Piater LA, Madala NE, Labuschagne N, Dubery IA (2018) The chemistry of plant-microbe interactions in the rhizosphere and the potential for metabolomics to reveal signaling related to defense priming and induced systemic resistance. Front Plant Sci 9:112

Mierziak J, Kostyn K, Kulma A (2014) Flavonoids as important molecules of plant interactions with the environment. Molecules 19:16240–16265

Mishra J, Kumar M, Sharan AK, Kumar J (2019) Detoxification assessment of Aflatoxin in *Aspergillus flavus* under the effect of *Calendula officinalis* Linn's methanolic extract. Agric Sci Dig 39:21–25

Moosa A, Farzand A, Sahi ST, Khan SA (2018) Transgenic expression of antifungal pathogenesis-related proteins against phytopathogenic fungi—15 years of success. Israel J Plant Sci 65:38–54

Nabavi SM, Šamec D, Tomczyk M, Milella L, Russo D, Habtemariam S, Suntar I, Rastrelli L, Daglia M, Xiao J, Giampieri F (2020) Flavonoid biosynthetic pathways in plants: versatile targets for metabolic engineering. Biotechnol Adv 38:107316

Nayaka HB, Londonkar RL, Umesh MK, Tukappa A (2014) Antibacterial attributes of apigenin, isolated from *Portulaca oleracea* L. Int J Bacteriol 2014:175851

Ohsugi T, Nishida R, Fukami H (1985) Oviposition stimulant of *Papilio xuthus*, a citrus-feeding-swallowtail butterfly. Agric Biol Chem 49:1897–1900

Pandey RP, Parajuli P, Koffas MA, Sohng JK (2016) Microbial production of natural and non-natural flavonoids: pathway engineering, directed evolution and systems/synthetic biology. Biotechnol Adv 34:634–662

Parmeggiani F, Lovelock SL, Weise NJ, Ahmed ST, Turner NJ (2015) Synthesis of d- and l-phenylalanine derivatives by phenylalanine ammonia lyases: a multienzymatic cascade process. Angew Chem 127:4691–4694

Pieterse CM, Poelman EH, Van Wees SC, Dicke M (2013) Induced plant responses to microbes and insects. Front Plant Sci 4:475

Pradhan A, Ghosh S, Sahoo D, Jha G (2021) Fungal effectors, the double edge sword of phytopathogens. Curr Genet 67:27–40

Prasad KVH (2022) Insect-plant interactions and role of secondary metabolites. In: Prasad KVH (ed) Insect ecology: concepts to management. Springer Nature, Singapore, pp 37–60

Qian Y, Lynch JH, Guo L, Rhodes D, Morgan JA, Dudareva N (2019) Completion of the cytosolic post-chorismate phenylalanine biosynthetic pathway in plants. Nat Commun 10:15

Radhakrishnan N (2022) Genesis, pathophysiology and management of venous and lymphatic disorders. Academic Press, Boca Raton

Raj VP, Thangavel S, Ponnusamy N, Prabakaran V (2022) Plant-microbe-insect interactions: concepts and applications for agricultural sustainability. In: Sayyed RZ, Singh A, Illyas N (eds) Antifungal metabolites of rhizobacteria for sustainable agriculture. Fungal biology. Springer, Cham, pp 335–349

Raji MR, Lotfi M, Tohidfar M, Ramshini H, Sahebani N, Alifar M, Baratian M, Mercati F, De Michele R, Carimi F (2022) Multiple fungal diseases resistance induction in *Cucumis melo* through co-transformation of different pathogenesis related (PR) protein genes. Sci Hortic 297:110924

Rocha MF, Sales JA, da Rocha MG, Galdino LM, de Aguiar L, Pereira-Neto WD, de Aguiar CR, Castelo-Branco DD, Sidrim JJ, Brilhante RS (2019) Antifungal effects of the flavonoids kaempferol and quercetin: a possible alternative for the control of fungal biofilms. Biofouling 35:320–328

Saini A, Panesar PS, Bera MB (2019) Valorization of fruits and vegetables waste through green extraction of bioactive compounds and their nanoemulsions-based delivery system. Bioresour Bioprocess 6:26

Sajid M, Channakesavula CN, Stone SR, Kaur P (2021) Synthetic biology towards improved flavonoid pharmacokinetics. Biomol Ther 11:754

Santisree P, Jalli LCL, Bhatnagar-Mathur P, Sharma KK (2020) Emerging roles of salicylic acid and jasmonates in plant abiotic stress responses. In: Roychoudhury A, Tripathi DK (eds) Protective chemical agents in the amelioration of plant abiotic stress: biochemical and molecular perspectives. Wiley, Hoboken, pp 342–373

Santos-Sánchez NF, Salas-Coronado R, Hernández-Carlos B, Villanueva-Cañongo C (2019) Shikimic acid pathway in biosynthesis of phenolic compounds. In: Soto-Hernández M, García-Mateos R, Palma-Tenango M (eds) Plant physiological aspects of phenolic compounds. IntechOpen, London. https://doi.org/10.5772/intechopen.83815

Shen N, Wang T, Gan Q, Liu S, Wang L, Jin B (2022) Plant flavonoids: classification, distribution, biosynthesis, and antioxidant activity. Food Chem 2022:132531

Shi J, Liu X, Wang N, Xu Q, Zhang Y, Yan G (2022) Comparative metabolomics analysis reveals the variation of metabolites associated with axillary bud development in upland cotton. J Plant Growth Regul 42:3071–3083

Siciliano I, Amaral Carneiro G, Spadaro D, Garibaldi A, Gullino ML (2015) Jasmonic acid, abscisic acid, and salicylic acid are involved in the phytoalexin responses of rice to Fusarium fujikuroi, a high gibberellin producer pathogen. J Agric Food Chem 63:8134–8142

Siddhardha B, Pandey U, Kaviyarasu K, Pala R, Syed A, Bahkali AH, Elgorban AM (2020) Chrysin-loaded chitosan nanoparticles potentiates antibiofilm activity against *Staphylococcus aureus*. Pathogens 9:115

Silva MS, Arraes FB, de Araújo CM, Grossi-de-Sa M, Fernandez D, de Souza Cândido E, Cardoso MH, Franco OL, Grossi-de-Sa MF (2018) Potential biotechnological assets related to plant immunity modulation applicable in engineering disease-resistant crops. Plant Sci 270:72–84

Singh BN, Upreti DK, Singh BR, Pandey G, Verma S, Roy S, Naqvi AH, Rawat AK (2015) Quercetin sensitizes fluconazole-resistant *Candida albicans* to induce apoptotic cell death by modulating quorum sensing. Antimicrob Agents Chemother 59:2153–2168

Swallah MS, Sun H, Affoh R, Fu H, Yu H (2020) Antioxidant potential overviews of secondary metabolites (polyphenols) in fruits. Int J Food Sci 2020:9081686

Tarfeen N, Nisa Q, Kahlief K (2022) Antioxidant defense system in plants against biotic stress. In: Aftab T, Hakeem KR (eds) Antioxidant defense in plants: molecular basis of regulation. Springer Nature, Singapore, pp 383–395

Thakur M, Bhattacharya S, Khosla PK, Puri S (2019) Improving production of plant secondary metabolites through biotic and abiotic elicitation. J Appl Res Med Arom Plants 12:1–12

Tiku AR (2020) Antimicrobial compounds (phytoanticipins and phytoalexins) and their role in plant defense. In: Mérillon JM, Ramawat K (eds) Co-evolution of secondary metabolites, Reference series in phytochemistry. Springer, Cham, pp 845–868

Waititu JK, Zhang C, Liu J, Wang H (2020) Plant non-coding RNAs: origin, biogenesis, mode of action and their roles in abiotic stress. Int J Mol Sci 21:8401

Wallis CM, Galarneau ER (2020) Phenolic compound induction in plant-microbe and plant-insect interactions: a meta-analysis. Front Plant Sci 11:580753

Wang Y, Lu C, Zhao X, Wang D, Liu Y, Sun S (2021) Antifungal activity and potential mechanism of asiatic acid alone and in combination with fluconazole against *Candida albicans*. Biomed Pharmacother 139:111568

Wang L, Chen M, Lam PY, Dini-Andreote F, Dai L, Wei Z (2022a) Multifaceted roles of flavonoids mediating plant-microbe interactions. Microbiome 10:233

Wang C, Yan Y, Huang M, Ma G, Wang L, Xie X, Xue W, Li X (2022b) Myricetin derivative LP11 targets cucumber mosaic virus 2b protein to achieve *in vivo* antiviral activity in plants. J Agric Food Chem 70:15360–15370

Zhang S, Li C, Si J, Han Z, Chen D (2022) Action mechanisms of effectors in plant-pathogen interaction. Int J Mol Sci 23:6758

Zhou H, Hua J, Li H, Song X, Luo S (2023) Structurally diverse specialized metabolites of maize and their extensive biological functions. J Cell Physiol. https://doi.org/10.1002/jcp.30955

Zhu Q, Sui S, Lei X, Yang Z, Lu K, Liu G, Liu YG, Li M (2015) Ectopic expression of the coleus R2R3 MYB-type proanthocyanidin regulator gene SMYB3 alters the flower color in transgenic tobacco. PLoS One 10:e0139392

Zribi I, Ghorbel M, Brini F (2021) Pathogenesis related proteins (PRs): from cellular mechanisms to plant defense. Curr Protein Peptide Sci 22:396–412

Mode of Interactions Between Opportunistic Fungi, Nematode, and Other Soil Microbiota

8

Anik Majumdar

8.1 Introduction

The functioning of ecosystem depends on a large scale upon the soil microbiota, and interactions between different soil microbes can have a big impact on the soil health as well as productivity. Due to its potential effect on plant growth and nutrient cycling, the interactions between fungus, nematodes, and other soil microorganisms has drawn a lot of study in particular. The immediate micro-environment around roots of the plant is known as rhizosphere, and it provides a unique environment to soil microbes because of changes in nutrient levels from time to time (Gupta et al. 2012; Yadav et al. 2015). Many different types of microorganisms colonize and reside in the rhizoplane and soil around the rhizosphere. Hence, the beneficial rhizospheric microbiota and the plant-parasitic nematode share a common ecological niche and have different effects on the growth of the plants and yield potential (Akhtar and Panwar 2011). Studies have also shown that opportunistic fungi residing in rhizosphere can have negative and positive impacts on nematodes. For example, fungi like as *Fusarium* spp. and *Rhizoctonia* spp. have been shown to enhance disease incidence on crops in association with nematodes, while on the other hand, nematophagous fungi can have negative effects on nematodes by producing toxic metabolites, which can lead to reduced nematode populations, thus aiding in biocontrol of nematodes. The objective of this chapter is to discuss the interaction among opportunistic fungi, nematodes, and soil microbiota in two halves, i.e., interaction between nematodes and soil microbiota; and interaction between nematodes and opportunistic fungi.

A. Majumdar (✉)
Division of Plant Pathology, ICAR-Indian Agricultural Research Institute, Pusa, New Delhi, India

M. S. Akhtar (ed.), *Opportunistic Fungi, Nematode and Plant Interactions*, https://doi.org/10.1007/978-981-97-2045-3_8

8.2 Interaction Between Nematodes and Soil Microbiota

The diverse group of microorganisms that inhabit soil is referred to as soil microbiota. These microbes, which include bacteria, fungus, archaea, protozoa, are crucial to soil functions like nitrogen cycling, decomposition, and interactions between plants and microbes. The chemical, physical, and biological soil characteristics, as well as plant growth, can all be significantly influenced by soil microbiota. For instance, some fungi and bacteria interact with plant roots in a positive way, providing them with nutrients and protection from pathogens, while others can produce toxic substances that inhibit plant's growth and cause diseases. The root exudates composition differs between plant genotypes and species, while the rhizospheric microbes are affected by plant genotype and soil type (Eisenhauer et al. 2016). When attacked by a pathogen, plants rely on the root microbiota that encourages the specific enrichment of microbes, which protect plants and microbial activity in the rhizosphere (Bakker et al. 2018). Different microbial processes are thought to regulate and mediate nematode-plant interaction. Before entering in plants, the infectious stages of plant-parasitic nematodes (PPNs) come into contact with a broad range of soil microbes, and this enables a diverse range of microorganisms to adhere to their cuticle and surface coat. Although nematodes may not be aware that, by accepting the attachment and transmitting the microbes to the plants, they may create self-deleterious effects; such as they can be antagonized by the transmitted microbiota or, even if the microbes remain in the soil, it can lead to the formation of nematode suppressive soil. On contrary, research has also shown that nematodes native microbiomes have a species-enriched population of bacteria (mostly predominated by the proteobacteria), which may help nematodes defend themselves against fungal infections. For instance, probiotic yeasts inhabiting the gut of *Caenorhabditis elegans* may shield them from being infected by non-infectious *Candida albicans* strain while decreasing their harmful impacts (Kunyeit et al. 2019). Dirksen et al. (2016) found that three *Pseudomonas* isolates showed antifungal activity in vitro which in turn protected *C. elegans* from fungal attack in vivo.

8.2.1 Nematode Suppressive Soil

Suppressive soils are characterized by the absence of soil-borne pathogens, either because they fail to establish, establish without causing disease, or initially establish to cause disease but quickly diminish (Baker and Cook 1974; Weller et al. 2002). Two types of soil suppressiveness are mainly categorized: specific and general. General soil suppressiveness is known to combat a broad spectrum of soil-borne diseases (Cook and Baker 1983). Consequently, soil steaming can reduce overall soil suppressiveness, but it is not transferable to conductive soils (Weller et al. 2002). Conversely, specific soil suppressiveness typically arises in field soils following a disease outbreak during crop monoculture. It relies on the antagonistic actions of specifically enriched microbial consortia that disrupt the life cycle of various plant pathogens (Raaijmakers and Mazzola 2016). Transfer of specific soil suppressiveness to conducive soil is possible with minimal amounts of suppressive soils (Borneman and Becker 2007). In 1962, Collingwood first observed the reduction of populations of the cyst nematode Heterodera avenae in the United Kingdom, attributing it to excessive cropping of cereal crops (Collingwood 1962). Subsequently, specific suppressive soil for various nematode species have been found globally. In soil, nematophagous fungi and bacteria use a variety of strategies to attack PPN, such as the formation of adhesive hyphal traps by nematode-trapping fungi, the use of spores by endoparasitic fungi and bacteria, the use of hyphal tips by fungi that prey on eggs and females, and the production of toxins by some fungi and bacteria to protect plant roots from nematodes. Thus the interaction between different soil microbiota with PPNs decides whether the soil would be either conductive or suppressive one for the PPNs to survive.

8.2.2 Interaction Between Nematodes and Rhizospheric Microbes

Numerous bacterial species present in the soil rhizosphere contribute to the growth of plants by enhancing the production of plant growth regulators and improving nutrient accessibility. These bacteria, known as plant growth-promoting rhizobacteria (PGPR), constitute a substantial population in the rhizosphere, actively fostering plant development and consequently enhancing crop yield. *Rhizobium* spp., *Xanthomonas* spp., *Arthrobacter* spp., *Bacillus* spp., *Bradyrhizobium* spp., *Frankia* spp., *Flavobacterium* spp., *Microbacterium* spp., *Pseudomonas* spp., *Serratia* spp., *Azospirillum* spp., *Agrobacterium* spp., *Enterobacter* spp., *Azotobacter* spp., *Burkholderia* spp., and *Erwinia* spp. are the bacterial genera most often detected in the rhizosphere (Tailor and Joshi 2014; Teymouri et al. 2016). Among these, the bacteria that are antagonistic to rhizosphere PPNs include *Pseudomonas* spp., *Bacillus* spp., *and Pasteuria* spp. To enhance plant growth, these PGPRs can engage in direct or indirect interactions with PPNs. Direct antagonism may manifest through predation, the release of toxic metabolites, nutrient competition, and niche exclusion, whereas indirect antagonism can be initiated through systemic acquired resistance (SAR) or induced systemic resistance (ISR). PGPRs that act as biocontrol agent for management of PPNs are listed below (Table 8.1).

8.2.3 Role of Nematode Surface Coat (SC) and Root Exudates in Nematode–Soil Microbiota Interactions

The surface coat (SC) constitutes the nematode's crucial body region in interactions with soil microorganisms. This protein layer, abundant in carbohydrates, is secreted onto the epicuticle by the hypodermis, excretory system, or nervous system. The SC likely plays a role in recognizing lectin-like protein molecules on the microbial surface through its glycohydrate epitopes (Davies and Curtis 2011). Studies indicate that root exudates and phytohormones like indole-acetic acid can alter the SC of

Table 8.1 PGPR acts as biocontrol agents for management of plant parasitic nematode

PGPR strain	Plant parasitic nematode	Mode of action	References
Pasteuria nishizwae	*Globodera* spp., *Heterodera* spp.	Predation	Sayre et al. (1991)
P. thornei	*Pratylenchus* spp.	Predation	Atibalentja et al. (2000)
Rhizobium etli	*Meloidogyne* spp.	ISR	Reitz et al. (2000)
Pseudomonas fluorescens CHA0	*M. incognita, M. javanica*	HCN, DAPG, pyoluteorin	Hamid et al. (2003), Siddiqui et al. (2005)
P. stutzeri	*M. incognita*	HCN	Khan et al. (2016)
P. aeruginosa	*M. javanica*	ISR and SAR	Fatima and Anjum (2017)
Bacillus subtilis	*M. incognita, M. javanica, Hekicotylenchus multicinctus, Rotylenchulus reniformis*	Secondary metabolites, hydrolytic enzymes	Basyony and Abo-Zaid (2018), Mazzuchelli et al. (2020)
P. penetrans	*Meloidogyne* spp.	Predation	Bhuiyan et al. (2018)
Bacillus megaterium	*M. incognita, H. glycines, M. javanica*	Protease, secondary metabolites	Mostafa et al. (2018)
B. coagulans	*M. incognita, H. glycines*	Hydrolytic enzymes	Xiang et al. (2018)
B. cereus	*H. avenae*	Protease, chitinase, antibiotic production	Ahmed (2019)
B. licheniformis	*M. incognita*	Protease, chitinase	El-Nagdi et al. (2019)
B. pumilus L1	*H. glycines, M. arenaria*	Protease, chitinase	Forghani and Hajihassani (2020)

plant-parasitic nematodes (PPN), influencing microorganism attachment to the nematode surface (Singh et al. 2014; Liu et al. 2017). For instance, diverse root exudates from different plant species exhibited varying effects on the attachment of *Pasteuria penetrans* endospores to J_2 of root-knot nematodes (Singh et al. 2014; Liu et al. 2017), revealing a correlation between nematode-microbe interactions in soil and host root exudates. Experiments demonstrated that when J_2 of *M. incognita* encountered soil microorganisms in the presence of root exudates, the adhesion of *P. penetrans* endospores to J_2 increased (Singh et al. 2014). While root exudates are pivotal for nematode attraction to roots, they also directly impact nematode interaction with soil microbes by modifying the PPN's surface. Depending on the nature of these interactions and the kinds of soil microorganisms involved, PPN may evade plant immune responses or fail to infect the plant.

8.2.4 Application of Nematode–Soil Microbiota Interactions in Agriculture: Management of Phytophagous Nematodes with PGPR

8.2.4.1 Direct Antagonism of PGPR Against PPNs

The prevention of egg hatch, as well as the growth and reproduction of PPNs, can be achieved through the discharge of toxins or hydrolytic enzymes such as hydrogen cyanide (HCN), 2,4-diacetylphloroglucinol (DAPG), glucanase chitinase, lipase, and protease along with the direct antagonism of PGPRs. The root knot nematode's (RKN) body cavity was directly attacked and colonized by *Pasteuria penetrans* (Sayre and Starr 1985), the most well-known (though not necessarily PGPR) predatory bacterium. *Pasteuria penetrans* spores initiate to manifest germination tubes after they are attached to the cuticle surface of RKN, and these tubes eventually pierce and destroy the PPN cell envelop (Bishop 2011). Similarly, *P. thornei* exhibits the capability to combat root-lesion nematodes (*Pratylenchus* spp.), while *P. nishizawae* is effective against cyst nematodes from two genera, *Heterodera* spp., and *Globodera* spp. (Atibalentja et al. 2000). In contrast, PGPR genera like *Bacillus* and *Pseudomonas* produce chemicals with nematicidal properties. For example, *P. fluorescens* strains that produce 2,4-diacetylphloroglucinol (DAPG) demonstrated reduced levels of *Globodera* spp. and *M. incognita* in both soil conditions and in vitro, respectively (Cronin et al. 1997). When compared to mutant strains with disrupted DAPG production, the wild type (WT) bacteria exhibited an 85% reduction in juvenile motility. The highly toxic substance hydrogen cyanide (HCN), produced by *P. fluorescens* strain CHA0 and *P. aeruginosa*, significantly impairs the proper functions and formation of mitochondria in several RKN species, including *M. incognita* and *M. javanica* (Siddiqui and Haque 2001; Siddiqui and Shaukat 2003). Notably, two secondary metabolites, DAPG and pyoluteorin, produced *by P. fluorescens* strain CHA0, were found to significantly decrease the population density of *M. javanica*, both in vitro and in soil (Hamid et al. 2003; Timper et al. 2009).

8.2.4.2 Indirect Antagonism Through Induced Systemic Resistance (ISR) of PGPR Against PPNs

The process of PGPR-mediated priming involves the indirect antagonistic action of PGPR against PPNs through ISR, enabling the entire plant to better cope with environmental challenges by triggering quicker and/or more potent defense responses. ISR, characterized by its nonspecific nature, bestows plants enduring defense and broad spectrum resistance against various pathogenic microorganisms, insects including PPNs such as *M. javanica*, *M. incognita*, *Globodera pallida*, and *Heterodera cajani* (Stone 1972). *Pseudomonas* spp. *and Bacillus* spp. including *P. fluorescens*, *B. amyloliquefaciens*, and *B. subtilis* inducing ISR, have shown promising outcomes in suppressing PPNs in both dicotyledonous plants (*Arabidopsis*, cucumber, bean, tobacco, radish, and tomato) and monocotyledonous plants (rice, sugarcane, maize).

8.3 Interactions Between Nematodes and Opportunistic Fungi

Owing to their predatory, antagonistic, or parasitic habits, opportunistic (OP) fungi are important among the several microorganisms that regulate the density of nematodes in soil. Certain OP fungi show potential for biocontrol and engage in a variety of aggressive activities. In addition to directly parasitizing nematodes, opportunistic fungi can also impact different stages of the nematode life cycle by secreting nematicidal metabolites (Lopez-Llorca and Jansson 2007). Opportunistic fungi have the capacity to invade and negatively impact the nematode reproductive structures. Some fungi can target the nematode's secondary stage during its life cycle when it is exposed to the soil and/or the host roots (Siddiqui and Mahmood 1996). The eggs and cysts of nematodes that are laid into the soil are very much exposed to colonization and deterioration. The OP fungi quickly multiply after coming into contact with eggs or cysts, eventually parasitizing all of them that are in the primary stages of embryonic emergence. Evidently, the parasitic activity of these juveniles decreases as they develop into adults. The different genera of opportunistic fungi include *Arthrobotrys, Paecilomyces, Pochonia, Trichoderma, Fusarium, Plectosphaerella, Cylindrocarpon, Rhizoctonia,* etc. The interaction between these opportunistic fungi and plant pathogenic nematodes can be sub-divided into parts, i.e., antagonistic interaction and synergistic interaction.

8.3.1 Antagonistic Interactions

Opportunistic fungi showing antagonistic interactions with PPNs are used as biocontrol agents. They use various mechanisms to kill PPNs in crops if applied as biocontrol agents (Table 8.2). Some of these fungi and their interactions with PPNs are discussed below.

8.3.1.1 *Arthrobotrys oligospora*

The initial nematode-trapping fungus identified was *Arthrobotrys oligospora* (Fresenius 1852), which stands out as the most frequently and widely encountered nematode-trapping fungus in the natural environment, as highlighted by Wachira et al. (2009). Significant insights into the realm of nematode toxic fungi have been garnered through the discovery that *A. oligospora* possesses the capability to immobilize plant-parasitic nematodes (PPNs) through the production of nematotoxins. *A. oligospora* has become a central focus of research, establishing itself as a popular model system for investigating various aspects of nematophagous fungal biology, spanning from pathogenesis to morphogenesis. In the context of agricultural ecosystems, the rhizosphere of crops and plants actively promotes the growth of *A. oligospora*, particularly within the top 30 cm of the soil, with minimal occurrences reported below a depth of 40 cm (Persmark et al. 1996). Originally discovered in Europe and characterized by Fresenius in 1852, *A. oligospora* has demonstrated a positive association with the rhizosphere environment. The structure of *A.*

Table 8.2 Opportunistic fungi used as biocontrol agent against plant parasitic nematodes

Opportunistic fungi	Mechanism of action	Target nematode and host plant	References
Trichoderma harzianum, *T. viride*, *T. virens*, *T. atroviridae*	Parasitism, antibiotic production, ISR and SAR, secretion of lytic enzymes, toxic metabolites	*M. incognita* (tomato), *M. javanica* (tomato), *M. graminicola* (rice), *M. arenaria* (maize)	Singh et al. (2017), Javeed et al. (2016), Abd-El-Kareem et al. (2019)
Pochonia chlamydosporia	Parasitism, toxic secretion	*H. zeae* (sweet corn), *M. javanica* (lettuce), *M. hapla* (lettuce)	Saxena (2018), Kumar (2020)
Paecilomyces lilacinus	Parasitism, antibiotics production, lytic enzymes	*H. zeae* (sweet corn), *M. incognita* (tomato), *M. javanica* (pigeon pea), *M. arenaria* (brinjal), *Tylenchulus semipenetrans* (Mandarin)	Hammam et al. (2016), Najafi et al. (2016), Moosavi and Zare (2020)

oligospora consists of an upright lateral branch that grows from a vegetative hypha, curves to join the parent hypha, and forms additional loops on the parent hypha or outside the initial loop. Furthermore, Woronin noted that *A. oligospora* conidiophores possess the capacity to directly produce a complex network structure which owes highly to its parasitic ability.

Interactions of *A. oligospora* with PPNs

Zopf (1888) provided a detailed account of the mechanism employed by *A. oligospora* to ensnare nematodes. He elucidated how the distinctive mycelial structure captured nematodes, penetrated the worm's cuticle, extended mycelia within the prey, and subsequently digested the internal contents of the worms. Following the death of the host, hyphae bearing conidia and conidiophores emerged. Notably, Zopf observed that the development of these trapping devices occurred exclusively in a nutrient-deficient culture, a finding that substantiated *A. oligospora's* predatory behavior against plant-parasitic nematodes (PPNs). It is widely accepted that enzymatic secretions play a crucial role in the invasion process. This belief stems from the implication of extracellular proteases in the penetration and digestion of host tissues by various plant pathogenic fungi. Growing evidence from ultrastructural and histochemical studies indicates the involvement of extracellular enzymes, including proteases, collagenases, and chitinases, in the penetration of nematode cuticles and the digestion of host cells (Yang et al. 2007). Given that collagen is identified as the primary component of nematode cuticles in most chemical composition studies, it has been proposed that collagenases produced by predatory fungi play a significant role in the infection of nematodes (Huang et al. 2004). Soliman et al. (2021) reported that *A. oligospora* effectively suppressed and preyed upon *M. incognita*. Aside from secreting toxic metabolites, the fungus also developed various trapping structures to capture and consume its prey. Hiep et al. (2019)

further noted that the nematode-trapping fungi *A. oligospora* and *A. guizhouensis* exhibited multi-trap abilities, proving effective against *M. incognita* and *P. coffea* on black pepper and coffee plants in Vietnam, respectively.

8.3.1.2 *Paecilomyces* spp.

The first report of genus *Paecilomyces* was made in 1907 (Bainier 1907). This genus is characterized by hyaline to yellowish septate hyphae, which frequently have smooth walls, conidiophores that are organized irregularly or verticillately, and phialides that have a wide base and an extended neck. Hyaline and unicellular conidia are formed in chains and show basipetal arrangement. *Paecilomyces* is an extensive genus encompassing numerous species with both pathogenic and saprophytic characteristics. These fungi can be encountered in diverse environments, including soil (He et al. 2011), decomposing plant material, and food sources. Upon recognizing a pathogen, *Paecilomyces* secretes enzyme complexes that instigate the growth of antagonists at the host's expense. The penetration mechanism employed by these fungi can be enzymatic, involving various lytic enzymes such as pectinases, cellulases, glucanases, laccases, lipases, proteases, and chitinases, or mechanical, accomplished through the development of appressoria.

Interactions of *P. lilacinus* with PPNs

Paecilomyces lilacinus exhibits the ability to penetrate the eggshells and structural build-ups of the adults and juveniles of numerous PPNs through the germination of its spores, followed by the development of hyphal branching and appressoria formation (Dong et al. 2007). The synthesis of amylase, lipase, protease, and chitinase in this species has been linked to the production of lytic enzymes with nematicidal properties, as reported by Morton et al. (2004), Gortari et al. (2008), and Gine and Sorribas (2017). In both in vitro and field studies, the application of *P. lilacinus* and *P. fumosoroseus* against *M. incognita* or *M. javanica* led to a significant reduction in their densities (Perveen and Shahzad 2013). Using *P. marquandii* to combat *M. hapla* resulted in increased weigh of the lettuce, a 25.7% reduction in formation of galls, and a 46.3% decrease in production of eggs, as reported by Chen et al. (2000). Additionally, *P. variotii* was found to be more effective than chemical substances in reducing gall numbers by over 90% (Khir Al-Assas and Naffaa 2011). Abdelrafaa (2019) demonstrated that a combination of *P. lilacinus* and *Trichoderma harzianum* significantly reduced the egg hatching of *Meloidogyne javanica* by 22.0–37.2% compared to the control.

8.3.1.3 *Pochonia chlamydosporia*

Batista and Fonseca first described the genus *Pochonia* (Batista and Fonsêca 1965). *Pochonia chlamydosporia* are naturally occurring facultative parasites of cyst and RKN that have been studied thoroughly as possible biological control agents. According to Zare and Gams (2004), *Pochonia* species may be distinguished from one another based on the conidial form, location, and quantity of chlamydospores as well as the development of conidia in either heads or chains. *P. chlamydosporia* exhibits a multitrophic existence, acting as an endophyte, parasite, and saprophyte simultaneously (Manzanilla-López et al. 2011).

Interactions of *P. chlamydosporia* with PPNs

Mostly PPN eggs and females are parasitized by *P. chlamydosporia*. According to Manzanilla-Lopez et al. (2013), the fungus preferentially colonizes fertilized eggs as well as those that have formed juveniles. The parasitic process of *Pochonia* involves adhesion, penetration, and colonization. The adherence of conidia and hyphae to the eggs is facilitated by glycoproteins produced by the fungus, providing protection against desiccation and aiding in anchoring the spore germ tube to the substrate. *Pochonia chlamydosporia* invades eggs of nematodes by forming appresoria at the hyphal tip or at lateral position, tightly encircling the surface of eggshells. Subsequently, the appresoria give rise to infection pegs that penetrate the eggshells (Holland et al. 1999). The development of mycelia within the eggs, stemming from a post-infection bulb, leads to the complete destruction of their contents (Esteves et al. 2009). *P. chlamydosporia* secretes serine proteases and chitinases, inducing significant changes within nematode eggs and potentially causing the breakdown of vitelline layer. According to Uddin et al. (2019), *P. chlamydosporia* extract PC-6 extract followed by extract PC-1 exhibited maximum suppression of *M. incognita* egg hatching. However, the PC1 isolate, followed by PC6 caused significant mortality in second-stage juveniles. Barbosa et al. (2019) reported that a formulation based on the *P. chlamydosporia* Pc-10 isolate reduced the quantity of *Meloidogyne javanica* eggs/gram of banana roots when applied to micro-propagated banana plants. These significant findings collectively underscore the biocontrol potential of *P. chlamydosporia* against PPNs.

8.3.1.4 *Trichoderma* spp.

The *Trichoderma* genus was proposed by Persoon (1794). *Trichoderma* spp. are characterized by conidiophores featuring a wide main axis that branches regularly, typically with successive branches at both apical and distal ends. These branches can be more or less divergent, occurring singly, in pairs, or forming verticils. The repeated verticillate branching can give rise to a highly branched pyramidal structure. Conidiogenous cells, known as phialides, are often arranged in divergent verticils at the tips of conidiophore branches. The conidia are single-celled, usually green but can also be colorless, or grayish. They vary in shape from smooth-walled to roughened, being subglobose, obovoid, oblong, and they aggregate in gloeoid heads (Gams and Bissett 2002). *Trichoderma* spp. have been extensively employed to enhance plant growth by suppressing PPNs (Sharon et al. 2011). They are widely distributed in plant root ecosystems, and capable of colonizing roots, thereby promoting growth and increasing disease resistance in host plants.

Interaction with PPNs

Commonly perceived mechanisms through which *Trichoderma* spp. suppress plant-parasitic nematodes (PPNs) include mycoparasitism, competition, induced resistance, and antibiosis (Nusaibah and Musa 2019). In direct interactions between nematodes and fungi, parasitism emerges as a crucial mechanism facilitating attachment to the nematode cuticle. For example, *T. longibrachiatum* TL6 has demonstrated successful control of *Heterodera avenae* by directly parasitizing both eggs and J2 (Zhang et al. 2017). *T. atroviride* and *Trichoderma asperellum*-203 stand out as highly effective parasites of *Meloidogyne javanica* juveniles, displaying a strong affinity for various life stages of the nematode both in vitro and within the plants (Sharon et al. 2007). The nematode eggshell, a significant barrier to external threats, consists of three different layers: an exterior vitelline layer, a middle chitinous layer, and an inner lipoprotein layer. The amino acid content of the eggshell suggests the presence of collagen-like proteins, imparting toughness and resilience to eggshells. *Trichoderma* lytic enzymes, including chitinase and protease, play a crucial role in disrupting the nematode eggshell, as explored by various researchers (Markovich and Kononova 2003; Steyaert et al. 2003). *T. harzianum* CECT 2413, for instance, produces PRA1 exhibiting nematicidal properties attributed to mycoparasitic interactions (Suarez et al. 2004). *Trichoderma* spp. are also recognized for promoting systemic acquired resistance (SAR) against nematode infections. In a notable instance, tomato plants primed with *Trichoderma harzianum* and *Trichoderma viride*, when challenged with RKNs, displayed stimulated expression of salicylic acid (SA) and jasmonic acid (JA) marker genes on tomato roots at various stages of *Meloidogyne* infection. This indicates priming of SA and JA responsive marker genes occur at different phases of nematode infection (Martínez-Medina et al. 2017).

8.3.1.5 Non-pathogenic *Fusarium* spp.

Fusarium species rank among the prevalent soil-borne plant pathogens responsible for substantial crop yield reductions, inducing root rot and wilts in numerous economically significant crops. However, it is noteworthy that *Fusarium* are also widely identified as the most common fungi isolated from the roots of crop plants without inducing any disease symptoms. Gordon and Martyn (1997) proposed that isolates not associated with production of disease symptoms in plants could be categorized as non-pathogenic isolates. These isolates excel at colonizing plant roots but do not elicit any disease symptoms (Benhamou and Garand 2001).

Interaction with PPNs

As an endophytic fungus, non-pathogenic *F. oxysporum* can activate host plant defense responses against PPNs, confer tolerance to environment related stresses, and stimulate the synthesis of auxins and gibberellins (Saikkonen et al. 1998). The defense mechanism activation and the initiation of resistance was observed, which in turn was complemented the formation of secondary metabolites such as hydrolytic enzymes, phytoalexins, pathogenesis-related (PR) proteins, and so on (Harish et al. 2009; Tonelli et al. 2011). One of the different forms of interaction of non-pathogenic *Fusarium* spp. with PPNs is competition (Kaur et al. 2011). Waweru et al. (2014) reported that endophytic non-pathogenic *F. oxysporum* provided effective management of *Pratylenchus goodeyi* and *Helicotylenchus multicinctus* in banana plants. Similarly, Kisaakye et al. (2022) demonstrated that endophytic *F. oxysporum* efficiently suppressed *Radopholus similis* on banana plants.

8.3.1.6 *Plectosphaerella cucumerina*

Plectosphaerella cucumerina is another opportunistic fungus that has been identified to interact with PPNs. These are ascomycetous fungi with a flat, slimy, appressed colony, and sparse aerial mycelium that varies in color from buff to salmon pink when produced on potato dextrose agar (PDA) media. Mycelia are hyaline, septate, and have many anastomosis that form coils of hypha. Conidiogenous cells developed on hyphal coils are phialidic, determinate, distinct, smooth, and solitary. Conidia are hyaline, aggregated in slimy heads, having rounded/tapering ends, with or without septa (Arzanlou et al. 2013).

Interaction with PPNs

Plectosphaerella cucumerina has demonstrated its efficacy as a successful biocontrol agent through interactions with various PPNs. Research by Elhady et al. (2017) identified an association between the bacterium *B. thuringiensis*, the fungus *P. cucumerina*, and *Meloidogyne incognita* J2 in the soil. Additionally, *Plectosphaerella cucumerina* has been discovered in egg masses of *Meloidogyne incognita* (Yu and Coosemans 1998), *Globodera pallida* eggs (Kooliyottil et al. 2017), and has also been observed infecting eggs of *Globodera rostochiensis* (Atkins et al. 2003).

8.3.2 Synergistic Interactions

The opportunistic fungi in the genera *Fusarium*, *Rhizoctonia,* and *Cylindrocarpon* aid PPNs to cause more intense disease on host plants. They may act as host modifiers, rhizosphere modifiers, host resistance breakers, and vector of plant pathogens.

8.3.2.1 *Fusarium* spp.

Atkinson (1892) first reported that the infection by PPNs could compromise host resistance or intensify the impact of fungi-infected plant diseases. He illustrated that the presence of the RKN *Meloidogyne* spp. disrupted the resistance of a cotton cultivar to *Fusarium oxysporum* f. sp. *vasinfectum*. Subsequently, numerous authors have reported several synergistic interactions between *Fusarium* spp. and PPNs. Ferraz and Lear (1976) noted that carnation plants infected with *Pratylenchus dianthus* exhibited exacerbated symptoms of *Fusarium* wilt. Griffin (1990) observed that the presence of *D. dipsaci* intensified *Fusarium* wilt in alfalfa. Likewise, *R. similis* increased the severity of wilt caused by *Fusarium oxysporum* f. sp. *cubense* in banana (Loos 1959). *M. incognita*, displaying a synergistic interaction with *Fusarium oxysporum* f. sp. *vasinfectum*, significantly heightened fusarium wilt severity in cotton (da Silva et al. 2019). Nematode infections can render eggplant more susceptible to wilt diseases like *Fusarium* wilt as well as bacterial wilt (Osman et al. 2018). Hua et al. (2019) identified a synergistic relationship between *M. incognita* and *Fusarium oxysporum* f. sp. *niveum*, exacerbating wilt symptoms in watermelon. Similarly, Wagner et al. (2022) reported that the synergistic interaction between *Fusarium oxysporum* f. sp. *vasinfectum* race 4 and RKNs significantly increased wilt-disease severity in cotton.

8.3.2.2 *Rhizoctonia* and *Cylindrocarpon* spp.

Rhizoctonia spp. (mostly *Rhizoctonia solani*) is notorious plant pathogens which are reported to be associated with many PPNs and cause elevated disease incidence on host plants. Juveniles of *Globodera rostochiensis* infecting roots of potato were linked with a greater magnitude of stolons infection by *Rhizoctonia solani* (Back et al. 2006). Bhattarai et al. (2010) discovered the connection between *Rhizoctonia solani* and *G. pallida* in potato plants under controlled environments and glasshouse conditions, which resulted in significant yield losses. Similarly, *Pratylenchus neglectus* increased the impact of *Verticillium dahliae* and *Rhizoctonia solani* infection of potato, resulting in a wilt-disease complex (Taheri et al. 2010). In soybean fields exhibiting pronounced disease symptoms caused by *Rhizoctonia solani*, elevated populations of stunt nematodes (*Tylemnchorhynchus* spp.) and root-lesion nematodes (*Pratylenchus* spp.) have been identified (Liu et al. 2016). Edin et al. (2019) found that temporal interaction of root-lesion nematode and *Rhizoctonia solani* resulted in decreased output of potato. Khan et al. (2020) also observed a synergistic relationship between *M. incognita* and *R. bataticola* that causes cotton root rot. Similarly, synergistic interaction between another opportunistic fungi *Cylindrocarpon* spp. with nematode *Pratylenchus* spp. has been reported to cause apple replant disease (ARD) in South Africa (Van Schoor et al. 2009).

8.4 Conclusions and Future Perspectives

The utilization of beneficial opportunistic fungi and PGPRs not only reduces the need for nemato-toxic compounds (nematicides) in agricultural practices but also enhances plant vigor by facilitating the absorption of essential nutrients and diminishing PPN populations in both the plant and soil through favorable interactions. Although the mechanisms by which these biocontrol agents safeguard plants from nematodes are intricate and require further investigation for a more comprehensive understanding. Conversely, increased exploration of synergic interactions between opportunistic fungal genera and PPNs can offer insights into disrupting these detrimental associations, identifying potential vulnerabilities in these interactions that can be targeted proactively. Consequently, advancing toward more sustainable production systems presents a future challenge that necessitates a deeper comprehension of the mechanism associated with the interactions among opportunistic fungi, nematodes, and soil microbiota.

References

Abd-El-Kareem F, Elshahawy IE, Abd-Elgawad MM (2019) Local *Trichoderma* strains as a control strategy of complex black root-rot disease of strawberry in Egypt. Bull Natl Res Cent 43:160

Abdelrafaa OAM (2019) Potential of *Trichoderma harzianum* and *Paecilomyces lilacinus* as biocontrol agents for *Meloidogyne javanica* on tomato. Doctoral dissertation, Sudan University of Science and Technology, Sudan

Ahmed S, Liu Q, Jian H (2019) *Bacillus cereus* a potential strain infested cereal cyst nematode (*Heterodera avenae*). Pak J Nematol 37:53–61

Akhtar M, Panwar J (2011) Arbuscular mycorrhizal fungi and opportunistic fungi: efficient root symbionts for the management of plant parasitic nematodes. Adv Sci Eng Med 3:165–175

Arzanlou M, Torbati M, Khodaei S (2013) Phenotypic and molecular characterization of *Plectosphaerella cucumerina* on bamboo from Iran. Mycosphere 4:647–651

Atibalentja N, Noel GR, Domier LL (2000) Phylogenetic position of the North American isolate of *Pasteuria* that parasitizes the soybean cyst nematode, *Heterodera glycines*, as inferred from 16S rDNA sequence analysis. Int J Syst Evol Microbiol 50:605–613

Atkins SD, Clark IM, Sosnowska D, Hirsch PR, Kerry BR (2003) Detection and quantification of *Plectosphaerella cucumerina*, a potential biological control agent of potato cyst nematodes, by using conventional PCR, real-time PCR, selective media, and baiting. Appl Environ Microbiol 69:4788–4793

Atkinson GF (1892) Some diseases of cotton. Frenching 41:19–29

Back M, Haydock P, Jenkinson P (2006) Interactions between the potato cyst nematode *Globodera rostochiensis* and diseases caused by *Rhizoctonia solani* AG3 in potatoes under field conditions. Eur J Plant Pathol 114:215–223

Bainier G (1907) Mycothéque de l'école de Pharmacie. *Paecilomyces* genre nouveau de Mucédinées. Bull Trimest Soc Mycol Franç 23:26–27

Baker KF, Cook RJ (1974) Biological control of plant pathogens. WH Freeman and Company, San Francisco, p 433

Bakker PA, Pieterse CM, de Jonge R, Berendsen RL (2018) The soil-borne legacy. Cell 172:1178–1180

Barbosa RT, Monteiro TSA, Coutinho RR, Silva JG, Freitas LG (2019) *Pochonia chlamydosporia* for controlling root-knot nematode in banana. Nematropica 49:99–106

Basyony AG, Abo-Zaid GA (2018) Biocontrol of the root-knot nematode *Meloidogyne incognita*, using an eco-friendly formulation from *Bacillus subtilis* lab and greenhouse studies. Egypt J Biol Pest Control 28:87

Batista AC, Fonsêca OJM (1965) *Pochonia humicola* n. gen. en. sp., uma curiosa entidade fúngica dos solos do Nordeste do Brasil. Publicações do Instituto de Micologia da Universidade do Recife, p 462

Benhamou N, Garand C (2001) Cytological analysis of defense-related mechanisms induced in pea root tissues in response to colonization by nonpathogenic *Fusarium oxysporum* Fo47. Phytopathology 91:730–740

Bhattarai S, Haydock PP, Back MA, Hare MC, Lankford WT (2010) Interactions between field populations of the potato cyst nematode *Globodera pallida* and *Rhizoctonia solani* diseases of potatoes under controlled environment and glasshouse conditions. Nematology 12:783–790

Bhuiyan SA, Garlick K, Anderson JM, Wickramasinghe P, Stirling GR (2018) Biological control of root-knot nematode on sugarcane in soil naturally or artificially infested with *Pasteuriapenetrans*. Austral Plant Pathol 47:45–52

Bishop AH (2011) *Pasteuria penetrans* and its parasitic interaction with plant parasitic nematodes. In: Logan N, Vos P (eds) Endospore-forming soil bacteria, Soil biology, vol 27. Springer, Berlin, pp 181–201

Borneman J, Becker JO (2007) Identifying microorganisms involved in specific pathogen suppression in soil. Annu Rev Phytopathol 45:153–172

Chen J, Abawi GS, Zuckerman BM (2000) Efficacy of *Bacillus thuringiensis*, *Paecilomyces marquandii*, and *Streptomyces costaricanus* with and without organic amendments against *Meloidogyne hapla* infecting lettuce. J Nematol 32:70–77

Collingwood CA (1962) Continuous corn growing and cereal root eelworm in the South-West. NAAS Q Rev 58:70–73

Cook RJ, Baker KF (1983) The nature and practice of biological control of plant pathogens. American Phytopathological Society, Saint Paul

Cronin D, Moenne-Loccoz Y, Fenton A, Dunne C, Dowling DN, O'Gara F (1997) Role of 2,4-diacetylphloroglucinol in the interactions of the biocontrol pseudomonad strain F113 with the potato cyst nematode *Globodera rostochiensis*. Appl Environ Microbiol 63:1357–1361

da Silva MB, Davis RF, Doan HK, Nichols RL, Kemerait RC, Halpern HC, Brewer MT, Jagdale G, Chee PW (2019) *Fusarium* wilt of cotton may commonly result from the interaction of *Fusarium oxysporum* f. sp. *vasinfectum* with *Belonolaimus longicaudatus*. J Nematol 51:1–10

Davies KG, Curtis RH (2011) Cuticle surface coat of plant-parasitic nematodes. Annu Rev Phytopathol 49:135–156

Dirksen P, Marsh SA, Braker I, Heitland N, Wagner S, Nakad R, Mader S, Petersen C, Kowallik V, Rosenstiel P, Félix MA, Schulenburg H (2016) The native microbiome of the nematode *Caenorhabditis elegans*: gateway to a new host-microbiome model. BMC Biol 14:38

Dong LQ, Yang JK, Zhang KQ (2007) Cloning and phylogenetic analysis of the chitinase gene from the facultative pathogen *Paecilomyces lilacinus*. J Appl Microbiol 103:2476–2488

Edin E, Gulsher M, Andersson Franko M, Englund JE, Flöhr A, Kardell J, Viketoft M (2019) Temporal interactions between root-lesion nematodes and the fungus *Rhizoctonia solani* lead to reduced potato yield. Agronomy 9:361

Eisenhauer N, Barnes AD, Cesarz S, Craven D, Ferlian O, Gottschall F, Türke M (2016) Biodiversity–ecosystem function experiments reveal the mechanisms underlying the consequences of biodiversity change in real world ecosystems. J Veg Sci 27:1061–1070

Elhady A, Giné A, Topalovic O, Jacquiod S, Sørensen SJ, Sorribas FJ, Heuer H (2017) Microbiomes associated with infective stages of root-knot and lesion nematodes in soil. PLoS One 12:e0177145

El-Nagdi WMA, Abd-El-Khair H, Soliman GM, Ameen HH, El-Sayed GM (2019) Application of protoplast fusants of *Bacillus licheniformis* and *Pseudomonas aeruginosa* on *Meloidogyne incognita* in tomato and eggplant. Middle East J Appl Sci 9:622–629

Esteves I, Peteira B, Atkins SD, Magan N, Kerry B (2009) Production of extracellular enzymes by different isolates of *Pochonia chlamydosporia*. Mycol Res 113:867–876

Fatima S, Anjum T (2017) Identification of a potential ISR determinant from *Pseudomonas aeruginosa* PM12 against *Fusarium* wilt in tomato. Front Plant Sci 8:848

Ferraz S, Lear B (1976) Interaction of four plant parasitic nematodes and *Fusarium oxysporum* f. sp. *dianthi* on carnation. Experientiae 22:272–277

Forghani F, Hajihassani A (2020) Recent advances in the development of environmentally benign treatments to control root-knot nematodes. Front Plant Sci 11:1125

Fresenius G (1852) Beitraege zur Mykologie. Heft 1–2:1–80

Gams W, Bissett J (2002) Morphology and identification of *Trichoderma and Gliocladium*. Taylor & Francis, Milton Park, pp 3–34

Gine A, Sorribas FJ (2017) Effect of plant resistance and BioAct WG (*Purpureocillium lilacinum* strain 251) on *Meloidogyne incognita* in a tomato–cucumber rotation in a greenhouse. Pest Manag Sci 73:880–887

Gordon TR, Martyn RD (1997) The evolutionary biology of *Fusarium oxysporum*. Annu Rev Phytopathol 35:111–128

Gortari MC, Galarza BC, Cazau MC, Hours RA (2008) Comparison of the biological properties of two strains of *Paecilomyces lilacinus* (Thom) Samson associated to their antagonistic effect onto *Toxocara canis* eggs. Malays J Microbiol 4:35–41

Griffin GD (1990) Pathological relationship of *Ditylenchus dipsaci* and *Fusarium oxysporum* f. sp. *medicaginis* on alfalfa. J Nematol 22:333–336

Gupta G, Panwar J, Akhtar MS, Jha PN (2012) Endophytic nitrogen-fixing bacteria as biofertilizer. Sustain Agric Rev 11:183–221

Hamid M, Siddiqui IA, Shahid Shaukat S (2003) Improvement of *Pseudomonas fluorescens* CHA0 biocontrol activity against root-knot nematode by the addition of ammonium molybdate. Lett Appl Microbiol 36:239–244

Hammam MMA, Wafaa MEN, Abd-Elgawad MMM (2016) Biological and chemical control of the citrus nematode *Tylenchulus semipenetrans* (Cobb 1913) on Mandarin in Egypt. Egypt J Biol Pest Control 26:345–349

Harish S, Kavino M, Kumar N, Samiyappan R (2009) Differential expression of pathogenesis-related proteins and defense enzymes in banana: interaction between endophytic bacteria, Banana bunchy top virus and *Pentalonianigro nervosa*. Biocontrol Sci Technol 119:843–857

He J, Kang J, Lei B, Wen T (2011) *Paecilomyces wawuensis*, a new species isolated from soil in China. Mycotaxon 115:303–310

Hiep NV, Ha NT, Thuy TTT, Van Toan P (2019) Isolation and selection of *Arthrobotrys* nematophagous fungi to control the nematodes on coffee and black pepper plants in Vietnam. Arch Phytopathol Plant Protect 52:825–843

Holland R, Williams K, Khan A (1999) Infection of *Meloidogyne javanica* by *Paecilomyces lilacinus*. Nematology 1:131–139

Hua GKH, Timper P, Ji P (2019) *Meloidogyne incognita* intensifies the severity of *Fusarium* wilt on watermelon caused by *Fusarium oxysporum* f. sp. *niveum*. Can J Plant Pathol 41:261–269

Huang X, Zhao N, Zhang K (2004) Extracellular enzymes serving as virulence factors in nematophagous fungi involved in infection of the host. Res Microbiol 155:811–816

Javeed MT, Al-Hazmi AS, Molan YY (2016) Antagonistic effects of some indigenous isolates of *Trichoderma* spp. against *Meloidogyne javanica*. Pak J Nematol 34:183–191

Kaur R, Kaur J, Singh RS (2011) Nonpathogenic *Fusarium* as a biological control agent. Plant Pathol J 9:79–91

Khan MR, Mohidin FA, Khan U, Ahamad F (2016) Native *Pseudomonas* spp. suppressed the root-knot nematode in *in vitro* and *in vivo*, and promoted the nodulation and grain yield in the field grown mungbean. Biol Control 101:159–168

Khan MA, Khan SA, Hasan Riaz NA, Binyamin R, Ashraf W, Ishtiaq M, Ikram RM (2020) Synergistic effect of *Meloidogyne incognita* and *Rhizoctonia bataticola* causing root rot diseases of cotton. Int J Agric Biol 24:1401–1408

Khir Al-Assas KM, Naffaa W (2011) Effectiveness of *Paecilomyces variotii*, plant extraction of hemlock *Conium maculatum* and some pesticides in controlling root-knot nematode *Meloidogyne incognita* on tomato. Arab J Arid Environ 4:48–54

Kisaakye J, Fourie H, Haukeland S, Kisitu J, Nakimera S, Cortada L, Coyne D (2022) Endophytic non-pathogenic *Fusarium oxysporum*-derived dual benefit for nematode management and improved banana (*Musa* spp.) productivity. Agriculture 12:125

Kooliyottil R, Dandurand LM, Knudsen GR (2017) Prospecting fungal parasites of the potato cyst nematode *Globodera pallida* using a rapid screening technique. J Basic Microbiol 57:386–392

Kumar KK (2020) Fungi: a bio-resource for the control of plant parasitic nematodes. In: Yadav A, Mishra S, Kour D, Yadav N, Kumar A (eds) Agriculturally important fungi for sustainable agriculture. Fungal biology. Springer, Cham, pp 285–311

Kunyeit L, Kurrey NK, Anu-Appaiah KA, Rao RP (2019) Probiotic yeasts inhibit virulence of non-albicans *Candida* species. MBio 10:e02307-19

Liu B, Shen W, Wei H, Smith H, Louws FJ, Steadman JR, Correll JC (2016) *Rhizoctonia* communities in soybean fields and their relation with other microbes and nematode communities. Eur J Plant Pathol 144:671–686

Liu C, Timper P, Ji P, Mekete T, Joseph S (2017) Influence of root exudates and soil on attachment of *Pasteuria penetrans* to *Meloidogyne arenaria*. J Nematol 49:304–310

Loos CA (1959) Symptom expression of Fusarium wilt disease of 'Gros Michel' banana in the presence of *Radopholus similis* (Cobb, 1893) Thorne, 1949, and *Meloidogyne incognita* acrita Chitwood, 1949, vol 26. In: Proceedings of the Helminthological Society, Washington, DC, pp 103–111

Lopez-Llorca LV, Jansson HB (2007) In: Robson GD, van West P, Gadd GM (eds) Exploitation of fungi. Cambridge University Press, Cambridge, pp 310–335

Manzanilla-López RH, Esteves I, Powers SJ (2011) Effects of crop plants on abundance of *Pochonia chlamydosporia* and other fungal parasites of root-knot and potato cyst nematodes. Ann Appl Biol 159:118–129

Manzanilla-Lopez RH, Esteves I, Finetti-Sialer MM, Hirsch PR, Ward E, Devonshire J, Hidalgo-Díaz L (2013) *Pochonia chlamydosporia*: advances and challenges to improve its performance as a biological control agent of sedentary endo-parasitic nematodes. J Nematol 45:1–7

Markovich NA, Kononova GL (2003) Lytic enzymes of *Trichoderma* and their role in plant defense from fungal diseases: a review. Appl Biochem Microbiol 39:341–351

Martínez-Medina A, Fernandez I, Lok GB, Pozo MJ, Pieterse CM, Van Wees SC (2017) Shifting from priming of salicylic acid-to jasmonic acid-regulated defences by *Trichoderma* protects tomato against the root knot nematode *Meloidogyne incognita*. New Phytol 213:1363–1377

Mazzuchelli RDCL, Mazzuchelli EHL, de Araujo FF (2020) Efficiency of *Bacillus subtilis* for root-knot and lesion nematodes management in sugarcane. Biol Control 143:104185

Moosavi MR, Zare R (2020) Fungi as biological control agents of plant-parasitic nematodes. In: Merillon JM, Ramawat KG (eds) Plant defence: biol control, 2nd edn. Springer, Cham, pp 333–384

Morton O, Hirsch P, Kerry B (2004) Infection of plant-parasitic nematodes by nematophagous fungi—a review of the application of molecular biology to understand infection processes and to improve biological control. Nematology 6:161–170

Mostafa FA, Khalil AE, Nour El-Deen AH, Ibrahim DS (2018) The role of *Bacillus megaterium* and other bio-agents in controlling root-knot nematodes infecting sugar beet under field conditions. Egypt J Biol Pest Control 28:66

Najafi F, Naddaf SR, Rezaie S, Kia EB, Mowlavi G (2016) In vitro ovicidal activity of nematophagous fungus *Paecilomyces lilacinus* on the eggs of parasitic helminths. J Med Microbiol Infect Dis 4:52–56

Nusaibah SA, Musa HA (2019) Review report on the mechanism of *Trichoderma* spp. as biological control agent of the basal stem rot (BSR) disease of *Elaeis guineensis*. In: Mohammad MS, Sharif U, Buhari TR (eds) *Trichoderma*—the most widely used fungicide. IntechOpen, London, pp 79–90

Osman HA, Ameen HH, Mohamed M, El-Mohamedy R, Elkelany US (2018) Field control of *Meloidogyne incognita* and root rot disease infecting eggplant using nematicide, fertilizers, and microbial agents. Egypt J Biol Pest Control 28:1–6

Persmark L, Banck A, Jansson HB (1996) Population dynamics of nematophagous fungi and nematodes in an arable soil: vertical and seasonal fluctuations. Soil Biol Biochem 28:1005–1014

Persoon CH (1794) Disposita methodical fungorum. Romers Neues Mag Bot 1:81–128

Perveen Z, Shahzad S (2013) A comparative study of the efficacy of *Paecilomyces* species against root-knot nematode *Meloidogyne incognita*. Pak J Nematol 31:125–131

Raaijmakers JM, Mazzola M (2016) Soil immune responses. Science 352:1392–1393

Reitz M, Rudolph K, Schroder I, Homann-Hergarten S, Hallmann J, Sikora RA (2000) Lipopolysaccharides of *Rhizobium etli* strain G12 act in potato roots as an inducing agent of systemic resistance to infection by the cyst nematode *Globodera pallida*. Appl Environ Microbiol 66:3515–3518

Saikkonen K, Faeth SH, Helander M, Sullivan TJ (1998) Fungal endophytes: a continuum of interactions with host plants. Annu Rev Ecol Syst 29:319–343

Saxena G (2018) Biological control of root-knot and cyst nematodes using nematophagous fungi. In: Giri B, Prasad R, Varma A (eds) Root biology, Soil biology, vol 52. Springer, Cham, pp 221–237

Sayre RM, Starr MP (1985) *Pasteuria penetrans* (ex Thorne, 1940) nom. rev., comb. n., sp. n., a mycelial and endospore-forming bacterium parasitic in plant-parasitic nematodes. In: Proceedings of the Helminthological Society of Washington, USA, vol 52, pp 149–165

Sayre RM, Wergin WP, Schmidt JM, Starr MP (1991) *Pasteuria nishizawae* sp. *nov*., a mycelial and endospore-forming bacterium parasitic on cyst nematodes of genera *Heterodera* and *Globodera*. Res Microbiol 142:551–564

Sharon E, Chet I, Viterbo A, Bar-Eyal M, Nagan H, Samuels GJ, Spiegel Y (2007) Parasitism of Trichoderma on *Meloidogyne javanica* and role of the gelatinous matrix. Eur J Plant Pathol 118:247–258

Sharon E, Chet I, Spiegel Y (2011) Trichoderma as a biological control agent. In: Davies K, Spiegel Y (eds) Biological control of plant-parasitic nematodes: building coherence between microbial ecology and molecular mechanisms. Progress in biological control, vol 11. Springer, Dordrecht, pp 183–201

Siddiqui IA, Haque SE (2001) Suppression of the root rot–root knot disease complex by *Pseudomonas aeruginosa* in tomato: the influence of inoculum density, nematode populations, moisture and other plant-associated bacteria. Plant Soil 237:81–89

Siddiqui ZA, Mahmood I (1996) Biological control of plant parasitic nematodes by fungi: a review. Bioresour Technol 58:229–239

Siddiqui IA, Shaukat SS (2003) Suppression of root-knot disease by *Pseudomonas fluorescens* CHA0 in tomato: importance of bacterial secondary metabolite, 2,4-diacetylpholoroglucinol. Soil Biol Biochem 35:1615–1623

Siddiqui IA, Haas D, Heeb S (2005) Extracellular protease of *Pseudomonas fluorescens* CHA0, a biocontrol factor with activity against the root-knot nematode *Meloidogyne incognita*. Appl Environ Microbiol 71:5646–5649

Singh J, Kumar MU, Walia RK (2014) Influence of plant root exudates on the adherence of *Pasteuria penetrans* endospores. Nematology 16:121–124

Singh J, Rajput RS, Kartikay Bisen K, Singh S, Singh HB (2017) Role of *Trichoderma* secondary metabolites in plant growth promotion and biological control. In: Singh HB, Sarma BK, Keswani C (eds) Advances in PGPR research. CABI, Wallingford, pp 411–426

Soliman MS, El-Deriny MM, Ibrahim DSS, Zakaria H, Ahmed Y (2021) Suppression of root-knot nematode *Meloidogyne incognita* on tomato plants using the nematode trapping fungus *Arthrobotrys oligospora* Fresenius. J Appl Microbiol 131:2402–2415

Steyaert JM, Ridgway HJ, Elad Y, Stewart A (2003) Genetic basis of mycoparasitism: a mechanism of biological control by species of *Trichoderma*. N Z J Crop Hortic Sci 31:281–291

Stone AR (1972) Heterodera pallida N. Sp. (Nematoda: Heteroderidae), a second species of potato cyst nematode 1. Nematologica 18:591–606

Suarez B, Rey M, Castillo P, Monte E, Llobell A (2004) Isolation and characterization of PRA1, a trypsin-like protease from the biocontrol agent *Trichoderma harzianum* CECT 2413 displaying nematicidal activity. Appl Microbiol Biotechnol 65:46–55

Taheri A, Hollamby GJ, Vanstone VA (2010) Interaction between root lesion nematode, *Pratylenchus neglectus* (Rensch 1924) Chitwood and Oteifa 1952, and root rotting fungi of wheat. N Z J Crop Hortic Sci 22:181–185

Tailor AJ, Joshi BH (2014) Harnessing plant growth promoting rhizobacteria beyond nature: a review. J Plant Nutr 37:1534–1571

Teymouri M, Akhtari J, Karkhane M, Marzban A (2016) Assessment of phosphate solubilization activity of rhizobacteria in mangrove forest. Biocatal Agric Biotechnol 5:168–172

Timper P, Kone D, Yin J, Ji P, Gardener BBM (2009) Evaluation of an antibiotic-producing strain of *Pseudomonas fluorescens* for suppression of plant-parasitic nematodes. J Nematol 41:234–240

Tonelli ML, Furlan A, Taurian T, Castro S, Fabra A (2011) Peanut priming induced by biocontrol agents. Physiol Mol Plant Pathol 75:100–105

Uddin M, Saifullah AM, Khan W, Khan B (2019) Evaluation of *Pochonia chlamydosporia* (Goddard) isolates for suppression of *Meloidogyne incognita*, root-knot nematode of tomato. J Agric Sci 11:70–81

Van Schoor L, Denman S, Cook NC (2009) Characterisation of apple replant disease under South African conditions and potential biological management strategies. Sci Hortic 119:153–162

Wachira P, Mibey R, Okoth S, Kimenju J, Kiarie J (2009) Diversity of nematode destroying fungi in Taita Taveta, Kenya. Fungal Ecol 2:60–65

Wagner TA, Duke SE, Davie SM, Magill C, Liu J (2022) Interaction of *Fusarium* wilt race 4 with root-knot nematode increases disease severity in cotton. Plant Dis 106:2558–2562

Waweru B, Turoop L, Kahangi E, Coyne D, Dubois T (2014) Non-pathogenic *Fusarium oxysporum* endophytes provide field control of nematodes, improving yield of banana (*Musa* sp.). Biol Control 74:82–88

Weller DM, Raaijmakers JM, Gardener BBM, Thomashow LS (2002) Microbial populations responsible for specific soil suppressiveness to plant pathogens. Annu Rev Phytopathol 40:309–348

Xiang N, Lawrence KS, Donald PA (2018) Biological control potential of plant growth-promoting rhizobacteria suppression of *Meloidogyne incognita* on cotton and *Heterodera glycines* on soybean: a review. J Phytopathol 166:449–458

Yadav BK, Akhtar MS, Panwar J (2015) Rhizospheric plant-microbe interactions: key factors to soil fertility and plant nutrition. In: Arora N (ed) Plant microbes symbiosis: applied facets. Springer, New Delhi, pp 127–145

Yang J, Tian B, Liang L, Zhang KQ (2007) Extracellular enzymes and the pathogenesis of nematophagous fungi. Appl Microbiol Biotechnol 75:21–31

Yu Q, Coosemans J (1998) Fungi associated with cysts of *Globodera rostochiensis*, *G. pallida*, and *Heterodera schachtii*; and egg masses and females of *Meloidogyne hapla* in Belgium. Phytoprotection 79:63–69

Zare R, Gams W (2004) A monograph of *Verticillium* section Prostrata. Rostaniha 3(S):188

Zhang S, Gan Y, Ji W, Xu B, Hou B, Liu J (2017) Mechanisms and characterization of *Trichoderma longibrachiatum* T6 in suppressing nematodes (*Heterodera avenae*) in wheat. Front Plant Sci 8:1491

Zopf W (1888) Zur Kenntnis der Infections-Krankheitenniederer Tiere. Nova Acta Leop Acad Naturf Halle 52:7

Molecular Interactions Between Opportunistic Fungi, Nematodes, and Plants

9

Arkadeb Chatterjee, Rashmita Saikia, Dikshita Saikia, and Pranjal Kumar Kaman

9.1 Introduction

Numerous creatures, such as nematodes, bacteria, fungus, plants, and arthropod herbivores, are found in the realm of soil (Gans et al. 2005; Bonkowski et al. 2009). Because there is typically little organic matter in the soil, organisms compete with one another, which leads to the development of specific relationships. Consequently, organisms evolved a set of specific signals that serve their community in order to increase fitness in dynamically fluctuating soil circumstances. One of the main mechanisms via which plants interact with soil organisms is chemical transmission that takes place close to the roots. Additionally, soil organisms, including plants, use chemicals (i.e., metabolites) to maintain the physical characteristics of the soil and other physical properties that allow them to endure abiotic stress. Therefore, a better knowledge of rhizosphere interactions is likely to be highly beneficial for strengthening plant resilience to biotic and abiotic stressors, which will ultimately lead to an increase in agricultural yield.

Plants are a major target for soil-dwelling creatures because they are essential for regulating the conditions beneath the surface of the earth and are the most prolific source of organic chemicals in the soil environment. Plants and soil animals interact in ways that have developed to be either beneficial or detrimental. Plants use root exudates to improve their fitness in a given soil environment by changing the physical and chemical features of the soil and boosting nutrient intake (Parniske 2008;

A. Chatterjee
Department of Nematology, Assam Agricultural University, Jorhat, Assam, India

R. Saikia · D. Saikia
Department of Plant Pathology, Punjab Agricultural University, Ludhiana, Punjab, India

P. K. Kaman (✉)
Department of Plant Pathology, Assam Agricultural University, Jorhat, Assam, India
e-mail: pranjal.k.kaman@aau.ac.in

M. S. Akhtar (ed.), *Opportunistic Fungi, Nematode and Plant Interactions*,
https://doi.org/10.1007/978-981-97-2045-3_9

Haichar et al. 2014). These exudates are rich in low-molecular-weight compounds that map concentration gradients in the rhizosphere due to their solubility and stability, which are dictated by their chemical properties (Bais et al. 2006). In contrast to exudation, rhizodeposition releases compounds from dead plant tissues, including leaves, roots, and detached root cells, which impacts soil microorganisms (Dennis et al. 2010). Roots emit molecules in the form of soluble exudates in addition to more hydrophobic volatile organic compounds (VOCs) that can permeate further into the rhizosphere and soil environment (Peñuelas et al. 2014). Studies on specialised metabolites in the soil matrix are scarce, which might be because it is challenging to analyse the composition of soil exudates and volatile organic compounds (VOCs). Soil microorganisms are predicted to have a major impact on the chemical composition. Because of microbial activity, the turnover of metabolites varies throughout soil types. Furthermore, different soil particles have different affinities to different chemical groups, which influences the methodology used for quantifying metabolites (Quastel 1965). While collecting root exudates is a common reason for using hydroponic and aeroponic systems, the chemical combinations produced by these methods may not be the same as those released by plants growing naturally (Kuijken et al. 2015). Owing to these limitations, there have been more efforts made to develop analytical methods for root exudate and VOC in-situ analysis (Phillips et al. 2011; Crespo et al. 2012; Kai et al. 2016).

Specialised metabolites produced and secreted by plants are vital to many different activities. Some of these specialised metabolites are generated in response to environmental cues, whereas others are constitutively released into the rhizosphere (Graham 1991; Dixon and Paiva 1995; Baetz and Martinoia 2014). The bulk of chemical groups' worth of metabolites, including phenolics, benzenoids, terpenoids, alkaloids, and sulphur-containing chemicals, have been found in exudates or as volatile organic compounds (VOCs) thus far. According to Schenkel et al. (2015), meta-analysis provides a comprehensive comparison of volatile organic compounds (VOCs) derived from roots and those produced by soil-borne bacteria. While water and solvent soluble molecules often allow for short-distance interaction, volatile organic compounds (VOCs) are more likely to be employed as long-distance communicators (Hiltpold and Turlings 2008; Insam and Seewald 2010). Furthermore, the latter may serve as indicators in trophic level interactions at various depths (Rasmann et al. 2005). Growing body of work (Kai et al. 2009; Wenke et al. 2010; Peñuelas et al. 2014) confirms the importance of volatile organic compounds (VOCs) generated by soil-borne bacteria in the chemical interactions between plants and microbes in the rhizosphere.

Over time, the composition of root exudate and volatile organic compounds (VOCs) varies in response to different cues from the plant's above- and below-ground environments (Haichar et al. 2008; De Coninck et al. 2015). Plant roots cause the rhizosphere to contain a greater number of molecules, although little is known about the functions of these materials there. Here, we provide an overview of recent studies on the specialised root exudate and volatile organic compounds (VOC) metabolites that are implicated in biotic interactions. We also discuss the formation, release, and role of these metabolites in facilitating the development of the optimal soil conditions for root system uptake of nutrients. These metabolites are produced especially for certain cell and tissue types. Although this overview might have been expanded to include many more specialised metabolites, its primary focus is on well-known chemical classes and the most current mechanistic findings associated with them.

Since the beginning of human civilisation, plant diseases have wreaked devastation on harvests and populations of people. For example, the Irish potato famine of the 1840s was caused by an outbreak caused by the fungus *Phytophthora infestans*. Infectious plant diseases continue to cause significant financial losses as well as human misery. Because of the expanding human population and the decreasing amount of land available for agriculture, every attempt to ensure the world's food supply needs to be treated seriously. Avoiding crop disease can lead to a large boost in agricultural productivity. Although pesticides have proven to be an efficient means of preventing and treating disease, their ongoing and widespread usage will be harmful to both the environment and human health. Another way to boost productivity is to employ high-yield crop varieties, but there's a catch: these genetically standardised kinds cultivated across large areas carry a risk of catastrophic epidemics. Understanding the complex ecosystem of the real disease processes and carefully implementing the plant's defensive mechanisms can help achieve effective protection against plant infections. This chapter looks at the ways that major host responses and microbial pathogenesis strategies affect plant–pathogen interactions. Our focus is on studies utilising the model plant Arabidopsis thaliana. There has been discussion of symbiotic interactions, such as the one between a bean and rhizobitim, elsewhere (Long 1996).

The fact that nematodes, both plant-parasitic and non-parasitic, congregate around plant roots as well as bacterial and fungal communities is widely recognised. A few of the factors that have been thought to be important are temperature, electrical potential, carbon dioxide, and various organic and inorganic chemicals (Croll 1970). To illustrate how these nematodes locate their hosts and eat fungus, researchers have mostly examined nematode attraction to fungi (Townshend 1964; Klink et al. 1970). The only research on nematodes' attraction to the live mycelium of nematophagous fungi was presented by Field and Webster (1977). It was discovered that culture filtrates of a nematophagous fungus attracted Aphelenchus avenue, a nematode that feeds on fungi (Monoson et al. 1973). Balan and Gerber (1972) and Balan et al. (1974, 1976) demonstrated that the bacteriovorous nematode *Panagrellus redivivus* was drawn to the culture filtrates of several nematophagous fungi by a variety of trapping methods.

9.2 Specialised Metabolites Secreted from Plants and Mycorrhizal Fungi Interactions

Plants and soil-borne fungi, or oomycetes, have evolved symbiotic, parasitic, and destructive associations through adaptation and co-evolution. Many plants get along well with AM fungus, but not much is known about the chemical communication between the two species at this time. The chemical signals most extensively studied in relation to interactions between roots and AM with root-AM are strigolactones. Nanomolar concentrations of the sesquiterpenes 5-deoxy-strigol, sorgolactone, and strigol, which were isolated from the root exudates of Lotus japonicas, significantly increased hyphal branching in the AM fungus Gigaspora margarita (Akiyama et al. 2005). This effect was also observed in GR24, the synthetic equivalent of sorgolactone (Dor et al. 2011). The similar effect of sorgolactones was also seen in two other AM strains, *Gigaspora rosea* and *Glomus intraradices* (Akiyama and Hayashi 2006). It was found that by activating ISR, plants colonised by AM changed their response to pathogens (Watts-Williams et al. 2015). Additional studies have revealed that exudates from AM-colonised plants have altered metabolic profiles (Scheffknecht et al. 2006), suggesting that the AM fungus may be indirectly responsible for notable changes in the composition of the soil microbiome. There is growing evidence that mycorrhizal fungi, like root-colonising bacteria, have an effect on the specialised metabolites that affect root architecture. Volatile sesquiterpenes released by the ectomycorrhizal fungus *Laccaria bicolor* stimulate the growth of lateral roots in Arabidopsis and poplar, increasing the root surface area and enabling improved fungal access and nutrient absorption (Ditengou et al. 2015). Similarly, 6-pentyl-2H-pyran-2-one, a volatile chemical generated by a fungus in the genus Trichoderma that colonises roots, suppresses the growth of primary roots and encourages the establishment of lateral roots (Garnica-Vergara et al. 2016).

9.3 Specialised Metabolites Secreted from Plants and Pathogenic Fungi/Oomycetes Interactions

A common target of oomycetes, or soil-borne fungi, is plant roots. Before colonising a plant, fungal diseases can use their hyphae to sense chemical cues and guide their development in that way. *Fusarium oxysporum*, a plant disease, is assumed to be directed towards tomato (*Solanum lycopersicum*) roots due in part to the catalytic activity of class III peroxidases released by roots (Turra et al. 2015). The substrates for these enzymes are yet unknown, even though it is believed that they are secreted from plant roots. Conversely, it was shown that *F. oxysporum* chemotrophic sensing is controlled by a distinct MAPK signalling cascade that is started by a protein that binds to α-pheromone peptide and resembles *Saccharomyces cerevisiae* Ste2 sex pheromone receptor. According to Turra et al. (2015), this finding suggests that the fungus may also be able to detect peptides in the root signals.

A similar eavesdropping study in legumes focused on three root-secreted isoflavones, genistein, and daidzein, which are essential for the *Rhizobium*-legume association as chemical lures, promoters of nodulation genes, and regulators of phytoalexin resistance (Hassan and Mathesius 2012). The chemotactic oomycete pathogens *Phytophthora niederhauserii, P. pisi, P. sojae,* and *P. vignae* have evolved receptors to identify the same chemical signals produced by legume roots, indicating an evolutionary shift to undermine a positive plant–microbe relationship (Hosseini et al. 2014).

Plants that are good at protecting themselves often release specialised chemicals with rapid defensive uses when they are attacked by a root fungal disease. After infection with the fungus *Verticillium longisporum*, phytoalexins such as camalexin, which are compounds produced from tryptophan, accumulate in Arabidopsis (Iven et al. 2012). Double mutants with impaired synthesis of these chemicals show increased susceptibility to the virus (Iven et al. 2012). Triterpenoids are among the most potent antifungal defence compounds produced by plant roots. Saponins, glycosylated triterpenoids produced by cereals, are commonly used as constitutive defence chemicals against fungal infections, as demonstrated by the antifungal activities of avenacin released by oat roots (Papadopoulou et al. 1999; Osbourn 2003). Two more terpenoids with phytoalexin-like characteristics that were only recently identified are the sesquiterpene zealexins and the diterpene kauralexins (Huffaker et al. 2011; Schmelz et al. 2011). These compounds accumulate in the roots of maize due to abiotic factors such drought and salt stress, as well as fungal pathogen infection (Vaughan et al. 2015). Mutants with decreased kauralexin production are more sensitive to drought. To determine whether these diterpenoids offer direct or indirect drought protection, more investigation is required. Additionally, some research studies have shown that volatile organic compounds (VOCs) produced by roots may suppress the spread of infections. One such study examined the impact of derivatives of oxygenated fatty acids on Fusarium (Cruz et al. 2012). Similar activities have been associated with volatile terpenoids derived from roots, such as the common belowground volatile organic compound 1,8-cineole (Wenke et al. 2010). Furthermore, studies conducted on the Arabidopsis plant have shown that the widely distributed unstable homoterpene (E)-4,8-dimethyl-1,3,7-nonatriene (DMNT) functions defensively to ward off infection by the root-rot-causing Pythium irregular disease (Sohrabi et al. 2015; Sohrabi et al. 2017). The triterpenoid precursor arabidiol is broken down into DMNT by physical contact between Pythium oospores and Arabidopsis roots. According to research conducted in vitro, it may reduce oospore germination in the early phases of infection (Sohrabi et al. 2015).

9.3.1 Nematophagous Fungi

Currently, nematophagous fungi are known to exist in roughly 700 species. These fungi belong to several evolutionary branches, such as Ascomycota, Basidiomycota, Zygomycota, and Chytridiomycota. Some species of Oomycetes, which belong to the fungus kingdom but are not true fungi, are capable of wiping out nematodes. Based on the strategies they use to combat nematodes, nematode-trapping fungi, endoparasitic fungi, egg-parasitic fungi, and toxin-producing fungi are the four main types of nematophagous fungi (Zhang and Hyde 2014). Many fungi can be easily distinguished from one another. For example, the order Orbiliales (Ascomycota) contains a monophyletic group of nematode-trapping fungi. These fungi can grow saprophytically in the soil. However, these fungi can become carnivorous in the presence of nematode prey by developing specific traps, such as nonconstricting rings, constricting rings, adhesive knobs, adhesive networks, and adhesive columns (Zhang and Hyde 2014).

The trapping methods may serve as trustworthy markers for the general delimitation of this fungus, based on phylogenetic analyses utilising ribosomal RNAs (rRNAs) and protein-coding genes (Ahrén et al. 1998; Scholler et al. 1999; Li et al. 2005). Consequently, three novel general hypotheses concerning nematode-trapping fungi have been proposed: Drechslerella is identified by constricting rings, whereas *Dactylellina* is identified by stalked sticky knobs and/or nonconstricting rings. Adhesive networks are what set *Arthrobotrys* apart (Li et al. 2005). Two basidiomycete fungi, *Coprinus comatus* and *Strophariarugoso annulata*, can produce special nematode-attacking instruments called spiny balls and acanthocytes, respectively (Luo et al. 2006, 2007). According to Luo et al. (2006, 2007), the distinct sharp characteristics rip the worm cuticle and cause nematode core components to leak, indicating that mechanical forces can also function as a virulence factor. In addition to the four kinds of fungi described above, a wide variety of funguses can also eradicate nematodes. For example, the mycoparasitic fungus *Trichoderma* spp. has been reported to be resistant to BCAs against PPNs; however, the underlying mechanism is still mostly unknown (Affokpon et al. 2011). Trichoderma secretes extracellular hydrolytic enzymes that can parasitize worm eggs and larvae, such as the chitinolytic enzymes chi18–5 and chi18-12 (Szabó et al. 2012), serine protease SprT (Chen et al. 2014), and trypsin-like protease PRA1 (Suarez et al. 2004). During in vitro nematode egg parasitism in *Trichoderma harzianum,* 13 peptidase-encoding genes are coexpressed, indicating that these genes probably have important functions in the egg-infection process. According to Szabó et al. (2013), these genes include the metalloendopeptidase gene P7455, the acidic serine protease gene PRA1, the aspartic protease genes P6281 and P9438, and the sedolisin serine protease. Furthermore, trichodermin (Yang et al. 2010), β-vinylcyclopentane-1α, 3α-diol, 6-pentyl-2H-pyran-2-one, and 4-(2-hydroxyethyl) phenol (Yang et al. 2012) are among the nematicidal compounds that have been identified from *Trichoderma* spp. According to Veresoglou and Rillig (2012) and Vos et al. (2012), certain endophytic and arbuscular mycorrhizal (AM) fungi have the ability to reduce nematode infestations and stimulate the growth of nematode-infected plants, such as *Acremonium* spp. and *Neotyphodium* spp. *Fusarium oxysporum*, a nonpathogenic fungus that has also been proven to boost the yield of tissue-cultured banana plants, has been demonstrated to significantly lower the populations of the PPNs *Pratylenchus goodeyi* and *Helicotylenchus multicinctus* (Waweru et al. 2014).

9.4 Specialised Metabolites Secreted from Plants and Plant-Parasitic Nematodes Interaction

Many specific metabolites have been connected to defensive roles, such as protecting plant roots against arthropods or nematodes. Soil nematodes are among the most diverse and diversified groups of organisms found in the soil. According to, a variety of soil nematodes feed mostly on the roots of subterranean plant tissues. Nematodes may be drawn to plant roots by specific signals produced as a result of mechanical root damage or insect herbivory (Rasmann and Turlings 2008; Laznik and Trdan 2013). There is a noticeable selectivity in the interactions between nematodes and plants, which is most likely caused by chemical cues. For example, in in vitro tests, *Ditylenchus destructor* was attracted to the crude root exudates of the sweet potato (Xu et al. 2015). Water-soluble compounds emitted from the root tips of green pea and maize caused nematodes to become dormant and lose their ability to spread infection (Zhao et al. 2000). Additionally, water-soluble substances were released by the ends of the green pea and maize roots, which put nematodes into a state akin to dormancy and reduced their vulnerability to infection (Zhao et al. 2000).

Plant-exuded specialised metabolites can affect several stages of the nematode life cycle, such as egg hatching, chemotaxis towards root hosts, and death. The root-knot nematode *Meloidogyne incognita* is well known for causing severe damage to tomato plants. An analysis of the composition of the infected tomato cultivars with varying susceptibilities to *M. incognita* showed a significant increase in L-ascorbyl 2,6-dipalmitate, dibutyl phthalate, dimethyl phthalate, and 2,6-ditertbutyl-p-cresol. According to Yang et al. (2016), all of these compounds increase the mortality rate of juvenile worms in their second stage and prevent *M. incognita* eggs from hatching. However, it was shown that the nortriterpene solanoeclepin A, which is released by potato and tomato roots, encourages the hatching of cyst nematode eggs from the genera *Globodera* and *Heterodera* (Schenk et al. 1999). Similar to this, kidney beans (*Phaseolus vulgaris*) and soybeans (*Glycine max*) release eclepins called glycinoeclepin A and glycinoeclepin B/C (Masamune et al. 1982; Fukuzawa et al. 1985). Intercropping tomato and crown daisy plants resulted in a considerable reduction in the amount of *M. incognita* infection of tomato roots. The lauric acid released by the crown daisy plant (*Chrysanthemum coronarium*) decreased the M. incognita root infection by changing the transcript level of Mi-flp-18, a neuromodulator that regulates nematode chemotaxis (Dong et al. 2014). Additionally, chemicals that affect root parasitic nematodes might be released by the action of microorganisms in the soil. According to Vos et al. (2012), the presence of AM fungus alters the root exudate profile of the host plant, preventing *M. incognita* from penetrating.

9.5 *Trichoderma*—The Game Changer

The molecular physiology of mycotrophy has been greatly advanced by recent advances in transcriptomics (Seidl et al. 2009; Lorito et al. 2010) and the genomic sequencing of three *Hypocrea/Trichoderma* spp. (*H. jecorina*, *H. virens*, and *H. atroviridis*; Kubicek et al. 2011). Numerous genes encoding oligopeptide transporters and proteases are expressed in numerous *Hypocrea/Trichoderma* spp. both before and during interaction with the prey (Suárez et al. 2007; Seidl et al. 2009). ESTs produced from *T. harzianum* CECT 2413 grown under biocontrol conditions greatly over-represent genes encoding these enzymes, and the majority of these proteases belong to the subtilisin-like serine protease group (Suárez et al. 2007). An analysis of the ESTs that accumulated at the time of *H. atroviridis* initial encounter with its fungal prey species *Thanatephorus cucumeris* (anamorph *Rhizoctonia solani*) and *S. sclerotiorum* showed a high abundance of genes encoding subtilisin-like serine proteases (Seidl et al. 2009). Strains of *H. atroviridis* that over expressed one of these proteases (encoded by the gene prb1) showed increased mycoparasitic activity (Flores et al. 1997). The release of oligopeptides from the action of any of these proteases on the prey fungus may then be absorbed by receptors on *H. atroviridis* that detect nitrogen deficiency (Seidl et al. 2009). Similar mechanisms in which the oligopeptides of the nematode cause the prey to be captured have also been discovered in nematophagous fungus (Dijksterhuis et al. 1994). It has been suggested that the class IV G protein-coupled receptors (GPCRs) identified in *H. atroviridis* (Seidl et al. 2009) serve as sensors for these oligopeptides (Kubicek et al. 2011). There are two class IV GPCR paralogues for each of the three species of *Haemophilus atroviridis, Haemophilus virens,* and *Haemophilus jecorina* (Kubicek et al. 2011).

The detection of the prey may involve additional GPCRs. For example, mycoparasitism in *H. atroviridis* requires the cyclic AMP receptor-like GPCR Gpr1 (protein ID 160995 in the JGI *T. atroviride* v2.0 genome of the Joint Genome Institute (JGI) Genome Portal) (Omann et al. 2009). Any extra signal transduction from any of these receptors is achieved through a common G protein signalling cascade consisting of three Gα subunits, one Gβ subunit, and one Gγ subunit. According to studies conducted by Rocha-Ramírez et al. (2002) and Reithner et al. (2005), the Gα subunit Tga1 of *H. atroviridis* loss-of-function mutants demonstrated a significant decrease in chitinase activity, a total loss of mycoparasitic proliferation on three hosts, and a reduction in the synthesis of the antifungal compound 6-pentyl pyrone. The loss of tgaA, a tga1 homologue, in *H. virens* is the cause of decreased mycoparasitic activity on *Athelia rolfsii*, an anamorph of *Sclerotium rolfsii* (Mukherjee et al. 2004). The mitogen-activated protein kinase (MAPK) pathway is one of the most widely used signal transduction routes in fungus (Schmoll 2008). The genomes of *Hypocrea/Trichoderma* spp. contain the genes encoding three MAPKs: the osmoregulatory MAPK (Hog1), the cell integrity kinase (TmkB), and the so-called pathogenicity MAPK (TmkA; also known as Tvk1 and Tmk1). When tmkA was deleted, it was demonstrated that a "P" strain of *H. virens*, which is effective against *Pythium* spp. and generates gliovirin, lost its antagonistic action against *S. rolfsii* but not *R. solani* (Mukherjee et al. 2003; Viterbo et al. 2005). Nevertheless, tmkA deletion resulted in significantly improved biocontrol against *Pythium ultimum* and *R. solani* in an *H. virens* "Q" strain ("Q" strains release copious amounts of gliotoxin and are effective against *R. solani*) (Mendoza-Mendoza et al. 2007). The different secondary metabolite profiles of the "P" and "Q" strains may account for these discrepant results; this notion will be investigated once the "P" strains' genomes are more understood. The deletion of the tmk1 homologue in *H. atroviridis* resulted in decreased mycoparasitism against R. solani and increased production of chitinase and antifungal compounds (Reithner et al. 2007). The functions of the other two MAPKs, TmkB and Hog1, are less evident since mutants for these genes show poor growth, which inhibits effective antagonism. For example, mycoparasitic capacity was absent in *H. atroviridis* mutants for Hog1 (a protein involved in tolerance to osmotic and oxidative stress) and inefficient in *H. virens* mutants for TmkB when it came to mycoparasitism on *S. rolfsii* (Delgado-Jarana et al. 2006; Kumar et al. 2010).

While mycotrophy only requires an attachment to the fungal substrate, mycoparasitism, which is frequently characterised by coiling around the prey mycelium and the production of helix-shaped hyphae (Harman et al. 2004; Harman 2011), depends on the identification of lectins from the fungal prey (Inbar and Chet 1996). Plant lectins also promote coiling to the same amount, suggesting that lectins are not responsible for determining the specificity of *Hypocrea/Trichoderma* spp. attachment to their respective prey species (Rocha-Ramírez et al. 2002). Furthermore, some *Hypocrea/Trichoderma* spp. hyphae have the capacity to coil around themselves in the absence of prey, hence coiling is not always linked to mycoparasitism (Lu et al. 2004). Hyphal elongations, whether spiral or helical, are in fact suggestive of the identity of various species, including *T. spirale* and *T. helicum.*

The creation of papilla-like structures and proliferation along the host hyphae are common initiators of *Hypocrea/Trichoderma* spp. mycoparasitic attacks, and they happen independently of the type of prey (Rocha-Ramírez et al. 2002; Chacón et al. 2007).The cell membrane starts to break down and the lumen starts to pierce when papilla-like structures appear (Harman et al. 2004; Chacón et al. 2007; Harman 2011). These structures resemble the appressorium of plant-pathogenic fungi and those generated by tomato (*Solanum lycopersicum*) in *T. harzianum* (Chacón et al. 2007). The rice blast fungus *Magnaporthe grisea* produces glycerol from store lipids, which helps to increase the turgor necessary for mechanical pressure that permits penetration of the plant cell wall (de Jong et al. 1997). According to Seidl et al. (2009), during the contact stage of mycoparasitism in *H. atroviridis*, transcription of genes related to lipid catabolism and osmoregulation increases. This suggests that the papilla-like structures of *Hypocrea/Trichoderma* spp. may also store glycerol for a comparable purpose.

Before germinating on *P. ultimum* hyphae, *H. atroviridis* spores adhere to them, indicating that parasite hyphae are not the only structures that can make contact with and bind to potential victims (Lu et al. 2004). According to genomic sequences, hydrophobins tiny, amphiphilic proteins with eight cysteine residues are most concentrated in *Hypocrea/Trichoderma* spp. among the ascomycetes. These proteins may play a role in the mechanism of conidial affinity for the host mycelium (Kubicek et al. 2008).

Genes encoding ABC efflux transporters, pleiotropic and multidrug resistance transporters, and the heat shock and oxidative stress responses are among the detoxification processes that *Hypocrea/Trichoderma* spp. induce in the presence of prey (Seidl et al. 2009; Lorito et al. 2010). The fungal prey *R. solani* uses radical oxygen species as signalling molecules during the formation of sclerotia and excretes antifungal metabolites (Aliferis and Jabaji 2010). Both of these substances have the potential to trigger the stress response seen in *Hypocrea/Trichoderma* spp. Deletion of one of the ABC transporter genes (Abc2) in *H. atroviridis* led to a decrease in

R. solani biocontrol, highlighting the significance of detoxification in mycoparasitism (Ruocco et al. 2009). The prey is killed by a combination of cell wall-hydrolytic enzymes produced by *Hypocrea/Trichoderma* spp. and antifungal secondary metabolites (BOX 2). The genomes of *Hypocrea/Trichoderma* spp., which contain a significant number of genes encoding enzymes for their production, mirror the relevance of these substances to mycoparasite lifestyle (Kubicek et al. 2011). For instance, among all the fungi, *H. virens* has the greatest number of nonribosomal-peptide synthetases. Furthermore, orthologous genes that are absent from *H. jecorina* but present in *H. atroviridis* and *H. virens* appear to encode proteins involved in the synthesis of secondary metabolites (Kubicek et al. 2011), indicating that these genes may function as precursors to the synthesis of antifungal compounds that are not yet known.

The cell wall of a fungal cell comprises around 30% of its dry weight and is mostly composed of chitin, β-1,3-glucans, α-1,3-glucans, and α-1,4-glucans (Latgé 2007). Interestingly, *H. atroviridis* and *H. virens* have 29 and 36 chitinases, respectively (Seidl 2008; Kubicek et al. 2011). By combining a cellulose-binding module with the chitinases Chit33 and Chit42, *T. harzianum* was able to more effectively mycoparasite (Limón et al. 2004). These chitinases' inclusion of cellulose-binding modules allowed them to bind to insoluble chitin substrates more securely. Certain chitinases from *Hypocrea/Trichoderma* spp. underwent positive selection during their evolution; this is indicative of a coevolutionary arms race between the pathogen and the host (Ihrmark et al. 2010). However, after deleting particular chitinase genes, gene redundancy probably kept numerous *Hypocrea/Trichoderma* species from losing their mycoparasitism or biocontrol capacities (Benítez et al. 2004; Harman et al. 2004). *Hypocrea and Trichoderma* spp. also have a larger group of GH family 75 chitosanases, which hydrolyse chitosan, a partially deacetylated form of chitin (Kubicek et al. 2011).

Fungal cell walls contain β-1,3-glucan (Latgé 2007), which is the second most prevalent polymer and has β-1,6-branches. β-1,3-glucanases hydrolyse this polymer. Genes encoding this specific type of enzyme appear to be over-represented in the genomes of *Hypocrea/Trichoderma* spp. as compared to other closely related fungi (Kubicek et al. 2011). β-1,6-glucanases have been detected in the area where *Hypocrea/Trichoderma* spp. and their prey species come into contact. By overexpressing the β-1,6-glucanase Bgn16.3 in *T. harzianum* CECT 2413, a more potent biocontrol strain was created that prevented the growth of *B. cinerea, R. solani*, and *Phytophthora citrophthora* (Montero et al. 2007). Moreover, strains of *H. virens* and *T. harzianum* that secrete an excessive amount of β-1,6-glucanases shown enhanced biocontrol of *P. ultimum* (Djonovic et al. 2006), *B. cinereal* (Ihrmark et al. 2010).

9.5.1 *Trichoderma* against Plant-Parasitic Nematodes

Regarding mycotrophy, a number of traits that seem to have evolved in *Hypocrea/Trichoderma* spp. could have functioned as pre-adaptations to allow animal predation or parasitism. In the rhizosphere, a variety of *Hypocrea/Trichoderma* species effectively oppose and eliminate plant-parasitic nematodes (Kubicek and Druzhinina 2007). Due to their diverse range of hosts, crop rotation is unable to control nematode pests that are economically important to agriculture, such as the root-knot nematodes (*Meloidogyne* spp.) and the cyst nematodes (*Heterodera* spp. and *Globodera* spp.) (Kubicek and Druzhinina 2007). But it is incredible how some *Hypocrea/Trichoderma* species, like *T. harzianum*, can colonise the nematode's eggs and second-stage juveniles to protect plants against *Meloidogyne incognita* assaults.

Nematode eggs must be parasitized through penetration of the eggshell, which consists of multiple layers (including a thick chitinous layer) and is believed to be a major barrier for infection (Bird and Bird 2012). For opportunistic nematophagy, the wide variety of chitinases seen in *Hypocrea/Trichoderma* spp. may therefore be beneficial. The presence of S8 proteases, which resemble subtilisin, in *Hypocrea/Trichoderma* spp. may also be essential for breaking through the adult worm cuticle, which is composed of proteins that resemble keratin and collagen. The ability of subtilisins and chemotrypsins from various *Hypocrea/Trichoderma* spp. to be cloned has been reported (Sharon et al. 2001; Suarez et al. 2004; Chen et al. 2009); examples of these include the chemotrypsin-like protease Pra1 from *T. harzianum* and the alkaline subtilisin Prb1 from *H. atroviridis*, both of which have significant roles in mycoparasitism (Seidl et al. 2009; Bird and Bird 2012). These cloned proteins can help the fungi to enter nematode eggs.

9.5.2 Activity in the Rhizosphere

One of the common ecological niches for *Hypocrea/Trichoderma* spp. is the rhizosphere, which provides opportunities for both saprotrophy and biotrophy on plant root exudates. In Ethiopian highland forests, the rhizosphere of the coffee plant, *Coffea arabica*, was found to harbour the most diverse species of fungi in a single habitat (Belayneh Mulaw et al. 2010); conversely, a similar survey conducted in non-rhizosphere soil on Sardinia (Italy) revealed very low diversity (Migheli et al. 2009).The reason *Hypocrea/Trichoderma* spp. are drawn to the rhizosphere can be explained by two of their food preferences. First, a mycorrhizal fungus that lives inside the roots of 92% of terrestrial plants may act as a potential food source for mycotrophs. While some research indicates that the two kinds of fungi work in concert, other investigations have discovered that Arbuscular mycorrhizal fungi are attacked by *Hypocrea/Trichoderma* spp. and their ability to colonise plant roots is suppressed. However, our understanding of the interactions between several species of *Hypocrea/Trichoderma* and mycorrhizal fungi remains incomplete (Calvet et al. 1993; McAllister et al. 1994; Datnoff et al. 1995; Green et al. 1999; Nemec et al. 1996; Siddiqui and Mahmood 1996). Furthermore, it has been noted that the density of *Hypocrea/Trichoderma* spp. is negatively impacted by arbuscular mycorrhizal fungus (Green et al. 1999).

Few investigations have been conducted on ectomycorrhizal fungus and *Hypocrea/Trichoderma* spp. (Werner and Zadworny 2003). Second, the plant roots, especially the tips, are covered in mucigel, a slimy capsule that resembles gel and is secreted by the outermost cells of the root cap. Mucigel is composed of highly hydrated polysaccharides like pectins and hemicelluloses, mainly rhamnogalacturonans and arabinoxylans. *Hypocrea/Trichoderma* spp. hemicellulases may have evolved to ingest polysaccharides generated from partially destroyed wood by potential fungal prey; these substances are easily biodegradable targets for them. According to Morán-Diez et al. (2009), *T. harzianum* CECT 2413 cannot thrive in the tomato rhizosphere without an endopolygalacturonase.

The development of *H. virens* in the rhizosphere is influenced by both sucrose and the monosaccharides and disaccharides released into the rhizosphere by plant roots. These substances are vital carbon substrates for mycorrhizae, as reported by Vargas et al. (2009) and Nehls et al. (2010). Before being hydrolysed, sucrose must be absorbed by a sucrose permease because the genomes of *H. atroviridis, H. virens*, and *H. jecorina* contain the genes for intracellular (but not extracellular) invertases. It is possible that sucrose is actively transferred from plant to fungal by a highly specialised sucrose transporter present in *H. virens*. This transporter is activated during the early stages of root colonisation and displays biochemical properties similar to those of plant sucrose carriers (Vargas et al. 2011). Moreover, the genomes of *H. atroviridis* and *H. virens* express a significant number of important facilitator solute transporters (Kubicek et al. 2011); nevertheless, it is unclear how these transporters contribute to the acquisition of extra root exudates. In conclusion, the main factors that may have attracted the ancestors of *Hypocrea/Trichoderma* spp. to colonise the rhizosphere and establish connections with plant roots were the availability of fungal prey and the accessibility of nutrients derived from plant roots (Kubicek et al. 2011).

9.5.3 Interaction between *Trichoderma* and Plants

Similar to fungi and mammals, plants can respond to the presence of other species by activating potential defence mechanisms. Different plant-pathogenic species that initiate a two-branched innate immune defence provide the most convincing explanation for this (Jones and Dangl 2006). The initial step, known as pamp-triggered immunity (PTI), is characterised by its ability to recognise and respond to molecules commonly found in bacteria, such as pathogen-associated molecular patterns (PAMPs) or microorganism-associated molecular patterns (MAMPs). Conversely, effector-triggered immunity, or ETI, responds to the pathogen's virulence factors. *Hypocrea/Trichoderma* spp., like other non-plant pathogenic microbes, induce systemic resistance (ISR) that results in the accumulation of enzymes involved in the ethylene and jasmonate signalling pathways, such as peroxidase, hydroperoxide lyase, and phenylalanine ammonia lyase (which promotes lignification) (Shoresh et al. 2010). For example, fungal endopectinases that interact with mucigel to activate plant defence systems provide oligogalacturonides (Morán-Diez et al. 2009).

Recognising MAMPs and/or molecules released during the early stages of the contact, the plant releases phenolic compounds and deposits more callose and cellulose in its cell walls. As was seen in the early phases of *T. asperelloides* root colonisation on *Cucumis sativus* (cucumber) (Yedidia et al. 1999; Segarra et al. 2007), both of these mechanisms halt additional colonisation. It is not expected that *Hypocrea/Trichoderma* spp. will cause the second stage of the plant innate immune response, in contrast to other plant infections. *T. asperellum* may do this in the early stages of the fungal interactions with roots, causing systemic acquired resistance (SAR) in cucumber plants in a concentration-dependent way (Segarra et al. 2007). SAR and the second phase of the plant immune response are frequently associated. In particular, *T. asperelloides* T203, *H. virens*, and the protoplast fusion hybrid *T. "harzianum"* T-22 are among the few *Hypocrea/Trichoderma* species and strains that are particularly good at inducing plant defences.

Several classes of *Hypocrea/Trichoderma* spp. chemicals, such as xylanases, peptaibols, swollenin, and cerato-platanins, are included in MAMPs. Endoxylanase Eix, sometimes referred to as Xyn2, from *"T. viride"* ATCC 52438 was the first protein from *Hypocrea/Trichoderma* spp. that was documented to cause the formation of ethylene in tobacco (*Nicotiana tabacum*) and tomato (Dean and Anderson 1991). Since the species identity of this strain has never been confirmed through the use of molecular tools, caution must be exercised when handling it. An effective biocontrol strain of *H. virens* secretes an endoxylanase that is similar to Eix (Hanson and Howell 2004). Genes encoding homologous enzymes are found in the genomes of *H. jecorina* (protein ID 123818 in the JGI *T. reesei* v2.0 genome) and *H. virens* (protein ID 72838 in the JGI *T. virens* Gv29-8 v2.0 genome). It is interesting to note that Eix's catalytic activity does not seem to be necessary for eliciting defensive mechanisms in plants (Enkerli et al. 1999; Sharon et al. 1993). This means that the MAMP must be the enzyme itself, rather than its reaction product. Actually, Eix binds to the plant Eix receptor 2 (Eix2; also called LeEix) to initiate the plant response. This protein is a member of the superfamily of leucine-rich repeat receptor-like proteins in plants and carries a signal for receptor-mediated endocytosis, which is required for the proper induction of defence responses (Ron and Avni 2004; Bar and Avni 2009). Additionally, the plant defence response is triggered by the binding of Eix to the receptors in plants, which leads to changes in membrane function (Bailey et al. 1992).

Restricting the synthesis of peptaibol, a family of non-ribosomal peptides, disrupts the peptaibol synthase gene Tex1 in *H. virens*, leading to strains that do not produce ISR in cucumber, though this may be corrected by adding peptaibol mixtures (Viterbo et al. 2007). While the exact mechanism of peptaibol's induction of ISR remains unknown, it is possible that these peptides, like Eix, alter the way membranes function (Bailey et al. 1992). The crystalline cellulose structure of plant cell walls can be harmed by a protein known as swollenin, which has a cellulose-binding module (Saloheimo et al. 2002). It encourages root colonisation and local defence mechanisms in *T. asperellum*, with the exception of ISR (Brotman et al. 2008). Given that the amino acid sequences of swollenin and expansins, two plant proteins that help the plant cell wall expand in roots and root hairs (Guo et al. 2011) are similar, *Hypocrea/Trichoderma* spp. may profit from the increased root surface area that swollenin causes when they settle in the plant rhizosphere. The unique characteristics of the tiny secreted proteins known as cerato-platanins are attributed to four cysteines that combine to form two disulphide bonds. The *H. virens* cerato-platanin Sm1, also known as Epl1, is responsible for inducing ISR in maize (*Zea mays*) and cotton (*Gossypium hirsutum*) (Djonovic et al. 2007). Epl1 is the orthologue of Sm1 in *H. atroviridis* and is one of the key proteins generated by the fungus on a constitutive basis. Sm1 glycosylation, which maintains the protein in a monomeric state, initiates ISR87 (Vargas et al. 2008). Deglycosylation results in the production of a Sm1 dimer but not ISR. Deglycosylation by the plant is thought to alter the protein's aggregation state, which affects Sm1's ability to elicit defensive responses. Given that most other fungi from related genera only have one paralogue of sm1, *H. jecorina*, *H. virens,* and *H. atroviridis* each contain three, it is probable that cerato-platanins are significant for *Hypocrea/Trichoderma* spp. The genomes of *Hypocrea/Trichoderma* spp. encode additional small-secreted cysteine-rich proteins that may be involved in root colonisation, similar to the small-secreted proteins of the ectomycorrhizal basidiomycete *Laccaria bicolor*, which accumulate in the hyphae that colonise the plant root (Kubicek et al. 2011; Martin et al. 2008).

Investigating the metabolic profiles of native soil exudates in a sensitive, reproducible, and comprehensive manner is technically difficult, despite the fact that the metabolites released by plant roots to the rhizosphere are exceedingly complicated. In order to overcome these limitations, it is anticipated that new technologies for soil root exudate sampling will be employed in the next years. Using mass spectrometry imaging (MSI), a technology that permits high-resolution spatial mapping of metabolites, potentially in soil regions close to the roots, would be one such method (Dong et al. 2016). Furthermore, the root portions and cell types that manufacture and secrete particular chemicals are mostly unknown. Together with the process of particular metabolite secretion and the associated transporters, these constitute the main unanswered concerns for rhizosphere research in the years to come. The role of particular specialised metabolite classes or individuals in underground connections remains our main area of unknown, nevertheless. Several investigations have demonstrated the diverse effects that a single class of specialised metabolites, or even the same metabolite, can have on a variety of organisms. The situation of SLs is among the best examples of a specific metabolite class having many roles.

It is anticipated that genetic techniques based on variations in natural metabolic processes or mutations in metabolic pathways would supplement efforts to comprehend these hidden mechanisms. Therefore, the discovery of novel roles for specialised metabolites in soil may have an impact on our future capacity to develop innovative methods for managing soil pathogens and biostimulating crop growth and yield. Over the past 10 years, tremendous progress has been made in our understanding of the molecular mechanisms controlling the interactions between nematophagous bacteria and model nematodes. These realisations have provided intriguing new goals and guidance for developing effective PPN biological control methods. Determining the functions of those critical genes/factors in defining the mode of action of BCAs should lead to the improvement of BCAs nematoid potential by targeted genetic modifications, hence boosting the biological control effectiveness of PPN management.

9.6 Conclusions and Future Prospects

For decades, researchers studying the mechanics of microbe–nematode interactions have focused primarily on the interaction between two species. The primary producers are plants, but nematodes that feed on plants and fungi either directly consume plant materials or use them as fuel for their own metabolism. There are definitely many interactions between species, as opposed to the simple relationships between a disease and a single host. Numerous nematicidal metabolites have biosynthetic routes and processes that are explained by the biocontrol activity of microbes; some of these pathways have even been generated as BCAs in order to control PPNs. The availability of genomic and metabolomics data from several species (PKSs, NRPSs, and P450s) should facilitate the discovery of novel metabolites and the genes responsible for their manufacture. Furthermore, with the discovery of novel metabolites with high nematicidal activity and the understanding of the molecular interactions during infections, some of these metabolites with high nematicidal efficiency will be developed or improved in the near future.

Numerous studies that are now available suggest that employing a range of organic amendments and biological control species may suppress nematodes more than utilising a single biological control agent. Strategies that integrate biological control agents with other control techniques, such as chemical nematodes, nematode-resistant cultivars, and crop rotations, and combine various microbial control agents with complementary and synergistic modes of action could be very effective in controlling pest nematode populations. Gaining further insight into the molecular processes that underpin the interactions between nematodes and bacteria will facilitate the development of more effective strategies.

References

Affokpon A, Coyne DL, Htay CC, Agbèdè RD, Lawouin L, Coosemans J (2011) Biocontrol potential of native *Trichoderma* isolates against root-knot nematodes in west African vegetable production systems. Soil Biol Biochem 43:600–608

Ahrén D, Ursing BM, Tunlid A (1998) Phylogeny of nematode-trapping fungi based on 18S rDNA sequences. FEMS Microbiol Lett 158:179–184

Akiyama K, Hayashi H (2006) Strigolactones: chemical signals for fungal symbionts and parasitic weeds in plant roots. Ann of Bot 97:925–931

Akiyama K, Matsuzaki KI, Hayashi H (2005) Plant sesquiterpenes induce hyphal branching in arbuscular mycorrhizal fungi. Nature 435:824–827

Aliferis KA, Jabaji S (2010) Metabolite composition and bioactivity of *Rhizoctonia solani* sclerotial exudates. J Agric Food Chem 58:7604–7615

Baetz U, Martinoia E (2014) Root exudates: the hidden part of plant defense. Trends Plant Sci 19:90–98

Bailey BA, Korcak RF, Anderson JD (1992) Alterations in *Nicotiana tabacum* L. cv. *Xanthi* cell membrane function following treatment with an ethylene biosynthesis-inducing endoxylanase. Plant Physiol 100:749–755

Bais HP, Weir TL, Perry LG, Gilroy S, Vivanco JM (2006) The role of root exudates in rhizosphere interactions with plants and other organisms. Annu Rev Plant Biol 57:233–266

Balan J, Gerber NN (1972) Attraction and killing of the nematode *Panagrellus redivivus* by the predaceous fungus *Arthrobotrys dactyloides*. Nematologica 18:163–173

Balan J, Krizkova L, Nemec P, Kolozsvary A (1976) A qualitative method for detection of nematode attracting substances and proof of production of three different attractants by the fungus *Monacrosporium rutgeriensis*. Nematologica 22:306–311

Balan J, Križková L, Nemec P, Vollek V (1974) Production of nematode-attracting and nematicidal substances by predacious fungi. Folia Microbiol 19:512–519

Bar M, Avni A (2009) EHD2 inhibits ligand-induced endocytosis and signaling of the leucine-rich repeat receptor-like protein LeEix2. Plant J 59:600–611

Belayneh Mulaw T, Kubicek CP, Druzhinina IS (2010) The rhizosphere of Coffea arabica in its native highland forests of Ethiopia provides a niche for a distinguished diversity of *Trichoderma*. Diversity 2:527–549

Benítez T, Rincón AM, Limón MC, Codon AC (2004) Biocontrol mechanisms of *Trichoderma* strains. Int'l Microbiol 7:249–260

Bird AF, Bird J (2012) The structure of nematodes. Academic Press, Boca Raton, USA

Bonkowski M, Villenave C, Griffiths B (2009) Rhizosphere fauna: the functional and structural diversity of intimate interactions of soil fauna with plant roots. Plant Soil 321:213–233

Brotman Y, Briff E, Viterbo A, Chet I (2008) Role of swollenin, an expansin-like protein from *Trichoderma*, in plant root colonization. Plant Physiol 147:779–789

Calvet C, Pera J, Barea JM (1993) Growth response of marigold (*Tagetes erecta* L.) to inoculation with *Glomus mosseae*, *Trichoderma aureoviride* and *Pythium ultimum* in a peat-perlite mixture. Plant Soil 148:1–6

Chacón MR, Rodríguez-Galán O, Benítez T, Sousa S, Rey M, Llobell A, Delgado-Jarana J (2007) Microscopic and transcriptome analyses of early colonization of tomato roots by *Trichoderma harzianum*. Int Microbiol 10:19

Chen LL, Liu LJ, Shi M, Song XY, Zheng CY, Chen XL, Zhang YZ (2009) Characterization and gene cloning of a novel serine protease with nematicidal activity from *Trichoderma pseudokoningii* SMF2. FEMS Microbiol Lett 299:135–142

Chen Q, Yang B, Liu X, Chen F, Ge F (2014) Long-term cultivation of Bt rice expressing the Cry1Ab/1Ac gene reduced phytoparasitic nematode abundance but did not affect other nematode parameters in paddy fields. Science of The Total Environment 607:463–474

Crespo E, Hordijk CA, de Graaf RM, Samudrala D, Cristescu SM, Harren FJ, van Dam NM (2012) On-line detection of root-induced volatiles in *Brassica nigra* plants infested with *Delia radicum* L. root fly larvae. Phytochemistry 84:68–77

Croll NA (1970) The behaviour of nematodes: their activity, senses and responses. Edward Arnold, Ltd., London, p 117

Cruz AF, Hamel C, Yang C, Matsubara T, Gan Y, Singh AK, Ishii T (2012) Phytochemicals to suppress *fusarium* head blight in wheat–chickpea rotation. Phytochemistry 78:72–80

Datnoff LE, Nemec S, Pernezny K (1995) Biological control of *Fusarium* crown and root rot of tomato in Florida using *Trichoderma harzianum* and *Glomus intraradices*. Biol Control 5:427–431

De Coninck B, Timmermans P, Vos C, Cammue BP, Kazan K (2015) What lies beneath: belowground defense strategies in plants. Trends Plant Sci 20:91–101

Dean JF, Anderson JD (1991) Ethylene biosynthesis-inducing xylanase: II. Purification and physical characterization of the enzyme produced by *Trichoderma viride*. Plant Physiol 95:316–323

Delgado-Jarana J, Sousa S, Gonzalez F, Rey M, Llobell A (2006) ThHog1 controls the hyperosmotic stress response in *Trichoderma harzianum*. Microbiology 152:1687–1700

Dennis PG, Miller AJ, Hirsch PR (2010) Are root exudates more important than other sources of rhizodeposits in structuring rhizosphere bacterial communities. FEMS Microbiol Ecol 72:313–327

Dijksterhuis J, Veenhuis M, Harder W, Nordbring-Hertz B (1994) Nematophagous fungi: physiological aspects and structure-function relationships. Adv Microb Physiol 36:111–143

Ditengou FA, Müller A, Rosenkranz M, Felten J, Lasok H, Van Doorn MM, Polle A (2015) Volatile signalling by sesquiterpenes from ectomycorrhizal fungi reprogrammes root architecture. Nat Commun 6:6279

Dixon RA, Paiva NL (1995) Stress-induced phenylpropanoid metabolism. Plant Cell 7:1085

Djonovic S, Pozo MJ, Kenerley CM (2006) Tvbgn3, a β-1, 6-glucanase from the biocontrol fungus *Trichoderma virens*, is involved in mycoparasitism and control of *Pythium ultimum*. Appl Environ Microbiol 72:7661–7670

Djonovic S, Vargas WA, Kolomiets MV, Horndeski M, Wiest A, Kenerley CM (2007) A proteinaceous elicitor Sm1 from the beneficial fungus *Trichoderma virens* is required for induced systemic resistance in maize. Plant Physiol 145:875–889

Dong L, Li X, Huang L, Gao Y, Zhong L, Zheng Y, Zuo Y (2014) Lauric acid in crown daisy root exudate potently regulates root-knot nematode chemotaxis and disrupts Mi-flp-18 expression to block infection. J Exp Bot 65:131–141

Dong Y, Li B, Aharoni A (2016) More than pictures: when MS imaging meets histology. Trends Plant Sci 21:686–698

Dor E, Joel DM, Kapulnik Y, Koltai H, Hershenhorn J (2011) The synthetic strigolactone GR24 influences the growth pattern of phytopathogenic fungi. Planta 234:419–427

Enkerli J, Felix G, Boller T (1999) The enzymatic activity of fungal xylanase is not necessary for its elicitor activity. Plant Physiol 121:391–398

Field JI, Webster J (1977) Traps of predacious fungi attract nematodes. Trans Br Mycol Soc 68:467–469

Flores A, Chet I, Herrera-Estrella A (1997) Improved biocontrol activity of *Trichoderma harzianum* by over-expression of the proteinase-encoding gene prb1. Curr Genet 31:30–37

Fukuzawa A, Matsue H, Ikura M, Masamune T (1985) Glycinoeclepins B and C, nortriterpenes related to glycinoeclepin A. Tetrahedron Lett 26:5539–5542

Gans J, Wolinsky M, Dunbar J (2005) Computational improvements reveal great bacterial diversity and high metal toxicity in soil. Science 309:1387–1390

Garnica-Vergara A, Barrera-Ortiz S, Muñoz-Parra E, Raya-González J, Méndez-Bravo A, Macías-Rodríguez L, Ruis-Herrera LF, López-Bucio J (2016) The volatile 6-pentyl- 2H-pyran-2-one from *Trichoderma atroviride* regulates *Arabidopsis thaliana* root morphogenesis via auxin signaling and ethylene insensitive 2 functioning. New Phytol 209:1496–1512

Graham TL (1991) Flavonoid and isoflavonoid distribution in developing soybean seedling tissues and in seed and root exudates. Plant Physiol 95:594–603

Green H, Larsen J, Olsson PA, Jensen DF, Jakobsen I (1999) Suppression of the biocontrol agent *Trichoderma harzianum* by mycelium of the arbuscular mycorrhizal fungus *Glomus intraradices* in root-free soil. Appl Environ Microbiol 65:1428–1434

Guo W, Zhao J, Li X, Qin L, Yan X, Liao H (2011) A soybean β-expansin gene GmEXPB2 intrinsically involved in root system architecture responses to abiotic stresses. Plant J 66:541–552

Haichar FZ, Marol C, Berge O, Rangel-Castro JI, Prosser JI, Balesdent J, Heulin T, Achouak W (2008) Plant host habitat and root exudates shape soil bacterial community structure. ISME J 2:1221–1230

Haichar FZ, Santaella C, Heulin T, Achouak W (2014) Root exudates mediated interactions belowground. Soil Biol Biochem 77:69–80

Hanson LE, Howell CR (2004) Elicitors of plant defense responses from biocontrol strains of *Trichoderma virens*. Phytopathology 94:171–176

Harman GE (2011) Multifunctional fungal plant symbionts: new tools to enhance plant growth and productivity. New Phytol 189:647–649

Harman GE, Howell CR, Viterbo A, Chet I, Lorito M (2004) *Trichoderma* species-opportunistic, avirulent plant symbionts. Nat Rev Microbiol 2:43–56

Hassan S, Mathesius U (2012) The role of flavonoids in root-rhizosphere signalling: opportunities and challenges for improving plant–microbe interactions. J Exp Bot 63:3429–3444

Hiltpold I, Turlings TC (2008) Belowground chemical signaling in maize: when simplicity rhymes with efficiency. J Chem Ecol 34:628–635

Hosseini P, Matthews BF (2014) Regulatory interplay between soybean root and soybean cyst nematode during a resistant and susceptible reaction. BMC plant biology 14:1–10

Huffaker A, Kaplan F, Vaughan MM, Dafoe NJ, Ni X, Rocca JR, Alborn HT, Teal PEA, Schmelz EA (2011) Novel acidic sesquiterpenoids constitute a dominant class of pathogen-induced phytoalexins in maize. Plant Physiol 156:2082–2097

Ihrmark K, Asmail N, Ubhayasekera W, Melin P, Stenlid J, Karlsson M (2010) Comparative molecular evolution of *Trichoderma* chitinases in response to mycoparasitic interactions. Evol Bioinforma 6:S4198

Inbar J, Chet I (1996) The role of lectins in recognition and adhesion of the mycoparasitic fungus *Trichoderma* spp. to its host. Adv Exp Med Biol 408:229–231

Insam H, Seewald M (2010) Volatile organic compounds (VOCs) in soils. Biol Fertil Soils 46:199–213

Iven T, König S, Singh S, Braus-Stromeyer SA, Bischoff M, Tietze LF, Braus GH, Lipka V, Feussner I, Dröge-Laser W (2012) Transcriptional activation and production of tryptophan-derived secondary metabolites in *Arabidopsis* roots contributes to the defense against the fungal vascular pathogen *Verticillium longisporum*. Mol Plant 5:1389–1402

Jones JD, Dangl JL (2006) The plant immune system. Nature 444:323–329

de Jong JC, McCormack BJ, Smirnoff N, Talbot NJ (1997) Glycerol generates turgor in rice blast. Nature 389:244–244

Kai M, Effmert U, Piechulla B (2016) Bacterial-plant-interactions: approaches to unravel the biological function of bacterial volatiles in the rhizosphere. Front Microbiol 7:108

Kai M, Haustein M, Molina F, Petri A, Scholz B, Piechulla B (2009) Bacterial volatiles and their action potential. Appl Microbiol Biotechnol 81:1001–1012

Klink JW, Dropkin VH, Mitchell JE (1970) Studies on the host-finding mechanisms of *Neotylenchus linfordi*. J Nematol 2:106

Kubicek CP, Baker S, Gamauf C, Kenerley CM, Druzhinina IS (2008) Purifying selection and birth and death evolution in the class II hydrophobin gene families of the ascomycete *Trichoderma/Hypocrea*. BMC Evol Biol 8:1–16

Kubicek CP, Druzhinina IS (2007) Environmental and microbial relationships. Springer, Berlin Heidelberg

Kubicek CP, Herrera-Estrella A, Seidl-Seiboth V, Martinez DA, Druzhinina IS, Thon M, Zeilinger S, Casas-Flores S, Horwitz BA, Mukherjee PK, Mukherjee M, Kredics L, Alcaraz LD, Aerts A, Antal Z, Atanasova L, Cervantes-Badillo MG, Challacombe J, Chertkov O, McCluskey K, Coulpier F, Deshpande N, von Döhren H, Ebbole DJ, Grigoriev IV (2011) Comparative genome sequence analysis underscores mycoparasitism as the ancestral life style of *Trichoderma*. Genome Biol 12:R40

Kuijken RC, Snel JF, Heddes MM, Bouwmeester HJ, Marcelis LF (2015) The importance of a sterile rhizosphere when phenotyping for root exudation. Plant Soil 387:131–142

Kumar A, Scher K, Mukherjee M, Pardovitz-Kedmi E, Sible GV, Singh US, Kale SP, Mukherjee PK, Horwitz BA (2010) Overlapping and distinct functions of two *Trichoderma virens* MAP kinases in cell-wall integrity, antagonistic properties and repression of conidiation. Biochem Biophys Res Commun 398:765–770

Latgé JP (2007) The cell wall: a carbohydrate armour for the fungal cell. Mol Microbiol 66:279–290

Laznik Ž, Trdan S (2013) An investigation on the chemotactic responses of different entomopathogenic nematode strains to mechanically damaged maize root volatile compounds. Exp Parasitol 134:349–355

Li Y, Hyde KD, Jeewon R, Cai L, Vijaykrishna D, Zhang K (2005) Phylogenetics and evolution of nematode-trapping fungi (Orbiliales) estimated from nuclear and protein coding genes. Mycologia 97:1034–1046

Limón MC, Chacón MR, Mejías R, Delgado-Jarana J, Rincón AM, Codón AC, Benítez T (2004) Increased antifungal and chitinase specific activities of *Trichoderma harzianum* CECT 2413 by addition of a cellulose binding domain. Appl Microbiol Biotechnol 64:675–685

Long SR (1996) *Rhizobium* symbiosis: nod factors in perspective. Plant Cell 8:1885–1898

Lorito M, Woo SL, Harman GE, Monte E (2010) Translational research on *Trichoderma*: from'omics to the field. Annu Rev Phytopathol 48:395–417
Lu Z, Tombolini R, Woo S, Zeilinger S, Lorito M, Jansson JK (2004) *In vivo* study of *Trichoderma*-pathogen-plant interactions, using constitutive and inducible green fluorescent protein reporter systems. Appl Environ Microbiol 70:3073–3081
Luo H, Li X, Li G, Pan Y, Zhang K (2006) Acanthocytes of *Stropharia rugosoannulata* function as a nematode-attacking device. Appl Environ Microbiol 72:2982–2987
Luo H, Liu Y, Fang L, Li X, Tang N, Zhang K (2007) Coprinus comatus damages nematode cuticles mechanically with spiny balls and produces potent toxins to immobilize nematodes. Appl Environ Microbiol 73:3916–3923
Martin F, Aerts A, Ahrén D, Brun A, Danchin EGJ, Duchaussoy F, Gibon J, Kohler A, Lindquist E, Pereda V, Salamov A, Shapiro HJ, Wuyts J, Blaudez D, Buée M, Brokstein P, Canbäck B, Cohen D, Courty PE, Coutinho PM, Delaruelle C, Detter JC, Deveau A, DiFazio S, Grigoriev IV (2008) The genome of *Laccaria bicolor* provides insights into mycorrhizal symbiosis. Nature 452:88–92
Masamune T, Anetai M, Takasugi M, Katsui N (1982) Isolation of a natural hatching stimulus, glycinoeclepin A, for the soybean cyst nematode. Nature 297:495–496
McAllister CÁ, Garcia-Romera I, Godeas A, Ocampo JA (1994) Interactions between *Trichoderma koningii, Fusarium solani* and *Glomus mosseae*: effects on plant growth, arbuscular mycorrhizas and the saprophyte inoculants. Soil Biol Biochem 26:1363–1367
Mendoza-Mendoza A, Rosales-Saavedra T, Cortes C, Castellanos-Juarez V, Martínez P, Herrera-Estrella A (2007) The MAP kinase TVK1 regulates conidiation, hydrophobicity and the expression of genes encoding cell wall proteins in the fungus *Trichoderma virens*. Microbiology 153:2137–2147
Migheli Q, Balmas V, Komoñ-Zelazowska M, Scherm B, Fiori S, Kopchinskiy AG, Kubicek CP, Druzhinina IS (2009) Soils of a Mediterranean hot spot of biodiversity and endemism (Sardinia, Tyrrhenian Islands) are inhabited by pan-European, invasive species of *Hypocrea/Trichoderma*. Environ Microbiol 11:35–46
Monoson HL, Galsky AG, Griffin JA, McGrath EJ (1973) Evidence for and partial characterization of a nematode attraction substance. Mycologia 65:78–86
Montero M, Sanz L, Rey M, Llobell A, Monte E (2007) Cloning and characterization of bgn163, coding for a β-1, 6-glucanase expressed during *Trichoderma harzianum* mycoparasitism. J Appl Microbiol 103:1291–1300
Morán-Diez E, Hermosa R, Ambrosino P, Cardoza RE, Gutiérrez S, Lorito M, Monte E (2009) The ThPG1 endopolygalacturonase is required for the *Trichoderma harzianum* plant beneficial interaction. Mol Plant-Microbe Interact 22:1021–1031
Mukherjee PK, Latha J, Hadar R, Horwitz BA (2003) TmkA, a mitogen-activated protein kinase of *Trichodermavirens*, is involved in biocontrol properties and repression of conidiation in the dark. Eukaryot Cell 2:446–455
Mukherjee PK, Latha J, Hadar R, Horwitz BA (2004) Role of two G-protein alpha subunits, TgaA and TgaB, in the antagonism of plant pathogens by *Trichoderma virens*. Appl Environ Microbiol 70:542–549
Nehls U, Göhringer F, Wittulsky S, Dietz S (2010) Fungal carbohydrate support in the ectomycorrhizal symbiosis: a review. Plant Biol 12:292–301
Nemec S, Datnoff LE, Strandberg J (1996) Efficacy of biocontrol agents in planting mixes to colonize plant roots and control root diseases of vegetables and citrus. Crop Prot 15:735–742
Omann M, Lehner S, Brunner K, Delic M, Stoppacher N, Schuhmacher R, Pucher M, Zeilinger S (2009) A cAMP receptor-like GPCR is involved in *Trichoderma atroviride* mycoparasitism. In: Elad Y, Maurhofer M, Keel C, Gessler C, Duffy B (eds) , vol 43. IOBC/WPRS Bull, pp 105–108
Osbourn AE (2003) Saponins in cereals. Phytochemistry 62:1–4
Papadopoulou K, Melton RE, Leggett M, Daniels MJ, Osbourn AE (1999) Compromised disease resistance in saponin-deficient plants. Proc Natl Acad Sci 96:12923–12928

Parniske M (2008) Arbuscular mycorrhiza: the mother of plant root endosymbioses. Nat Rev Microbiol 6:763–775
Peñuelas J, Asensio D, Tholl D, Wenke K, Rosenkranz M, Piechulla B, Schnitzler J (2014) Biogenic volatile emissions from the soil. Plant Cell Environ 37:1866–1891
Phillips RP, Finzi AC, Bernhardt ES (2011) Enhanced root exudation induces microbial feedbacks to N cycling in a pine forest under long-term CO_2 fumigation. Ecol Lett 14:187–194
Quastel JH (1965) Soil metabolism. Annu Rev Plant Physiol 16:217–240
Rasmann S, Köllner TG, Degenhardt J, Hiltpold I, Toepfer S, Kuhlmann U, Gershenzon J, Turlings TCJ (2005) Recruitment of entomopathogenic nematodes by insect-damaged maize roots. Nature 434:732–737
Rasmann S, Turlings TC (2008) First insights into specificity of belowground tritrophic interactions. Oikos 117:362–369
Reithner B, Brunner K, Schuhmacher R, Peissl I, Seidl V, Krska R, Zeilinger S (2005) The G protein α subunit Tga1 of *Trichoderma atroviride* is involved in chitinase formation and differential production of antifungal metabolites. Fungal Genet Biol 42:749–760
Reithner B, Schuhmacher R, Stoppacher N, Pucher M, Brunner K, Zeilinger S (2007) Signaling via the *Trichoderma atroviride* mitogen-activated protein kinase Tmk1 differentially affects mycoparasitism and plant protection. Fungal Genet Biol 44:1123–1133
Rocha-Ramírez V, Omero C, Chet I, Horwitz BA, Herrera-Estrella A (2002) *Trichoderma atroviride* G-protein α-subunit gene tga1 is involved in mycoparasitic coiling and conidiation. Eukaryot Cell 1:594–605
Ron M, Avni A (2004) The receptor for the fungal elicitor ethylene-inducing xylanase is a member of a resistance-like gene family in tomato. Plant Cell 16:1604–1615
Ruocco M, Lanzuise S, Vinale F, Marra R, Turrà D, Woo SL, Lorito M (2009) Identification of a new biocontrol gene in *Trichoderma atroviride*: the role of an ABC transporter membrane pump in the interaction with different plant-pathogenic fungi. Mol Plant-Microbe Interact 22:291–301
Saloheimo M, Paloheimo M, Hakola S, Pere J, Swanson B, Nyyssönen E, Bhatia A, Ward M, Penttilä M (2002) Swollenin, a *Trichodermareesei* protein with sequence similarity to the plant expansins, exhibits disruption activity on cellulosic materials. Eur J Biochem 269:4202–4211
Scheffknecht S, Mammerler R, Steinkellner S, Vierheilig H (2006) Root exudates of mycorrhizal tomato plants exhibit a different effect on microconidia germination of *Fusarium oxysporum* f. sp. *lycopersici* than root exudates from non-mycorrhizal tomato plants. Mycorrhiza 16:365–370
Schenk H, Driessen RA, de Gelder R, Goubitz K, Nieboer H, Brüggemann-Rotgans IE, Diepenhorst P (1999) Elucidation of the structure of Solanoeclepin A, a natural hatching factor of potato and tomato cyst nematodes, by single-crystal x-ray diffraction. Croat Chem Acta 72:593–606
Schenkel D, Lemfack MC, Piechulla B, Splivallo R (2015) A meta-analysis approach for assessing the diversity and specificity of belowground root and microbial volatiles. Front Plant Sci 6:707
Schmelz EA, Kaplan F, Huffaker A, Dafoe NJ, Vaughan MM, Ni X, Rocca JR, Alborn HT, Teal PE (2011) Identity, regulation, and activity of inducible diterpenoid phytoalexins in maize. Proc Natl Acad Sci USA 108:5455–5460
Schmoll M (2008) The information highways of a biotechnological work horse–signal transduction in *Hypocrea jecorina*. BMC Genomics 9:430
Scholler M, Hagedorn G, Rubner A (1999) A re-evaluation of predatory orbiliaceous fungi. II. A new generic concept. Sydowia 51:89–113
Segarra G, Casanova E, Bellido D, Odena MA, Oliveira E, Trillas I (2007) Proteome, salicylic acid, and jasmonic acid changes in cucumber plants inoculated with *Trichoderma asperellum* strain T34. Proteomics 7:3943–3952
Seidl V (2008) Chitinases of filamentous fungi: a large group of diverse proteins with multiple physiological functions. Fungal Biol Rev 22:36–42
Seidl V, Song L, Lindquist E, Gruber S, Koptchinskiy A, Zeilinger S, Schmoll M, Martínez P, Sun J, Grigoriev I, Herrera-Estrella A, Baker SE, Kubicek CP (2009) Transcriptomic response

of the mycoparasitic fungus *Trichoderma atroviride* to the presence of a fungal prey. BMC Genomics 10:567

Sharon A, Fuchs Y, Anderson JD (1993) The elicitation of ethylene biosynthesis by a *Trichoderma* xylanase is not related to the cell wall degradation activity of the enzyme. Plant Physiol 102:1325–1329

Sharon E, Bar-Eyal M, Chet I, Herrera-Estrella A, Kleifeld O, Spiegel Y (2001) Biological control of the root-knot nematode *Meloidogyne javanica* by *Trichoderma harzianum*. Phytopathology 91:687–693

Shoresh M, Harman GE, Mastouri F (2010) Induced systemic resistance and plant responses to fungal biocontrol agents. Annu Rev Phytopathol 48:21–43

Siddiqui ZA, Mahmood I (1996) Biological control of *Heterodera cajani* and *Fusarium udum* on pigeonpea by *Glomus mosseae*, *Trichoderma harzianum*, and *Verticillium chlamydosporium*. Israel J Plant Sci 44:49–56

Sohrabi R, Huh JH, Badieyan S, Rakotondraibe LH, Kliebenstein DJ, Sobrado P, Tholl D (2015) In planta variation of volatile biosynthesis: an alternative biosynthetic route to the formation of the pathogen-induced volatile homoterpene DMNT via triterpene degradation in *Arabidopsis* roots. Plant Cell 27:874–890

Sohrabi F, Sheikholeslami M, Heydari R, Rezaee S, Sharifi R (2017) Evaluation of four rhizobacteria on tomato growth and suppression of root-knot nematode, *Meloidogyne javanica* under greenhouse conditions, a pilot study. Egypt J Biol Pest Control 28:1–5

Suarez B, Rey M, Castillo P, Monte E, Llobell A (2004) Isolation and characterization of PRA1, a trypsin-like protease from the biocontrol agent *Trichoderma harzianum* CECT 2413 displaying nematicidal activity. Appl Microbiol Biotechnol 65:46–55

Suárez MB, Vizcaíno JA, Llobell A, Monte E (2007) Characterization of genes encoding novel peptidases in the biocontrol fungus *Trichoderma harzianum* CECT2413 using the TrichoEST functional genomics approach. Curr Genet 51:331–342

Szabó M, Csepregi K, Gálber M, Virányi F, Fekete C (2012) Control plant-parasitic nematodes with *Trichoderma* species and nematode-trapping fungi: the role of chi18-5 and chi18-12 genes in nematode egg-parasitism. Biol Control 63:121–128

Szabó M, Urbán P, Virányi F, Kredics L, Fekete C (2013) Comparative gene expression profiles of *Trichoderma harzianum* proteases during *in vitro* nematode egg-parasitism. Biol Control 67:337–343

Townshend JL (1964) Fungus hosts of *Aphelenchus avenae* Bastian, 1865 and *Bursaphelenchus* fungivorus Franklin & Hooper, 1962 and their attractiveness to these nematode species. Can J Microbiol 10:727–737

Turra D, El Ghalid M, Rossi F, Di Pietro A (2015) Fungal pathogen uses sex pheromone receptor for chemotropic sensing of host plant signals. Nature 527:521–524

Vargas WA, Crutcher FK, Kenerley CM (2011) Functional characterization of a plant-like sucrose transporter from the beneficial fungus *Trichoderma virens*. Regulation of the symbiotic association with plants by sucrose metabolism inside the fungal cells. New Phytol 189:777–789

Vargas WA, Djonovic S, Sukno SA, Kenerley CM (2008) Dimerization controls the activity of fungal elicitors that trigger systemic resistance in plants. J Biol Chem 283:19804–19815

Vargas WA, Mandawe JC, Kenerley CM (2009) Plant-derived sucrose is a key element in the symbiotic association between *Trichoderma virens* and maize plants. Plant Physiol 151:792–808

Vaughan MM, Christensen S, Schmelz EA, Huffaker A, Mcauslane HJ, Alborn HT, Romero M, Allen LH, Teal PEA (2015) Accumulation of terpenoid phytoalexins in maize roots is associated with drought tolerance. Plant Cell Environ 38:2195–2207

Veresoglou SD, Rillig MC (2012) Suppression of fungal and nematode plant pathogens through arbuscular mycorrhizal fungi. Biol Lett 8:214–217

Viterbo A, Harel M, Horwitz BA, Chet I, Mukherjee PK (2005) *Trichoderma* mitogen-activated protein kinase signaling is involved in induction of plant systemic resistance. Appl Environ Microbiol 71:6241–6246

Viterbo ADA, Wiest ARIC, Brotman Y, Chet ILAN, Kenerley C (2007) The 18mer peptaibols from *Trichoderma virens* elicit plant defence responses. Mol Plant Pathol 8:737–746

Vos C, Claerhout S, Mkandawire R, Panis B, De Waele D, Elsen A (2012) Arbuscular mycorrhizal fungi reduce root-knot nematode penetration through altered root exudation of their host. Plant Soil 354:335–345

Watts-Williams SJ, Jakobsen I, Cavagnaro TR, Grønlund M (2015) Local and distal effects of arbuscular mycorrhizal colonization on direct pathway Pi uptake and root growth in *Medicago truncatula*. J Exp Bot 66:4061–4073

Waweru B, Turoop L, Kahangi E, Coyne D, Dubois T (2014) Non-pathogenic *Fusarium oxysporum* endophytes provide field control of nematodes, improving yield of banana (*Musa* sp.). Biol Control 74:82–88

Wenke K, Kai M, Piechulla B (2010) Belowground volatiles facilitate interactions between plant roots and soil organisms. Planta 231:499–506

Werner A, Zadworny M (2003) *In vitro* evidence of mycoparasitism of the ectomycorrhizal fungus *Laccaria laccata* against *Mucor hiemalis* in the rhizosphere of *Pinus sylvestris*. Mycorrhiza 13:41–47

Xu Z, Zhao YQ, Yang DJ, Sun HJ, Zhang CL, Xie YP (2015) Attractant and repellent effects of sweet potato root exudates on the potato rot nematode, *Ditylenchus destructor*. Nematology 17:117–124

Yang G, Zhou B, Zhang X, Zhang Z, Wu Y, Zhang Y, Lu S, Zou Q, Gao Y, Teng L (2016) Effects of tomato root exudates on *Meloidogyne incognita*. PLoS One 11:e0154675

Yang Z, Yu Z, Lei L, Xia Z, Shao L, Zhang K, Li G (2012) Nematicidal effect of volatiles produced by *Trichoderma* sp. J Asia Pac Entomol 15:647–650

Yang ZS, Li GH, Zhao PJ, Zheng X, Luo SL, Li L, Niu XM, Zhang KQ (2010) Nematicidal activity of *Trichoderma* spp. and isolation of an active compound. World J Microbiol Biotechnol 26:2297–2302

Yedidia I, Benhamou N, Chet I (1999) Induction of defense responses in cucumber plants (*Cucumis sativus* L.) by the biocontrol agent *Trichoderma harzianum*. Appl Environ Microbiol 65:1061–1070

Zhang KQ, Hyde KD (eds) (2014) Nematode-trapping fungi, vol 23. Springer Science & Business

Zhao X, Schmitt M, Hawes MC (2000) Species-dependent effects of border cell and root tip exudates on nematode behavior. Phytopathology 90:1239–1245

Involvement of Parasitism Proteins in Plant Nematode and Opportunistic Fungi Connections

10

Dikshita Saikia, Anurag Kashyap, Rashmita Saikia, Arkadeb Chatterjee, and Mohd. Sayeed Akhtar

10.1 Introduction

Plants are often under attack to a wide range of pathogens that use them as a source of nourishment and shelter (Rashid et al. 2010). Fungi and nematodes have been associated with parasitism of plants since ages that has led to catastrophic destruction of major plants causing significant economic loss. This encompasses diverse diseases such as cereal smuts and rusts caused by fungi, apple scab, brown rot affecting stone fruits, formation of root-knot galls, and cysts by various plant-parasitic nematodes and various instances of wilts, leaf spots and necrotic blights. A range of parasitism proteins are deployed by these pathogens such as cell wall modifying enzymes, plant mimic molecules, toxins, host cell regulators, effector molecules that can breach the protective boundaries of the plant and lead to events of pathogenesis, or in another scenario activate a series of immune responses (Davis et al. 2008). Because of the long co-evolutionary relationship between the host and pathogens, a highly effective defence system has evolved, successfully protecting against possible microbial threats (Allard et al. 2008).

Plant cell walls vary in structure across plant lineages due to their complex combination of polysaccharides, proteins, and aromatic polymers. A matrix of hemicellulose, pectin, lignin, and structural proteins surround the cellulose microfibrils. To

D. Saikia · R. Saikia
Department of Plant Pathology, Punjab Agricultural University, Ludhiana, Punjab, India

A. Kashyap (✉)
Department of Plant Pathology, Assam Agricultural University, Jorhat, Assam, India
e-mail: anurag.kashyap@aau.ac.in

A. Chatterjee
Department of Nematology, Assam Agricultural University, Jorhat, Assam, India

M. S. Akhtar
Department of Botany, Gandhi Faiz-e-Aam College, Shahjahanpur, Uttar Pradesh, India

M. S. Akhtar (ed.), *Opportunistic Fungi, Nematode and Plant Interactions*,
https://doi.org/10.1007/978-981-97-2045-3_10

breach the plant cell wall barrier, phytopathogenic fungi and nematodes create cell wall-degrading enzymes (CWDEs) that degrade cellulose, pectin, and xylan. Enzymatic activity assists these organisms in entering the plant cell wall, which is a critical method for infections lacking particular penetration structures. This necessity becomes more apparent as the invasion progresses (Gibson et al. 2011). It is seen that many plant-pathogenic fungi, to sustain, deliberately break down plant tissue to obtain carbon from the freed mono and oligosaccharides. On the other hand, effector molecules produced by pathogens aid in causing infection of plant species. These compounds can alter various plant processes, which is critical in understanding the complex interplay between plants and diseases. Plants have evolved effective mechanisms to respond to environmental changes, encompassing both long and short defence reactions through various biochemical and molecular pathways (Nurnberger et al. 2004).

10.2 Parasitism Proteins in Nematodes

Among the nematodes, cyst and root-knot nematodes are bringing catastrophe to most of the important crops around the globe and setting noteworthy examples of parasitic evolution. Effector molecules are anticipated to be predominant in the region where feeding cell development and defence responses are localised around the nematode's head when resistance genes are present (Davis et al. 2008). The majority of potential effector molecules discovered are proteins expressed by nematode parasitism genes that are released into plant tissues through the stylet, except for an unknown factor generated from second juvenile of root-knot nematodes that can stimulate Nod factor pathways in legumes and a candidate avirulence protein exuded from the amphids of root-knot nematodes (Baum et al. 2007; Mitchum et al. 2007).

10.2.1 Origin of Parasitism Proteins

Hussey (1989) reported that three enlarged oesophageal gland secretory cells, i.e., two subventral and one dorsal, produce the proteins involved in nematode parasitism. With golgi derived membrane bound secretory granules and subsequent exocytosis of granule contents into the oesophagus lumen via specialised valves, these gland cells exhibit characteristics similar to those of true animal secretory cells. Similar to plant-pathogenic bacterium type III secretion systems, parasitism protein secretion in nematodes occurs via the stylet (Alfano and Collmer 2004). The identification of parasitism genes producing endoglucanases produced by phytoparasitic cyst nematodes was the first finding of endogenous cellulases in animals (Smant et al. 1998) and one of the earliest studies to corroborate the current proof of horizontal gene transfer to eukaryotes from prokaryotes (Hotopp et al. 2007). An arsenal of cell wall modifying proteins have been discovered as parasitism genes such as galactosidases, pectinases, expansins, xylanases and that seem to be active during movement of the nematodes inside plant tissues (Mitchum et al. 2007). Recent evidence points to prokaryotes as an evolutionary source for numerous phytonematode cell-membrane-modifying enzymes (Ledger et al. 2006) and some other family members that have a role in various functions in modifications of the cell wall during nematode parasitism of plants (Gao et al. 2004; Ledger et al. 2006; Kudla et al. 2007; Vanholme et al. 2007).

10.2.2 How Do Parasitism Proteins Function

These proteins begin as initial proteins having N-terminal signal peptides that direct the developing protein chain towards the endoplasmic reticulum during parasitism protein mRNA translation. The signal peptide is broken at this point, allowing the mature protein to follow the secretory route. Despite this, there are a few cases when putative parasitism proteins use an alternate secretion mechanism that does not rely on a signal peptide. Despite these anomalies, secretory parasitism proteins that depend on the presence of a signal peptide are encoded by parasitism proteins, which are mostly active in the oesophageal glands (Baum et al. 2007).

Plant-parasitic nematodes only later develop into diseases, depending on how severe the parasitism is seen by people, i.e., whether it results in obvious, economically detrimental symptoms. Therefore, the molecular processes that enable nematodes to infect plants are those processes that distinguish nematodes as parasites rather than pathogens. Thus, parasitism genes are suitable names for the genetic factors that allow nematodes to infect plants. Determining what causes a nematode to become a plant parasite, or which help to parasitize plants, is of utmost importance. A more in-depth look at nematode parasitism genes focuses on nematode proteins that directly interact with host molecules, increasing parasite interaction.

10.2.3 Roles of Parasitism Proteins in Feeding Tube Formation in Nematodes

A variety of potential roles for parasitism genes in the life cycle of nematodes might be hypothesised when examining parasitism genes such as the genes that are expressed in the oesophageal gland and encode secretory proteins secreted by the stylet. At first, nematodes have to make their way to the roots by passing through the host tissues. Despite the moderate size of cyst nematode-infective J_2s and root-knot nematodes, plant cell walls present considerable constraints. Nematodes utilise a combination of enzymes that can degrade the cell walls. The development of the extensive feeding cells referred to as giant cells and syncytia, appears to be the most amazing accomplishment, in addition to these crucial roles. Also, the nematodes must interact with the majority of differentiated root cells in order to encourage the growth of the parasitized cells into various feeding cell types. Additionally, in order to retain these cells, the nematodes must probably block plant defences and/or cell death mechanisms that might be engaged during parasitism. After secretions are released via the dorsal gland valve, emergence of a tube-like structure, within the cytoplasm at the stylet opening (Wyss and Zunke 1986; Wyss 1992) was revealed (Hussey and Mims 1991). Therefore, parasitism genes may play the roles of feeding tube formation and feeding cell maintenance during food absorption.

10.2.4 Current Hypotheses of Putative Parasitism Protein in Nematodes and their Functions

Similarities between parasitism proteins and functionally characterised proteins from other organisms allow the formulation of plausible working hypotheses about the parasitic mechanisms used by the nematodes. These are present in the form of enzymes, proteins, cell function regulators, etc.

10.2.4.1 Cell Wall Digesting Enzymes

The finding of cellulase genes in cyst nematodes constituted the first important success in parasitism gene discovery and was especially significant because there were no known cellulase genes from vertebrates at the time (Smant et al. 1998). It was discovered that nematode cellulases and bacterial proteins are rather similar, leading to the intriguing theory that it is possible that some parasitism genes were obtained by lateral gene transfer (Smant et al. 1998; Yan et al. 1998). Likewise, the pectate lyase proteins present in bacteria and fungi, as well as cyst and root-knot nematodes, were also discovered in nematode pectinase proteins. Outside of the world of sedentary nematodes, the lesion nematode *Pratylenchus penetrans* was shown to include β-1,4-endoglucanase genes (Uehara et al. 2001). This parasite is migratory and necessitates effective methods to penetrate plant cell walls. Recently, β-1,3-endoglucanase cellulase was found in *Bursaphelenchus xylophilus*, a pinewood nematode (Kikuchi et al. 2005). It is thought that this cellulase is involved in nematode feeding on fungal mycelium.

10.2.4.2 Expansins

It has been demonstrated that the potato cyst nematode secretes a protein that resembles expansin with the potential to breakdown not just the covalent bonds present in plant cell walls but also the non-covalent bonds (Qin et al. 2004). By breaking non-covalent bonds that stop cell-wall fibrils from gliding past one another, this group of proteins weaken the cell walls. One way to explain the resulting weakening of plant cell walls is to use the potato cyst nematode expansin parasitism gene (Qin et al. 2004).

10.2.4.3 Ubiquitination/Proteasome Functions (Regulator of Gene Expression)

One last and effective way to control gene expression is by targeted and scheduled protein break down. Given that cyst nematodes seem to release polyubiquitination-related proteins, which is a procedure that adds ubiquitin protein molecules precisely to proteins and targets the former for degradation, it is apparent that using this method, these pathogens change the expression of certain genes in plant cells they have parasitized. This theory is supported by the finding that those secretory isotypes of cyst nematodes are normally purely cytoplasmic ubiquitination pathway proteins, such as proteins and ubiquitin itself and the ones present in the E3 ubiquitin protein ligase complex also resemble the former proteins (Gao et al. 2003; Tytgat et al. 2004).

10.2.4.4 Small Bioactive Peptides (Mimic Molecule)

It now appears that small peptides are vital for development of the plant according to recent scientific advancements (Lindsey et al. 2002). For instance, it has been demonstrated that the tiny extracellular ligand CLAVATA3 present in *Arabidopsis* has a critical role to play in the differentiation of shoot meristems (Fletcher et al. 1999). The discovery that the cyst nematode of soybean produces a brief parasitism peptide seen in CLAVATA3-like ligand peptides that has a conserved C-terminal motif was of particular interest (Wang et al. 2001; Olsen and Skriver 2003). The restoration of the wild-type phenotype in the clavata3 (clv3) Arabidopsis mutant through the expression of this soybean cyst nematode CLAVATA3-like peptide was the first occurrence of ligand mimicry in phytonematology (Wang et al. 2005). In a nutshell, an endogenous plant receptor ligand is secreted by the nematode in order to effectively parasitize the host plant.

10.2.4.5 Nuclear Localised Parasitism Proteins

Considerable fraction of putative parasite proteins containing predicted nuclear localisation signals (NLS) that enable active absorption into the nucleus were found when parasitism proteins were analysed using computational methods to predict the location of proteins and destiny. After being secreted into a plant cell, NLS-containing nematode parasitism proteins enter the endoplasmic reticulum and secretory pathway before being absorbed by the plant nucleus. In order to test this theory, a limited set of cyst nematode parasitism proteins have been demonstrated to actively enter plant nuclei (Tytgat et al. 2004). Deciphering protein functions inside the plant nucleus will be of highest importance presently.

10.2.4.6 RanBPM

A collection of parasitism gene candidates was discovered in a project that examined the patterns of gene expression in the potato cyst nematode during several developmental stages (Qin et al. 2000). Further examination of these genes disclosed the existence of a small gene family that codes for secretory proteins that resemble the ones binding to the G-protein Ran, known as RanBPMs (Ran-Binding Protein in the Microtubule Organising Centre). Several of these genes (Qin et al. 2000) were discovered to express themselves in the dorsal gland. The precise activities of RanBPMs are unknown, but they seem to be complicated and varied, including cell cycle regulation. As a result, it is intriguing that similar to RanBPM, proteins found in potato cyst nematodes may be involved in controlling cell cycle activity as seen in growing syncytia (Qin et al. 2000). It is yet unknown if these nematode peptides can bind to Ran or have any effect on the phenotype of plant cells.

10.2.4.7 Venom-Allergen Proteins

A few potential parasite proteins are comparable to existing proteins, but their activities are currently unknown or too diversified. Members of this intriguing class of parasitism genes, called venom-allergen proteins (vaps), come from root-knot nematodes and cyst nematodes (Ding et al. 2000; Gao et al. 2001, 2003). In the parasitic worm *Ancylostoma caninum,* vaps were identified as secreted proteins (ASP) (Zhan et al. 2003). Since then, there are genes for vaps found in a range of nematodes, including parasites and free-living *Caenorhabditis elegans*. Even though it was discovered that some of these proteins are released or proteins secreted by soybean cyst nematodes express themselves in the subventral glands (Gao et al. 2001), their exact role is still unknown.

10.2.4.8 Calcirecticulin

In a related occurrence, the subventral glands of a root-knot nematode were seen to synthesise a protein comparable to calreticulin that was preceded by a signal peptide (Jaubert et al. 2002). Other parasitic nematodes release proteins that are similar to calreticulin, making them strong candidates to play a role in host–parasite interactions (Pritchard et al. 1999). However, it is challenging to hypothesise a role for calreticulin in root-knot nematode plant parasitism given the perplexing variety of potential or proven calreticulin functions documented (Nakhasi et al. 1998).

10.2.4.9 Chitinase

Additionally, there is an example of a parasitism protein with a well-defined function, however, the protein did not appear to have any purpose when it was created. Within the subventral glands of the soybean cyst nematode was a chitinase protein that is believed to represent the same (Gao et al. 2002). Further investigation is required to determine whether chitinase synthesis plays a role in this stage of parasitism, as it does with many other parasitism proteins.

10.2.4.10 Peroxidase

As a kind of defence against nematode attack, it appears likely that nematodes will use defence mechanisms against the reactive oxygen species (ROS) the host plant produces (Waetzig et al. 1999). As reported by Robertson et al. 2000 and Jones et al. (2004), ROS-detoxifying enzymes have been identified as peroxidases derived from the potato cyst nematode. Peroxidase genes are expressed in the hypodermis of potato cyst nematodes, and peroxidase proteins accumulate on their body surface, possibly aiding in ROS detoxification.

10.2.4.11 Fatty Acid-Binding Protein (FAR)

Another protein is retinol- and fatty acid-binding (FAR) protein which is a surface associated protein that may play a part in undermining plant defences. It is present in the potato cyst nematode *Globodera pallida*. The precursor lipids of the jasmonic acid signalling pathway and plant defence substances were both discovered to bind to this protein (Prior et al. 2001). This protein is a great match for a protein that might obstruct plant defence systems due to the detected concentration of this protein at the nematode surface.

10.2.4.12 Avr Protein

It was discovered that the amphids of a root-knot worm expressed nematode avirulence protein, which is a protein whose presence starts efficient plant resistance mechanisms induced by the tomato Mi resistance gene (Semblat et al. 2001). It is crucial to determine this protein's primary function and how it seems to induce an immune response.

10.2.4.13 14-3-3 Proteins

A final protein that may play a part in nematode parasitism has been found in the nematode *Meloidogyne incognita*. Despite lacking an N-terminal signal peptide, the dorsal gland-expressed gene produces a 14-3-3 family protein, which appears to be released (Jaubert et al. 2004). It is still unknown what role, if any, 14-3-3 proteins have in nematode parasitism. These proteins have a wide range of potential roles and are highly conserved in eukaryotes.

10.3 Parasitism Proteins of Fungi

When a pathogen comes in contact with a suitable host, plants use their cell walls as an essential defence mechanism to stop the infection (Hematy et al. 2009; Underwood 2012). Likewise, a specific group of enzymes is used by pathogenic fungi to compromise the structural integrity of the cell wall of the host plant. The word "glycoside hydrolases" (GHs) is used to refer to enzymes that may hydrolytically cleave glycosidic bonds and are spread among numerous, genetically and structurally distinct gene families. Some of the important fungal enzymes in cell wall degradation have been listed in Table 10.1.

Table 10.1 List of enzymes produced by phytopathogenic fungi (Kubicek et al. 2014)

Enzymes	Target component	Function	GH family
β-Glucosidase	Cellulose	Produces glucose by hydrolysing the soluble cellodextrin oligomers	GH1, GH3
Endo-β-(1 → 4)-glucanases	Hemicellulose	Hydrolyses the backbone chain of xyloglucan by cleaving the glycosidic bonds	GH5, GH12, GH16, GH44
Xylo-β-1,4-xylanase	Xylan	Hydrolyses xylan and produces xylose as its main product	GH10, GH11
β-Mannanase	Mannan	Hydrolyses mannan and produces mannose	GH5, subfamilies 7 and 8, and GH26
Polygalacturonase	Pectin	Cleaves glycosidic bond by hydrolysis	GH28
Rhamnogalacturonan hydrolase	Pectin	Cleaves the α-1,2-glycosidic bonds between D-galacturonic acid and L-rhamnose residues	GH28

10.3.1 Cell Wall-Degrading Enzymes and their Role in Plant Pathogenesis

To mimic plant environmental conditions, fungi were cultivated in plant substrate or in media where several studies showed a spike of genes coding for secretory proteins, such as effectors, which affect the host environment in the intracellular matrix, and genes expressing CWDE (Kubicek et al. 2014). Genes involved in the degradation of cellulose and hemicellulose, such as endoglucanase and xylanase were found to be increased upon the growth of *Ustilago maydis*, the causative agent of smut disease, in an infection model of *Arabidopsis thaliana* (Martinez-Soto et al. 2013). Likewise, according to Mathioni et al. (2011), the rice blast fungus *Magnaporthe oryzae* has elevated levels of gene expression for cellulases and hemicellulases after infection in comparison to a minimal medium control.

The GH28 Polygalacturonases (PG) are CWDEs for which there is some indication that they operate as virulence factors. When phytopathogens come into contact with plant cell walls, these are the first enzymes that are released (De Lorenzo and Ferrari 2002). Inhibition of *Aspergillus flavus* lesion growth in cotton bolls was observed with PG pecA gene deletion, on the other hand, in an *A. flavus* strain lacking in PG, expression of the same gene increases lesion size (Shieh et al. 1997). Some other pathogens such as *Botrytis cinerea* and *Alternaria citri* (ten Have et al. 1998; Isshiki et al. 2001) require a PG gene to express virulence. It is significant to note that a strain of *Claviceps purpurea* lacking the genes cppg1 and cppg2 makes the pathogen almost nonpathogenic without compromising its vegetative characteristics (Oeser et al. 2002).

Further instances come from methods that suppress entire sets of GH enzymes by using RNA interference (RNAi). GH6 and GH7 family cellulases, as well as xylanases from the GH10 and GH11 families, were all knocked down in *Magnaporthe oryzae* using RNA interference (Nguyen et al. 2011; VanVu et al. 2012). There was also a parallel decrease in virulence with a decrease in the synthesis of several enzymes. Knockdowns of the GH10 and GH11 families resulted in larger virulence impairments than knockdowns of the GH6 and GH7 families, while it was unclear if this was due to a greater relevance of xylanases during plant infection (VanVu et al. 2012).

10.3.2 Effector Molecules/Avr Genes of Fungal Plant Pathogens and Their Role in Host Plant Defence

Apart from breaking the cellular integrity of the host cell, parasitism proteins have made their way to induce defence responses. According to Flor's proposed "gene-for-gene" theory, there is a cognate resistance (R) gene in the host for every dominant avirulence (Avr) gene in the pathogen, and their interaction activates host defence mechanisms like the hypersensitive response (HR), which stops the growth of biotrophic fungi (Flor 1942). van Kan et al. (1991) discovered the first fungal Avr gene and Shan et al. (2004) discovered the first oomycete Avr gene. Plants have developed at least two active defences in response to pathogen attacks where the initial line of defence provides the most important defence against any possible infection and is based on the activation of PAMP-triggered immunity (PTI) and the detection of conserved pathogen-associated molecular patterns (PAMPs) by the so-called PAMP recognition receptors (PRRs) (De Wit 2007). There is now evidence that certain Avr genes generate effectors that obstruct PTI, permitting a pathogen to cause damage to a plant. During evolution, based on effector perception by R proteins and the subsequent activation of effector-triggered immunity (ETI), plants evolved a second recognition mechanism. This causes plants' defensive mechanisms, particularly HR, to respond more quickly and effectively, after the pathogens had defeated the basal defence system. In response, pathogens mutated their effectors or created new ones that might prevent effector-triggered immunity, while plants created new R proteins which served as a mediator in identifying these new effectors. Hence, this set off a second phase of plant-pathogen co-evolutionary conflict (Jones and Dangl 2006). Reports suggest that a number of necrotrophic fungal diseases may create ribosomal proteins that trigger necrosis in plants by acting as effectors with sensitive genes, hence facilitating disease (Friesen et al. 2008). However, Ottmann et al. (2009) also reported that necrosis and ethylene-inducing peptide 1 (Nep1)-like proteins (NLPs), produced by a number of species, including fungi, may have the similar function.

10.3.2.1 *Cladosporium fulvum*

Four Avr genes from *C. fulvum* (Cf) have been cloned, namely, Avr2, Avr4, Avr4E, and Avr9, all of which generate tiny cysteine-rich proteins released during infection, and are recognised by their corresponding cognate receptor Cf proteins Cf-2, Cf-4, Cf-4E, and Cf-9, respectively (Thomma et al. 2005). It is noteworthy that an effector Avr4 with a functional chitin-binding domain protects chitinous fungus such as *C. fulvum* from plant chitinases (van den Burg et al. 2004, 2006; van Esse et al. 2007).

10.3.2.2 *Rhynchosporium secalis*

Nip1-Nip3 is three low-molecular-weight peptides secreted by *Rhynchosporium secalis*. In barley cultivars bearing the Nip1 resistance gene, which is still uncloned induces particular (non-HR) defensive responses (Hahn et al. 1993). Nip1 has been reported to interact using a single receptor on the plasma membrane that is crucial for both defence triggering and virulence mediation, although the specific receptor is still unknown (Van't Slot et al. 2003).

10.3.2.3 *Fusarium oxysporum* f. sp. *lycopersici*

Six1 (renamed Avr3) produced by *F. oxysporum* is essential for full virulence on tomato, although it also causes ETI in the presence of the homologous I-3 resistance gene (Huang and Lindhout 1997; Rep et al. 2004). Furthermore, renamed as Avr2, Six3 increases virulence but also induces ETI in the presence of the homologous I-2 resistance gene (Houterman et al. 2009).

10.3.2.4 *Magnaporthe oryzae*

Several *M. oryzae* Avr genes have been cloned and characterised, like Avr-Pita (Orbach et al. 2000), Ace1 (Collemare et al. 2008), and the Pwl effectors (Sweigard et al. 1995). According to Jia et al. (2000) and Orbach et al. (2000), the Avr-Pita effector has similarity to fungal zinc-dependent metalloproteases and is often avirulent on rice.

10.3.2.5 *Ustilago maydis*

According to Doehlemann et al. (2009), Pep1, a recently discovered effector protein from U. maydis, is essential for penetration. Pep1 is released into the apoplast and accumulates at the sites of cell-to-cell passages.

10.3.2.6 Necrotrophic Fungi

Two necrotrophic fungi, *Stagonospora nodorum* and *Pyrenophora tritici-repentis* causing Stagonospora nodorum blotch (SNB) and tan spot, respectively, in wheat, create a number of deadly host-specific peptide effectors that can be recognised by host susceptibility genes and result in disease. Friesen et al. (2007) and Liu et al. (2006) reported that SnTox1 was the first toxic peptide produced by *S. nodorum* that interacted with Snn1, a similar host susceptibility gene. The most prevalent races of *P. tritici-repentis* were the first to develop ToxA, a peptide toxin. Ciuffetti et al. (1997) reported that PtrToxA, a 13.2-kDa protein, causes necrosis in particular wheat genotypes.

10.4 Conclusions and Future Prospects

Numerous strategies have been successful in identifying multiple candidate parasitism proteins that are secreted by phytopathogens in the form of effectors, Avr genes, gene regulators, etc. The identification of newsuch candidates will undoubtedly rise with the generation of further genomic resources. The development of molecular in vivo tests in intact host cells and the validation of the role of potential parasitism genes in pathogen–plant interactions are still essential for comprehending the processes and evolution of parasitism. In addition to offering excellent tools for the functional studies of parasitism genes, the usefulness of model plant species and the validation of the effects of RNA interference on phytopathogens also show that these genes have the potential to be viable targets for the development of artificial transgenic resistance in plants.

The proposed models are partially theoretical, therefore much more study is needed to confirm or disprove them. Numerous pathogenic genomes of fungi and nematodes will be sequenced in the near future owing to the quick development of new sophisticated and effective sequencing platforms, which will speed up the search for novel parasitism genes through comparative genomics along with related improvements in bioinformatics tools. Though there is yet little homology among fungal effectors, it is predicted that a number of fungal effectors participate in the inhibition of PAMP Triggered Immunity induced by PAMPs of fungal pathogens.

By using various methods such as gene silencing and overexpression we can experimentally investigate a role in virulence for most. Hence, we anticipate to unravel more PAMPs, which will improve our knowledge of how the plant defence system has evolved and how PTI and ETI interact.

References

Alfano JR, Collmer A (2004) Type III secretion system effector proteins: double agents in bacterial disease and plant defense. Annu Rev Phytopathol 42:385–414

Allard A, Gourgues M, Galiana E, Panabieres F, Ponchet M, Keller H (2008) Strategies of attack and defensein plant-oomycete interactions, accentuated for *Phytophthora parasitica*. J Plant Physiol 165:83–94

Baum TJ, Hussey RS, Davis EL (2007) Root-knot and cyst nematode parasitism genes: the molecular basis of plant parasitism. Genet Eng 28:17–43

van den Burg HA, Harrison SJ, Joosten MHAJ, Vervoort J, De Wit PJGM (2006) *Cladosporium fulvum* Avr4 protects fungal cell walls against hydrolysis by plant chitinases accumulating during infection. Mol Plant-Microbe Interact 19:1420–1430

van den Burg HA, Spronk CAE, Boeren S, Kennedy MA, Vissers JPC, Vuister GW, DeWit PJGM, Vervoort J (2004) Binding of the avr4 elicitor of *Cladosporium fulvum* to chitotriose units is facilitated by positive allosteric protein-protein interactions: the chitin-binding site of Avr4 represents a novel binding site on the folding scaffold shared between the invertebrate and the plant chitin-binding domain. J Biol Chem 279:16786–16796

Ciuffetti LM, Tuori RP, Gaventa JM (1997) A single gene encodes a selective toxin causal to the development of tan spot of wheat. Plant Cell 9:135–144

Collemare J, Pianfetti M, Houlle AE, Morin D, Camborde L, Gagey MJ, Barbisan C, Fudal I, Lebrun MH, Bohnert HU (2008) *Magnaporthe grisea* avirulence gene ACE1 belongs to an infection-specific gene cluster involved in secondary metabolism. New Phytol 179:196–208

Davis EL, Hussey RS, Mitchum MG, Baum TJ (2008) Parasitism proteins in nematode–plant interactions. Curr Opin Plant Biol 11:360–366

De Lorenzo G, Ferrari S (2002) Polygalacturonase-inhibiting proteins in defense against phytopathogenic fungi. Curr Opin Plant Biol 5:295–299

De Wit PJG (2007) How plants recognize pathogens and defend themselves. Cell Mol Life Sci 64:2726–2732

Ding X, Shields J, Allen R, Hussey RS (2000) Molecular cloning and characterisation of a venom allergen AG5-like cDNA from *Meloidogyne incognita*. Int J Parasitol 30:77–81

Doehlemann G, van der Linde K, Amann D, Schwammbach D, Hof A, Mohanty A, Jackson D, Kahmann R (2009) Pep1, a secreted effector protein of *Ustilago maydis*, is required for successful invasion of plant cells. PLoS Pathog 5:e1000290

van Esse HP, Bolton MD, Stergiopoulos I, De Wit PJGM, Thomma BPHJ (2007) The chitin-binding *Cladosporium fulvum* effector protein Avr4 is a virulence factor. Mol Plant-Microbe Interact 20:1092–1101

Fletcher J, Brand U, Running M, Simon R, Meyerowitz E (1999) Science 283:1911–1914

Flor HH (1942) Inheritance of pathogenicity in *Melampsora lini*. Phytopathology 32:653–669

Friesen TL, Faris JD, Solomon PS, Oliver RP (2008) Hostspecific toxins: effectors of necrotrophic pathogenicity. Cell Microbiol 10:1421–1428

Friesen TL, Meinhardt SW, Faris JD (2007) The *Stagonospora nodorum*–wheat pathosystem involves multiple proteinaceous hostselective toxins and corresponding host sensitivity genes that interact in an inverse gene-for-gene manner. Plant J 51:681–692

Gao B, Allen R, Davis EL, Baum TJ, Hussey RS (2004) Developmental expression and biochemical properties of a β-1,4-endoglucanase family in the soybean cyst nematode, Heteroderaglycines. Mol Plant Pathol 5:93–104

Gao B, Allen R, Maier T, Davis E, Baum T, Hussey R (2001) Mol Plant-Microbe Interact 14:1247–1254
Gao B, Allen R, Maier T, Davis E, Baum T, Hussey R (2003) Mol Plant-Microbe Interact 16:720–726
Gao B, Allen R, Maier T, McDermott J, Davis E, Baum T, Hussey R (2002) Int J Parasitol 32:1293–1300
Gibson DM, King BC, Hayes ML, Bergstrom GC (2011) Plant pathogens as a source of diverse enzymes for lignocellulose digestion. Curr Opin Microbiol 14:264–270
Hahn M, Jungling S, Knogge W (1993) Cultivar-specific elicitation of barley defense reactions by the phytotoxic peptide NIP1 from *Rhynchosporium secalis*. Mol Plant-Microbe Interact 6:745–754
Have TA, Mulder W, Visser J, van Kan JA (1998) The endopolygalacturonase gene Bcpg1 is required for full virulence of *Botrytis cinerea*. Mol Plant-Microbe Interact 11:1009–1116
Hematy K, Cherk C, Somerville S (2009) Host-pathogen warfare at the plant cell wall. Curr Opin Plant Biol 12:406–413
Hotopp JC, Clark ME, Oliveira DC, Foster JM, Fischer P, Torres MC, Giebel JD, Kumar N, Ishmael N, Wang S, Ingram J (2007) Widespread lateral gene transfer from intracellular bacteria to multicellular eukaryotes. Science 317:1753–1756
Houterman PM, Ma L, van Ooijen G, de Vroomen MJ, Cornelissen BJC, Takken FLW, Rep M (2009) The effector protein Avr2 of the xylem-colonizing fungus *Fusarium oxysporum* activates the tomato resistance protein I-2 intracellularly. Plant J 58:970–978
Huang CC, Lindhout P (1997) Screening for resistance in wild *Lycopersicon* species to *Fusarium oxysporum* f. sp. *lycopersici* race 1 and race 2. Euphytica 93:145–153
Hussey RS (1989) Disease-inducing secretions of plant-parasitic nematodes. Annu Rev Phytopathol 27:123–141
Hussey RS, Mims CW (1991) Ultrastructure of feeding tubes formed in giant-cells induced in plants by the root-knot nematode *Meloidogyne incognita*. Protoplasma 162:99–107
Isshiki A, Akimitsu K, Yamamoto M, Yamamoto H (2001) Endopolygalacturonase is essential for citrus black rot caused by *Alternaria citri* but not brown spot caused by *Alternaria alternata*. Mol Plant-Microbe Interact 14:749–757
Jaubert S, Laffaire JB, Ledger TN, Escoubas P, Amri EZ, Abad P, Rosso MN (2004) Comparative analysis of two 14-3-3 homologues and their expression patterns in the root-knot nematode *Meloidogyne incognita*. Int J Parasitol 34:873–880
Jaubert S, Ledger TN, Laffaire JB, Piotte C, Abad P, Rosso MN (2002) Direct identification of stylet secreted proteins from root-knot nematodes by a proteomic approach. Mol Biochem Parasitol 121:205–211
Jia Y, McAdams SA, Bryan GT, Hershey HP, Valent B (2000) Direct interaction of resistance gene and avirulence gene products confers rice blast resistance. EMBO J 19:4004–4014
Jones JDG, Dangl JL (2006) The plant immune system. Nature 444:323–329
Jones JT, Reavy B, Smant G, Prior AE (2004) Glutathione peroxidase of potato cyst nematode *Globodera rostochiensis*. Gene 324:47–54
van Kan JAL, van den Ackerveken GFJM, De Wit PJGM (1991) Cloning and characterization of cDNA of avirulence gene avr9 of the fungal pathogen *Cladosporium fulvum*, causal agent of tomato leaf mold. Mol Plant-Microbe Interact 4:52–59
Kikuchi T, Shibuya H, Jones J (2005) Molecular and biochemical characterization of an endo-β-1,3 glucanase from the pinewood nematode *Bursaphelenchus xylophilus* acquired by horizontal gene transfer from bacteria. Biochem J 389:117–125
Kubicek CP, Starr TL, Glass NL (2014) Plant cell wall-degrading enzymes and their secretion in plant-pathogenic fungi. Annu Rev Phytopathol 52:427–451
Kudla U, Milac AL, Qin L, Overmars H, Roze E, Holterman M, Petrescu AJ, Goverse A, Bakker J, Helder J, Smant G (2007) Structural and functional characterization of a novel, host penetration related pectate lyase from the potato cyst nematode *Globodera rostochiensis*. Mol Plant Pathol 8:293–305

Ledger TN, Jaubert S, Bosselut N, Abad P, Rosso MN (2006) Characterization of a new β-1,4 endoglucanase gene from the root-knot nematode *Meloidogyne incognita* and evolutionary scheme for phytonematode family 5 glycosyl hydrolases. Gene 382:121–128

Lindsey K, Casson S, Chilley P (2002) Peptide: new signalling molecules in plants. Trends Plant Sci 7:78–83

Liu ZH, Friesen TL, Ling H, Meinhardt SW, Oliver RP, Rasmussen JB, Faris JD (2006) The Tsn1–ToxA interaction in the wheat-*Stagonospora nodorum* pathosystem parallels that of the wheat–tan spot system. Genome 49:1265–1273

Martinez-Soto D, Robledo-Briones AM, Estrada-Luna AA, Ruiz-Herrera J (2013) Transcriptomic analysis of *Ustilago maydis* infecting *Arabidopsis* reveals important aspects of the fungus pathogenic mechanisms. Plant Signal Behav 8:e25059

Mathioni SM, Belo A, Rizzo CJ, Dean RA, Donofrio NM (2011) Transcriptome profiling of the rice blast fungus during invasive plant infection and *in vitro* stresses. BMC Genomics 12:49

Mitchum MG, Hussey RS, Davis EL, Baum TJ (2007) Application of biotechnology to understand pathogenesis of nematode plant pathogens. In: Biotechnology and plant disease management. CAB International, Wallingford, UK, pp 58–86

Nakhasi HL, Pogue GP, Duncan RC, Joshi M, Atreya CD, Lee NS, Dwyer DM (1998) Implications of calreticulin function in parasite biology. Parasitol Today 14:157–160

Nguyen QB, Itoh K, Van Vu B, Tosa Y, Nakayashiki H (2011) Simultaneous silencing of endo-β-1,4 xylanase genes reveals their roles in the virulence of *Magnaporthe oryzae*. Mol Microbiol 81:1008–1119

Nurnberger T, Brunner F, Kemmerling B, Piater L (2004) Innate immunity in plants and animals: striking similarities and obvious differences. Immunol Rev 198:249–266

Oeser B, Heidrich PM, Muller U, Tudzynski P, Tenberge KB (2002) Polygalacturonase is a pathogenicity factor in the *Claviceps purpurea*/rye interaction. Fungal Genet Biol 36:176–186

Olsen AN, Skriver K (2003) Ligand mimicry? Plant-parasitic nematode polypeptide with similarity to CLAVATA3. Trends Plant Sci 82:55–57

Orbach MJ, Farrall L, Sweigard JA, Chumley FG, Valent B (2000) A telomeric avirulence gene determines efficacy for the rice blast resistance gene Pi-ta. Plant Cell 12:2019–2032

Ottmann C, Luberacki B, Kufner I, Koch W, Brunner F, Weyand M, Mattinen L, Pirhonen M, Anderluh G, Seitz HU, Nurnberger T, Oecking C (2009) A common toxin fold mediates microbial attack and plant defense. Proc Natl Acad Sci USA 106:10359–10364

Prior A, Jones J, Blok V, Beauchamp J, McDermott L, Cooper A, Kennedy M (2001) A surface-associated retinol- and fatty acid-binding protein (Gp-FAR-1) from the potato cyst nematode *Globodera pallida*: lipid binding activities, structural analysis and expression pattern. Biochem J 356:387–394

Pritchard D, Brown A, Kasper G, Mcelroy P, Loukas A, Hewitt C, Berry C, Füllkrug R, Beck E (1999) A hookworm allergen which strongly resembles calreticulin. Parasite Immunol 21:439–450

Qin L, Kudla U, Roze E, Goverse A, Popeijus H, Nieuwland A, Overmars H, Jones J, Schots A, Smant G, Bakker J, Helder J (2004) Plant degradation: a nematode expansin acting on plants. Nature 427:30

Qin L, Overmars H, Helder J, Popeijus H, van der Voort JR, Groenink W, van Koert P, Schots A, Bakker J, Smant G (2000) An efficient cDNA-AFLP-based strategy for the identification of putative pathogenicity factors from the potato cyst nematode *Globodera rostochiensis*. Mol Plant-Microbe Interact 13:830–836

Rashid R, Bhat JA, Malik A, Bhat ZA, Dar WA, Untoo SA, Ganaie MY (2010) Molecular basis of host pathogen interaction. Int J Appl Agric Res 5:35–45

Rep M, van der Does HC, Meijer M, van Wijk R, Houterman PM, Dekker HL, De Koster CG, Cornelissen BJC (2004) A small, cysteine-rich protein secreted by *Fusarium oxysporum* during colonization of xylem vessels is required for I-3-mediated resistance in tomato. Mol Microbiol 53:1373–1383

Robertson L, Robertson W, Sobczak M, Helder J, Tetaud E, Ariyanayagam M, Ferguson M, Fairlamp A, Jones J (2000) Cloning, expression and functional characterisation of a peroxiredoxin from the potato cyst nematode *Globodera rostochiensis*. Mol Biochem Parasitol 111:41–49

Semblat J, Rosso M, Hussey R, Abad P, Castagnone-Sereno P (2001) Molecular cloning of a cDNA encoding an amphid-secreted putative avirulence protein from the root-knot nematode *Meloidogyne incognita*. Mol Plant-Microbe Interact 14:72–79

Shan WX, Cao M, Dan LU, Tyler BM (2004) The Avr1b locus of *Phytophthora sojae* encodes an elicitor and a regulator required for avirulence on soybean plants carrying resistance gene Rps1b. Mol Plant-Microbe Interact 17:394–403

Shieh MT, Brown RL, Whitehead MP, Cary JW, Cotty PJ, Cleveland TE, Dean RA (1997) Molecular genetic evidence for the involvement of a specific polygalacturonase, P2c, in the invasion and spread of *Aspergillus flavus* in cotton bolls. Appl Environ Microbiol 63:3548–3552

Smant G, Stokkermans JPWG, Yan YT, de Boer JM, Baum TJ, Wang XH, Hussey RS, Gommers FJ, Henrissat B, Davis EL, Helder J, Schots A, Bakker J (1998) Endogenous cellulases in animals: isolation of β-1,4-endoglucanase genes from two species of plant-parasitic cyst nematodes. Proc Natl Acad Sci USA 95:4906–4911

Sweigard JA, Carroll AM, Kang S, Farrall L, Chumley FG, Valent B (1995) Identification, cloning, and characterization of PWL2, a gene for host species specificity in the rice blast fungus. Plant Cell 7:1221–1233

Thomma BPH, Van Esse HP, Crous PW, De Wit PJGM (2005) *Cladosporium fulvum* (syn. *Passalora fulva*), a highly specialized plant pathogen as a model for functional studies on plant pathogenic Mycosphaerellaceae. Mol Plant Pathol 6:379–393

Tytgat T, Vanholme B, De Meutter J, Claeys M, Couvreur M, Vanhoutte I, Gheysen G, Van Criekinge W, Borgonie G, Coomans A, Gheysen G (2004) Mol Plant-Microbe Interact 17:846–852

Uehara T, Kushida A, Momota Y (2001) PCR-based cloning of two beta-1,4-endoglucanases from the root-lesion nematode *Pratylenchus penetrans*. Nematology 3:335–341

Underwood W (2012) The plant cell wall: a dynamic barrier against pathogen invasion. Front Plant Sci 3:85

Van't Slot KAE, van den Burg HA, Kloks CPAM, Hilbers CW, Knogge W, Papavoine CHM (2003) Solution structure of the plant disease resistance-triggering protein NIP1 from the fungus *Rhynchosporium secalis* shows a novel b-sheet fold. J Biol Chem 278:45730–45736

Vanholme B, Van Thuyne W, Vanhouteghem K, De Meutter J, Cannoot B, Gheysen G (2007) Molecular characterization and functional importance of pectate lyase secreted by the cyst nematode *Heterodera schachtii*. Mol Plant Pathol 8:267–278

VanVu B, Itoh K, Nguyen QB, Tosa Y, Nakayashiki H (2012) Cellulases belonging to glycoside hydrolase families 6 and 7 contribute to the virulence of *Magnaporthe oryzae*. Mol Plant-Microbe Interact 25:1135–1141

Waetzig G, Sobczak M, Grundler F (1999) Localization of hydrogen peroxide during the defence response of *Arabidopsis thaliana* against the plant-parasitic nematode *Heterodera glycines*. Nematology 1:681–686

Wang X, Allen R, Ding X, Goellner M, Maier T, de Boer J, Baum T, Hussey R, Davis E (2001) Signal peptide-selection of cDNA cloned directly from the esophageal gland cells of the soybean cyst nematode *Heterodera glycines*. Mol Plant-Microbe Interact 14:536–544

Wang X, Mitchum M, Gao B, Li C, Diab H, Baum T, Hussey R, Davis E (2005) A parasitism gene from a plant-parasitic nematode with function similar to CLAVATA3/ESR (CLE) of *Arabidopsis thaliana*. Mol Plant Pathol 6:187–191

Wyss U (1992) Observations on the feeding behaviour of *Heterodera schachtii* throughout development, including events during moulting. Fundum Appl Nematol 15:75–89

Wyss U, Zunke U (1986) Observations on the behaviour of second stage juveniles of Hetero inside host roots. Rev Nematol 9:153–165

Yan YT, Smant G, Stokkermans J, Qin L, Helder J, Baum T, Schots A, Davis E (1998) Genomic organization of four beta-1,4-endoglucanase genes in plant-parasitic cyst nematodes and its evolutionary implications. Gene 220:61–70

Zhan B, Liu Y, Badamchian M, Williamson A, Feng J, Loukas A, Hawdon JW, Hotez PJ (2003) Molecular characterisation of the Ancylostoma-secreted protein family from the adult stage of *Ancylostoma caninum*. Int J Parasitol 33:897–907

Role of Lifestyle Pattern and Soil Microbiome Interaction of *Arthrobotrys Species* for the Biostimulant Production

11

Subbaiah Revathi Lakshmi, Madasamy Sundar, and Selvaraj Kunjiappan

11.1 Introduction

Advances in biotechnology and widespread concern over the negative side effects of pesticides have encouraged the hunt for new, environmentally acceptable insect control methods. One of the best replacements among them is biological control. The employment of biological agents to lessen the impact caused by an insect pest population is known as biological control. Since the beginning of the agricultural era, only a very small proportion of known insect species are regarded as pests, and their control mechanism is a significant task for the scientific community. Chemical pesticides are expensive, even though they work well, and only offer short-term respite since insects may rapidly reproduce and evolve, making them immune to these and other control measures. In addition to the risks they represent to the health of the population, they also have an impact on non-target organisms and disrupt the ecological balance, which results in permanent environmental harm. A few criteria must be achieved for biocontrol to be a key component of modern agriculture, including the identification and designing of improved biocontrol agents, the construction of fermentation systems for biomass generation, and the developing of formulation and delivery systems consistent with both microorganism needs and

S. Revathi Lakshmi (✉)
Centre for Research and Postgraduate Studies in Botany, Ayya Nadar Janaki Ammal College, Affiliated to Madurai Kamaraj University, Madurai, Sivakasi, Tamil Nadu, India

M. Sundar
Department of Biotechnology, Kalasalingam Academy of Research and Education, Krishnankoil, Tamil Nadu, India

S. Kunjiappan
Department of Biotechnology, Kalasalingam Academy of Research and Education, Krishnankoil, Tamil Nadu, India

M. S. Akhtar (ed.), *Opportunistic Fungi, Nematode and Plant Interactions*,
https://doi.org/10.1007/978-981-97-2045-3_11

conventional agricultural practices. Fungi have been recognized as potential biological control agents primarily owing to their tremendous generative potential, target-specific activity, rapid multiplication times, and ability to produce a resting stage or saprophytic phase that allows them to survive for prolonged periods of time in the absence of a host.

Nematodes that parasitize plants are major cause for crop loss and productivity of many plants. The developing novel strategies for the control of these pathogens is the major concern since the established methods rely on the practice of nematicides which cause negative impact on human health and environment. At this juncture, designing a methodology that involves the use of natural enemies of pathogens will be the better choice (Stirling 2018). The nematophagous fungus, a group of predominantly soil-dwelling fungi that are contagious to nematodes, have gathered the most attention among them.

Nematophagous fungus are a significant class of soil microbes that have the ability to control populations of nematodes that are parasitic on plants and animals (Nordbring-Hertz et al. 2011; Swe et al. 2011). From the tropics to Antarctica, these fungi are distributed widely in soil and water bodies (Pramer 1964; Nordbring-Hertz et al. 2011). Nematophagous fungus can be divided into four major groups based on the way of interaction with the nematode. Toxin-producing fungi immobilize nematodes prior to invasion, endoparasitic fungi use their spores, egg-parasitic fungi invade worm eggs with their hyphal tips, and nematode-trapping fungi use sticky or mechanical hyphal structures (e.g., Liu et al. 2009; Swe et al. 2011; Moosavi and Zare 2020). The group includes more than 200 taxonomically diverse species of fungi that can create specialized traps to capture nematodes (eggs, juveniles, and adults) and then use a variety of strategies to obtain essential minerals from the prey (Jansson and Lopez-Uorca 2001; Yang et al. 2007; Nordbring-Hertz et al. 2011). Often these nematode-trapping fungi can coexist as saprophytes and parasites (Nordbring-Hertz et al. 2011; Pramer 1964). In a variety of ecological contexts, they are essential for maintaining the density of nematode populations. However, many of the fungus that generate traps may survive in soil without nematodes as a saprophyte.

The predatory (trapping) fungi create intricate hyphal structures, viz., hyphal nets, knobs, branches, or rings, whereas the nematodes get trapped mechanically or by adhesion-bound mechanism (Nordbring-Hertz et al. 2001; Singh 2007; Singh et al. 2012a). One of the most thoroughly researched nematode-trapping fungi is *Arthrobotrys oligospora* Fres (Nordbring-Hertz 2004; Nordbring-Hertz et al. 2011). Pfister and Liftik (1995) and Mo et al. (2006) have isolated different strains of *A. oligospora* from various soil environments. *A. oligospora* attains the parasitic stage when nematode is present by building intricate 3-D networks to ensnare nematode. The nematodes adhere, penetrate, and get immobilized prior to the trapping, among other things (Nordbring-Hertz et al. 2011; Nordbring-Hertz 2004). The high frequency of nematode-trapping capacity of the fungus makes it a desirable choice as biological control. Though several reports focusing on the biocontrol potential of *A. oligospora,* there are several other species belongs to the genus *Arthrobotrys* with potential nematicidal activity are also available. Therefore, this chapter discusses the mechanisms and factors responsible for the trap formation of biocontrol fungus *Arthrobotrys* and also highlights the plant growth potential of this fungus.

11.2 Mechanism of Nematicidal Activity of *Arthrobotyrs* Sp.

The nematodes are being predated by *Arthrobotrys* and it occurs in four stages. They are as follows:

11.2.1 Recognition of *Arthrobotrys* sp. by Nematodes

Most of the NTFs transit their lifestyle from saprophyte to predator when there is a scarcity of available nutrients in the environment. In order to survive, they tend to lead a predatory lifestyle. *Arthrobotrys* sp. is one among such nematode-trapping fungi. It is a non-motile predator and hence it devises a suitable method to attract the nematodes. The volatile organic compounds (VOCs) being excreted by the *Arthrobotrys* act as a pheromone and attract the nematodes. The functional group and nature of the VOCs may vary depending on the species of *Arthrobotrys*. Studies have also shown that almost all of the metabolites produced by NTFs to attract nematodes are small, organic compounds that are liposoluble (Hsueh et al. 2017; Wang et al. 2018) as these chemicals can more easily pass through the nematode cell membrane and induce worms to exhibit an odor stimulation reaction.

11.2.2 Recognition of Nematodes by *Arthrobotrys*

The recognition of nematodes of *Arthrobotrys* occurs by several approaches. They are as follows:

(a) As the nematodes have olfactory nerves, the AWC neurons and other odor recognizing genes respond to the stimuli of VOCs (Volatile Organic Compounds) secreted by *Arthrobotyrs* and thus aid further trap formation and infection (Ma et al. 2021).

(b) Along the line of NTFs (nematode-trapping fungi), nematodes also release some metabolites that helps the *Arthrobotrys* to recognize the prey. They are nemin, a putative morphogenic signal from nematode (Liang et al. 2019) and ascarosides, discharged by the nematode attracts *Arthrobotrys* (Worthy et al. 2018). Meerupati et al. (2013) reported that the ascarosides mediated recognition of prey occurs when the NTFs are deprived of nutrients.

(c) Lectin, an extracellular protein formed by *Arthrobotrys* plays a key role in the recognizing its prey and the initial stages of trap formation (Wes and Bargmann 2001). Lectin can detect all sorts of cell membrane residues of the nematodes and thus lectin mediated nematode—NTF interactions has widely been accepted by several researchers (Nordbring-Hertz and Mattiasson 1979; Rosenzweig and Ackroyd 1983; Borrebaeck et al. 1984; Wes and Bargmann 2001; Niu and Zhang 2011).

(d) Molecular studies on trap formation of *Arthrobotrys oligospora* revealed that the ascarosides acts as a pheromone and attracts the NTFs by the activation of MAPK (mitogen-activated protein kinase) (signaling pathway (Chou et al. 2001; Chen et al. 2021).

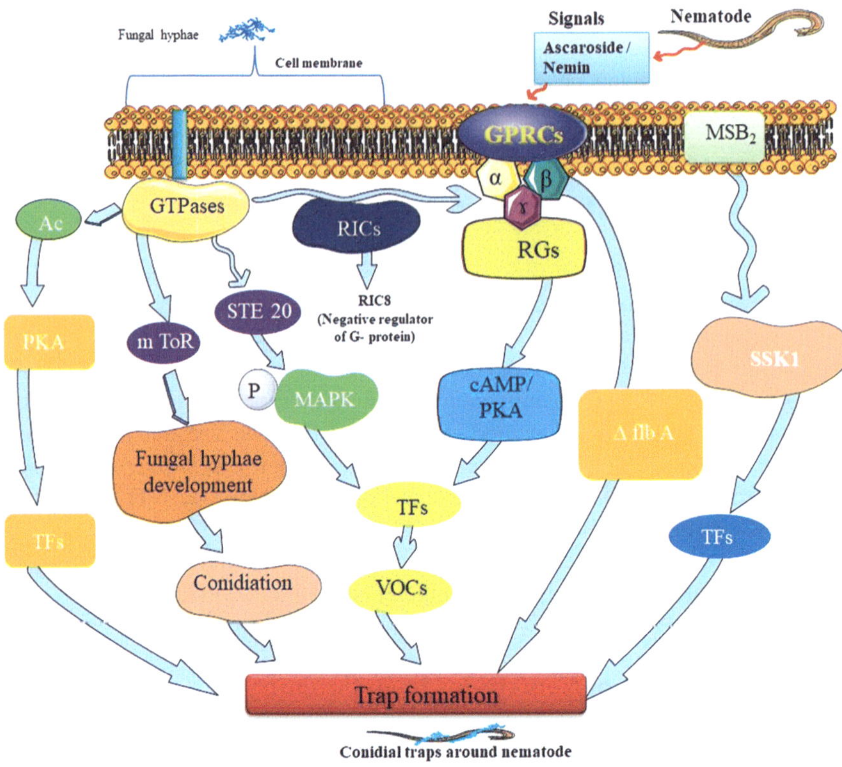

Fig. 11.1 Mechanism of trap formation by nematophagous fungi *Arthrobotrys* sp. based on the ideas derived from Yang et al. (2011) and Zhu et al. (2022)

11.2.3 Infection and Trap Formation

Infection and trap formation by *Arthrobotrys* sp. is mediated by several signaling pathways and metabolic reactions which have been evidenced through several reports (Fig. 11.1). They are summarized as follows:

(a) Ascarosides released by the nematodes have been received by GPCRs (G-protein coupled receptors) at the cell membrane of *Arthrobotrys* sp. Then the signal received from the ascarosides was being transduced by G-proteins, i.e., G-protein subunits (Gα, Gβ, and Gγ). Furthermore, the interaction of RIC8 with the α-subunit of G-Protein regulates the nematode predation by *Arthrobotrys* sp. (Li et al. 2007).

(b) The G-Protein regulates the synthesis of ΔflbA gene, a conserved gene sequence that is responsible for the trap formation against nematodes. Trap formation, i.e., concentric rings and adhesive bodies tends to be produced as a result of the transcription factors released by ΔflbA gene and RIC 8 (Resistance to Inhibitors of Cholinesterase) mediated G-Protein signaling pathway. In addition, inducer of meiosis 2 (IME2) and MAPK pathway was involved in the fungal hyphae growth and colonization (Ma et al. 2021; Bai et al. 2022).

(c) In the meantime, activation of SSK1 (response regulator) by MSB2 (mucin family signaling protein) initiates the MAPK cascades and binding of Phosphates to MAPK signaling pathway through the activation of STE 20 (serine/threonine protein kinase) by the GTPases (Guanine Triphospahtases) at the cell membrane of *Arthrobotrys* sp. also involved in trap formation (Xie et al. 2020; Fan et al. 2021).
(d) The α-subunit of the G-Protein binds with AC (Adenylate Cyclase) to regulate the synthesis of PKA (Protein Kinase A), which in turn regulates the infection or lysis of the nematode cell wall by secreting extracellular enzymes such as serine proteases, collagenases, and chitinases. These extracellular enzymes facilitate the entry of *Arthrobotrys sp*. into the nematode cuticle (Fan et al. 2021).
(e) GTPases also regulate the synthesis of mTOR, thereby activating the formation of fungal hyphae. Additionally, they contribute to conidiation, immobilization, and further autophagy of nematode cells (Yang et al. 2021).

11.2.4 Immobilization and Lysis of Nematodes

From the available literatures, it is evident that PII a serine protease enzyme that plays a vital role in immobilizing the nematodes (Huang et al. 2004). The PII and secretion of extracellular enzymes like collagenases form piercing tube followed by infection bulb. Inside the infection bulb, fungal hyphae mature and digest the nematodes (Veenhuis et al. 1985; Tunlid et al. 1994; Niu and Zhang 2011). In vegetative hyphae, lipid droplets, as observed by Rosén et al. (1997), both assimilate and store the nutrients obtained from infected nematodes, which are then utilized to sustain fungal growth.

11.3 Factors Influencing Trap Formation

Nematophagous fungi like *Arthrobotrys* lead saprophytic lifestyle and it became predacious when it comes in contact with the nematode and forms different forms of traps. The causes of the genesis of trap have been a topic of discussion for many years (Xie et al. 2010; Hsueh et al. 2013; Wang et al. 2014). The following are potential inducers (chemicals) or factors identified through the literature analysis:

11.3.1 Nematode as the Influencing Factor

(a) Induction of trap formation occurs effectively when there is direct contact between the living nematode and the hyphae of nematophagous fungi (Xie et al. 2010).
(b) The likelihood of trap formation increases along with the population of nematode juveniles (Xie et al. 2010).

(c) The nematode extract has been found to be responsible forming concentric rings by *A. dactyloides* (Lawton 1957) and vegetative fungal hyphae by *A. oilgospora* (Yang et al. 2011).
(d) Nemin, a low molecular weight amino acid, extracted from the culture filtrates found to be involved in the trap formation of *A. conoides* (Pramer and Stoll 1959; Pramer and Kuyama 1963) and nemin induced trap formation is carbon dioxide dependent (Balan and Lechevalier 1972).
(e) The chances of trap formation by *Arthrobotrys* is highly influenced by ascarosides, a highly conserved molecule found in nematodes under nutrient deficient circumstances (Hsueh et al. 2013).

11.3.2 Amino Acids as Inducing Factor

Several researchers found that amino acid supplementation to the in vitro culture medium facilitate the trap formation. The amino acids associated with the trap formation are as follows:

(a) Valine, aspartic acid, and glutamic acid (Nordbring-Hertz 1973).
(b) Low nutrient medium fortified with valine, phenylalanine, and aspartic acid (Rosenzweig 1984).
(c) Cornmeal agar with phenylalanine and valine (Friman et al. 1985).
(d) Incorporation of phosphate along with the amino acids like valine and phenyl alanine in the low nutrient medium showed negative influence over trap formation (Friman 1996).

11.3.3 Bacteria as Trap Inducing Factor

Co-culturing of *A. oligospora* with *Chryseobacterium* sp. TFB facilitates the mycelial traps and conidial traps around nematode (Li et al. 2011). It has also been reported that the trap formation is accelerated by bacteria mediated urea transport further the bacterium also mediate the defense mechanism of *A. oligospora* by aiding its mobility and lysis of nematodes (Wang et al. 2014). Diketopiperazines (DKPs) have the potential to promote *Stenotrophomonas*' chemotaxis towards fungus extract, which has been implicated in the development of bacterial biofilms on surfaces which in turn facilitate the trap forming potential of *A. oligospora* (Li et al. 2016).

11.3.4 Other Trap Inducing Factors

The chances of formation of adhesive networks by *A. oligospsora* were high in water agar medium supplemented with cow dung (Dackman and Nordbring-Hertz 1992). The prevalence of conidal traps by *A. dactyloides* enhanced on water agar surfaces in the soil environment (Persmark and Nordbring-Hertz 1997).

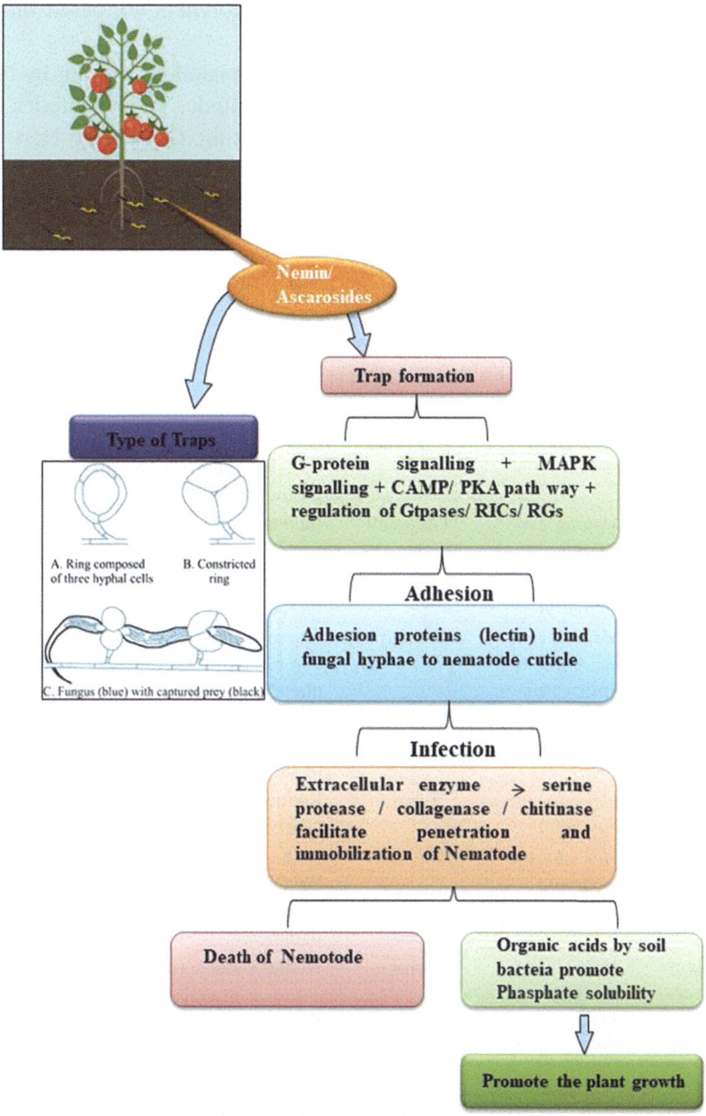

Fig. 11.2 Interaction of nematode, *Arthrobotrys* sp., and soil microbiome

11.4 *Arthrobotrys* sp. as Potential Biocontrol Agent

One of the most extensively researched nematode-trapping fungi is *Arthrobotrys oligospora* (Nordbring-Hertz et al. 2011; Nordbring-Hertz 2004). *A. oligospora* strains have been discovered in a variety of soil conditions (Pfister and Liftik 1995; Mo et al. 2006). Understanding the interaction of nematode, *Arthrobotrys* sp. and soil microbiome is quite important to develop *Arthrobotrys* sp. based biocontrol agents and was depicted in Fig. 11.2.

A. oligospora reaches the parasitic stage when nematode is present by building intricate three-dimensional networks to capture nematode. The nematodes adhere, penetrate, and get immobilized prior to the trapping (Nordbring-Hertz 2004; Nordbring-Hertz et al. 2011). The fungus is a desirable candidate for use in the management of plant-parasitic nematodes due to its high capacity to capture nematodes. *A. oligospora* builds three-dimensional adhesive nets in order to

Table 11.1 Biocontrol potential of *Arthrobotrys* species reported in crop plants

Name of the strain	Pathogen	Host plant	Responses	References
A. amerospora, Rhizoboium japanicum	*Heterodera glycines*	Soybean	Reduces nematodes but does not show any growth promoting activity	Niblack and Hussey (1986)
Arthrobotrys sp.	*Meloidogyne javanica*	Guldaudi	Increase in root length	Dávila and Hío (2005)
A.oligospora	*Meloidogyne incognita*	Okra	Nematicidal activity and increase in yield	Dhawan and Singh (2011)
A. oligospora VNS-1	*Meloidogyne incognita and Rhizoctonia solani*	Tomato	Increased level of chlorophyll, total phenolic content, phenylalanine lyase activity and accumulation of salicylic acid	Singh et al. (2012a)
A. oligospora	*Meloidogyne incognita*	Tomato	Reduced nematode egg count	Jamshidnejad et al. (2013)
A. oligospora	*Meloidogyne incognita*	Tomato	Reduce nematode at the larva stage and positive effect on growth parameters	Bakr et al. (2014)
A. oligospora	*Meloidogyne incognita*	Carrot	Nematocidal activity	Hussain et al. (2017)
A. oligospora strain C-2197	*Meloidogyne incognita*	Tomato	Reduce nematode population and increased leaf area and well developed root system	Quevedo et al. (2021)
A. conoides *A. musiformis*	*Rhabditis* spp.	In vitro culture filtrates	Nematicidal activity	Tkalenko et al. (2021)
A. musiformis	*Haemonchus contortus*	In vitro fungal culture assay	Predatory activity	Pérez-Anzúrez et al. (2022)
A. thaumasia *A. musiformis*	*Meloidogyne halpa*	Tomato	Reduced root-knot infection rate and increase in root length, shoot length, and biomass	Purba et al. (2022)

mechanically capture nematodes. The trap development by the fungal hyphae is greatly influenced by the environmental factors especially the presence of nematodes.

Since plant-parasitic nematodes often target plant roots, the ability of nematophagous fungi to colonize roots should be a significant advantage if the fungus is utilized for biological control (Singh et al. 2012b). *Arthobotrys* species have long been employed in the elimination of nematodes of plants and animals. The significant result on biocontrol ability of different *Arthrobotrys* species have been reported by several researchers and was enlisted in Table 11.1. From the available literatures, it is apparent that the biocontrol potential of different *Arthrobotrys* species may subject to vary based on the nematode they infect, host plant, microbiome on the host plant rhizosphere, population of both nematode and *Arthrobotrys* species.

Apart from the above discussed factors, the nutrient medium selected for the in vitro culture of *Arthrobotrys* spp. showed a wide range of difference with respect to the species cultured. From the available data on the examination of biocontrol potential of different species of *Arthrobotrys* (Table 11.2) explicate the nutrient source (medium components) play vital role in the density of *Arthrobotrys* spp.,

Table 11.2 In vitro culture requirements of different species of *Arthrobotrys*

Name of the strain	Culture conditions	Reference
A. amerospora	Corn meal: Water agar on 1:3 at 25 °C	Niblack and Hussey (1986)
A. oligospora	Corn meal agar medium at 25 ° C and 3 weeks incubation	Sanyal (2000)
A. oligospora	TSB medium at 25 °C and 1 week incubation at dark	Duponnois et al. (2006)
A. oligospora	Corn meal agar medium at 25 °C and 1 week incubation at dark, pH 7.0	Singh et al. (2012a)
A. oligospora	PDA medium at 25 °C and 10 days incubation	Jamshidnejad et al. (2013)
A. oligospora	Corn meal agar medium at 25 °C	Bakr et al. (2014)
A. conoides (JX979095)	Corn meal agar medium at 28 °C and 1 week incubation	Pandit et al. (2014)
A. thaumasia	PDB Medium at 25 °C and 10 days incubation at 180 rpm in shaker	Kassam et al. (2021)
A. oligospora C-2197	Corn meal agar medium/oat meal agar medium/corn meal plus carrot agar medium/V8 juice agar medium at 25 °C	Quevedo et al. (2021)
A. oligospora Fres. 1852 *A. musiformis* *A. thaumasia* *A. spheroides* *A. flagrans* *A. amerospora*	PSU-20 at 200 rpm at 25–26 °C and 72 h incubation	Tkalenko et al. (2021)
A. thaumasia *A. musiformis*	PDA medium at 25 °C and 14 days incubation	Purba et al. (2022)

which in turn affect the hyphal growth, conidiation and trap formation followed by nematicidal activity. Furthermore, it is apparent that culture conditions such as temperature, incubation period, and incubation time also have impact on the density of *Arthrobotrys* spp.

Singh et al. (2013) observed the following changes in tomato plants while treated with *A. oligospora*—EAO-147 and RST-101 strains. They observed increased activity of enzymes such as superoxide dismutase, chitinase, catalase, β-1,3-glucanase, ascorbate peroxidase, phenylalanine lyases, guaiacol peroxidase, and polyphenol oxidase. Increased levels of total phenolic content in leaves and ethylene in roots have also been reported by Singh et al. (2013). These results provide information about the pathogenicity brought out by *A. oligospora*—EAO-147 and RST-101 strains. Superoxide dismutase and peroxidases scavenge reactive oxygen radicals generated as a result of abiotic and biotic stresses, similarly, polyphenol oxidases may induce the synthesis of lignin and other phenolics that gives protection to the cell wall against the pathogen attack. Increased activity of PAL (Phenylalanine lyases) improved the plant growth by reducing the disease severity.

The use of fungal inoculants from the nematode-trapping group, according to Hastuti and Faull (2018), can promote plant development by increasing auxin production as a result of high microbial activity in the host plant's root region. This can concurrently increase the biomass of the plant and alter the increase in organic and inorganic components in host plant cells. The quantity of pathogenic nematodes on plant roots can be decreased, according to Singh et al. (2013) and Marin-Bruzos and Grayston (2019). By assisting in the enhancement of other soil microflora and the synthesis of plant secondary metabolites that can obstruct reproduction, these fungi can also promote plant growth.

11.5 Plant Growth Promoting Potential of *Arthrobotrys* Species

Nutrient deficiency is the second major concern that affects the productivity of crops next to pathogenesis. Need for the utility of microbiomes as phytostimulants increasing day by day. In order to regulate plant growth and metabolism, ample quantity of phosphorus is required for the plants next to nitrogen. Irrespective of the surplus availability of phosphorus in the soil, it has not been utilized by the plants because of its binding nature with iron, aluminum oxides, hydroxides or calcium made it unavailable for the roots to absorb. Several microorganisms have widely been utilized for the phosphate solubilization. Nematophagous fungi are one such having the potential to solubilize phosphate much efficiently than bacteria because of its tolerance to acidic soils, siderophore, and organic acid synthesis.

The in vitro and in vivo ability of *Arthrobotyrys oligospora* to solubilize rock phosphate has been investigated by Duponnois et al. (2006). They used four concentrations of three types of rock phosphates, namely, Barkina Faso, Senegal, and Mali and all the three types of rock phosphates were being solubilized by *A. oligospora*. Maximum solubilization was observed with Senegal. They further investigated the effect of Senegal alone and in combination with *A. oligospora* on *Acacia holosericea*. Furthermore, it was conferred that combined effect of *A. oligospora and Senegal* was found to be significant in increasing the growth rate of *Acacia holosericea*.

The in vitro phosphate solubilizing potential of *Arthrobotrys conoides* (JX979095) has been studied by Pandit et al. (2014). They experimented with different phosphate sources, namely, tricalcium phosphate, zinc phosphate, rock phosphate, and aluminum phosphate. They conducted the in vitro phosphate solubility assay on different nutrient mediums, viz., PVK, AYG, NBRIP, and NBRIY to identify the most suitable medium for phosphate solubilization. HPLC profiling of fungal filtrates were made and reported the presence of citric acid and malic acid. Moreover, fall in pH level of the culture filtrates was also observed. These finds lead to a conclusion that the organic acid released by the fungi and fall in pH level of the culture filtrates might be the reason for the solubility of phosphorus and providing an opportunity to utilize it as a biofertilizer. Recently, Aophytase, an enzyme similar to the β-propeller phytases (BPP) found in bacteria, has been isolated from *A. oligospora*. The recombinant phytase, rAophytase, was generated and investigated for the release of phosphate from soymeal. Significant results were observed in the release of phosphate and other minerals (Hou et al. 2020).

Based on the literatures available on the phosphate solubilizing potential of *Arthrobotrys* sp., the possible reasons for the failure or lack of proper evidences for the growth promoting potential of this nematophagous fungus could be derived. The phosphate solubilization can be made possible with the synthesis of organic acids or siderophore formation or chelating with iron oxides but the ultra-performance liquid chromatography–quadrupole/time of flight mass spectrometry (QToF-MS) analysis on *A. oligospora* by Kassam et al. (2021) does not show the presence of organic acids related to phosphate solubilization. To date none of the available literatures on *Arthrobotrys* sp. provide details on the either siderophore formation or binding nature with iron oxides or organic acids synthesis. The data (phosphate solubilizing potential) derived from the findings of Hou et al. (2020), Pandit et al. (2014), and Duponnois et al. (2006) might possibly be due to the potential activity of phosphate solubilizing bacteria that are inhabited in the rhizosphere of the host plant where the nematophagous fungi was introduced. Moreover, it is very clear that *Arthrobotyrs*, being a nematophagous fungi, it could not be able to survive when there is a nonexistence of nematodes and it would possibly be lead a saprophytic lifestyle that too happens based on the availability of organic matters (Pramer 1964; Nordbring-Hertz et al. 2011). Ali (1990) and Kushwaha and Sarma (2015) evaluated the effect of organic amendments farm yard manure and pigeon droppings with *A. oligospora* and observed increased predatory activity and increased the plant growth by providing better root system.

11.6 Conclusions and Future Prospects

Regardless of the use of several species of *Arthrobotrys* as biological control agent against nematodes over several years, none of the species were commercialized since the fungi dwells in soil as a saprophyte consuming nutrients from the soil. However, there are limited commercialization options due to the expensive, inefficient, and difficult-to-regulate nature of solid substrate mass production techniques for ensuring product quality. The inclusion of a nutrient source might cause microbial contamination issues, and the bulkiness of the substrate makes these products costly to carry, store, and apply using traditional farm equipment. It is much evident to design a nutrient source that should satisfy the nutritional requirement of the fungi and it should also be free from contaminants. To achieve the above requirements, incipient fungal spores or fungal conidia or fungal hyphae with under developed concentric rings and adhesive traps shall be coated with kaolin or sodium alginate or gum arabic and prepared as granules might satisfy the need. Furthermore, the research on the standardizing a common nutrient substrate that keeps the incipient fungal spores viable as well as maintaining its pathogenicity against nematodes is much important. Moreover, it is evident from the literatures available that, in order to design a commercial biostimulant from *Arthrobotrys*, it is essential to include some other beneficial bacteria that dwell along with the fungi in native environment to exploit the plant growth promoting potential.

References

Ali AHH (1990) The use of nematode trapping fungi and organic amendments to control root-knot nematodes. Bull Facul Agric Univer Cairo 41:1001–1012

Bai N, Zhang G, Wang W, Feng H, Yang X, Zheng Y, Yang L, Xie M, Zhang KQ, Yang J (2022) Ric8 acts as a regulator of G-protein signalling required for nematode-trapping lifecycle of *Arthrobotrys oligospora*. Environ Microbiol 24:1714–1730

Bakr RA, Mahdy ME, Mousa E-SM (2014) Biological control of root-knot nematode Meloidogyne incognita by *Arthrobotrys oligospora*. Egypt J Crop Prot 9:239252

Balan J, Lechevalier HA (1972) The predaceous fungus *Arthrobotrys dactyloides*: induction of trap formation. Mycologia 64:919–922

Borrebaeck CA, Mattiasson B, Nordbring-Hertz B (1984) Isolation and partial characterization of a carbohydrate-binding protein from a nematode-trapping fungus. J Bacteriol 159:53–56

Chen SA, Lin HC, Schroeder FC, Hsueh YP (2021) Prey sensing and response in a nematode-trapping fungus is governed by the MAPK pheromone response pathway. Genetics 217:iyaa008

Chou JH, Bargmann CI, Sengupta P (2001) The *Caenorhabditis elegans* odr-2 gene encodes a novel Ly-6-related protein required for olfaction. Genet 157:211–224

Dackman C, Nordbring-Hertz B (1992) Conidial traps-a new survival structure of the nematode-trapping fungus *Arthrobotrys oligospora*. Mycol Res 96:194–198

Dávila L, Hío JC (2005) Evaluación de la actividad biocontroladora de *Arthrobotrys* sp. y *Paecilomyces* sp. sobre *Meloidogyne javanica in vitro* y bajo condiciones de invernadero en crisantemo (*Drendranthema grandiflora* Andernson). Agron Colomb 23:91–101

Dhawan SC, Singh S (2011) Biological control activity of *Arthrobotrys oligospora* against *Meloidogyne incognita* infesting okra. Indian J Nematol 41:70–78

Duponnois R, Kisa M, Plenchette C (2006) Phosphate-solubilizing potential of the nematophagous fungus *Arthrobotrys oligospora*. J Plant Nutr Soil Sci 169:280–282

Fan Y, Zhang W, Chen Y, Xiang M, Liu X (2021) DdaSTE12 is involved in trap formation, ring inflation, conidiation, and vegetative growth in the nematode-trapping fungus *Drechslerella dactyloides*. Appl Microbiol Biotechnol 105:7379–7393

Friman E (1996) Influence of phosphate on development of the nematode-trapping fungus *Arthrobotrys oligospora*. Microbiol Res 151:77–80

Friman E, Olsson S, Nordbring-Hertz B (1985) Heavy trap formation by *Arthrobotrys oligospora* in liquid culture. FEMS Microbiol Ecol 1:17–21

Hastuti LDS, Faull J (2018) An investigation on Sumateran *Arthrobotrys oligospora* and carbofuran Againts root-knot nematode (*Meloidogyne hapla*) on tomato (*Solanum lycopersicum* mill.). Int J Sci Technol Res 7:32–38

Hou X, Shen Z, Li N, Kong X, Sheng K, Wang J, Wang Y (2020) A novel fungal beta-propeller phytase from nematophagous *Arthrobotrys oligospora*: characterization and potential application in phosphorus and mineral release for feed processing. Microb Cell Factories 19:1–13

Hsueh YP, Gronquist MR, Schwarz EM, Nath RD, Lee CH, Gharib S, Schroeder FC, Sternberg PW (2017) Nematophagous fungus *Arthrobotrys oligospora* mimics olfactory cues of sex and food to lure its nematode prey. elife 6:e20023

Hsueh Y-P, Mahanti P, Schroeder FC, Sternberg PW (2013) Nematode-trapping fungi eavesdrop on nematode pheromones. Curr Biol 23:83–86

Huang X, Zhao N, Zhang K (2004) Extracellular enzymes serving as virulence factors in nematophagous fungi involved in infection of the host. Res Microbiol 155:811–816

Hussain M, Zouhar M, Ryšánek P (2017) Potential of some nematophagous fungi against *Meloidogyne hapla* infection in Czech Republic. Pak J Zool 49:35–43

Jamshidnejad V, Sahebani N, Etebarian H (2013) Potential biocontrol activity of *Arthrobotrys oligospora* and *Trichoderma harzianum* BI against *Meloidogyne javanica* on tomato in the greenhouse and laboratory studies. Arch Phytopathol Plant Protect 46:1632–1640

Jansson HB, Lopez-Uorca LV (2001) Biology of nematophagous fungi. In: Mishra JK, Hora BW (eds) *Trichomycetes* and other fungal groups. CRC Press, Boca Raton, USA, pp 145–174

Kassam R, Yadav J, Chawla G, Kundu A, Hada A, Jaiswal N, Bollinedi H, Kamil D, Devi P, Rao U (2021) Identification, characterization, and evaluation of nematophagous fungal species of *Arthrobotrys* and *Tolypocladium* for the management of *Meloidogyne incognita*. Front Microbiol 12:790223

Kushwaha RK, Sarma BK (2015) Studies on some nematophagous fungi for management of root knot diseases of tomato. J Biotechnol Crop Sci 4:56–60

Lawton JR (1957) The formation of constricting rings in nematode-catching Hyphomycetes grown in pure culture. J Exp Bot 8:50–54

Li L, Ma M, Liu Y, Zhou J, Qu Q, Lu K, Fu D, Zhang K (2011) Induction of trap formation in nematode-trapping fungi by a bacterium. FEMS Microbiol Lett 322:157–165

Li L, Wright SJ, Krystofova S, Park G, Borkovich KA (2007) Heterotrimeric G protein signaling in filamentous fungi. Ann Rev Microbiol 61:423–452

Li L, Yang M, Luo J, Qu Q, Chen Y, Liang L, Zhang K (2016) Nematode-trapping fungi and fungus-associated bacteria interactions: the role of bacterial diketopiperazines and biofilms on *Arthrobotrys oligospora* surface in hyphal morphogenesis. Environ Microbiol 18:3827–3839

Liang LM, Zou CG, Xu J, Zhang KQ (2019) Signal pathways involved in microbe–nematode interactions provide new insights into the biocontrol of plant-parasitic nematodes. Philos Tran Royal Soc B 374:20180317

Liu X, Xiang M, Che Y (2009) The living strategy of nematophagous fungi. Mycoscience 50:20–25

Ma N, Zhao Y, Wang Y, Yang L, Li D, Yang J, Jiang K, Zhang KQ, Yang J (2021) Functional analysis of seven regulators of G protein signaling (RGSs) in the nematode-trapping fungus *Arthrobotrys oligospora*. Virulence 12:1825–1840

Marin-Bruzos M, Grayston SJ (2019) Biological control of nematodes by plant growth promoting rhizobacteria: secondary metabolites involved and potential applications. In: Singh H, Keswani C, Reddy M, Sansinenea E, García-Estrada C (eds) Secondary metabolites of plant growth promoting Rhizomicroorganisms: discovery and applications. Springer, Singapore, pp 253–264

Meerupati T, Andersson K-M, Friman E, Kumar D, Tunlid A, Ahren D (2013) Genomic mechanisms accounting for the adaptation to parasitism in nematode-trapping fungi. PLoS Genet 9:e1003909

Mo MH, Chen WM, Su HY, Zhang KQ, Duan CQ, He DM (2006) Heavy metal tolerance of nematode-trapping fungi in lead-polluted soils. Appl Soil Ecol 31:11–19

Moosavi MR, Zare R (2020) Fungi as biological control agents of plant-parasitic nematodes. In: Mérillon JM, Ramawat KG (eds) Plant defence: biological control, vol 22. Springer, Cham, pp 333–384

Niblack TL, Hussey RS (1986) Evaluation of *Arthrobotrys amerospora* as a biocontrol agent for Heterodera glycines on soybean. Plant Dis 70:448–451

Niu X-M, Zhang K-Q (2011) *Arthrobotrys oligospora*: a model organism for understanding the interaction between fungi and nematodes. Mycology 2:59–78

Nordbring-Hertz B (1973) Peptide-induced morphogenesis in the nematode trapping fungus *Arthrobotrys oligospora*. Physiol Plant 29:223–233

Nordbring-Hertz B, Jansson HB, Tunlid A (2001) Nematophagous fungi. Encyclopedia of Life Sci:1–11

Nordbring-Hertz B (2004) Morphogenesis in the nematode-trapping fungus *Arthrobotrys oligospora*-an extensive plasticity of infection structures. Mycologist 18:125–133

Nordbring-Hertz B, Jansson HB, Tunlid A (2011) Nematophagous fungi. In: Encyclopedia of life sciences. John Wiley & Sons Ltd, Chichester, UK, pp 1–13

Nordbring-Hertz B, Mattiasson B (1979) Action of a nematode-trapping fungus shows lectin-mediated host–microorganism interaction. Nature 281:477–479

Pandit R, Kunjadia P, Mukhopadhyaya P, Kunjadia A (2014) Inorganic phosphate solubilizing potential of *Arthrobotrys conoides* and *Duddingtonia flagrans*, a nematode trapping fungi a potential biocontrol agent. J Agric Technol 10:559–570

Pérez-Anzúrez G, Olmedo-Juárez A, von-Son de Fernex E, Alonso-Díaz MÁ, Delgado-Núñez EJ, López-Arellano ME, González-Cortázar M, Zamilpa A, Ocampo-Gutierrez AY, Paz-Silva A, Mendoza-de Gives P (2022) *Arthrobotrys musiformis* (Orbiliales) kills *Haemonchus contortus* infective larvae (Trichostronylidae) through its predatory activity and its fungal culture filtrates. Pathogens 11:1068

Persmark L, Nordbring-Hertz B (1997) Conidial trap formation of nematode-trapping fungi in soil and soil extracts. FEMS Microbiol Ecol 22:313–323

Pfister DH, Liftik ME (1995) Two *Arthrobotrys* anamorphs from *Orbilia auricolor*. Mycologia 87:684–688

Pramer D (1964) Nematode-trapping fungi: an intriguing group of carnivorous plants inhabit the microbial world. Science 144:382–388

Pramer D, Kuyama S (1963) Nemin and the nematode-trapping fungi. Bacteriol Rev 27:282–292

Pramer D, Stoll NR (1959) Nemin: a morphogenic substance causing trap formation by predaceous fungi. Science 129:966–967

Purba RTT, Fauzi F, Sari RW, Naibaho DC, Putri QA, Maulana A, Hastuti LDS, Punnapayak H (2022) *Arthrobotrys thaumasia* and *Arthrobotrys musiformis* as biocontrol agents against *Meloidogyne hapla* on tomato plant. Biodivers J Biol Divers 23:3659–3666

Quevedo A, Vera-Morales M, Espinoza-Lozano F, Castañeda-Ruiz RF, Sosa D, Magdama F (2021) Assessing the predatory activity of *Arthrobotrys oligosporus* strain C-2197 as biocontrol of the root-knot nematode *Meloidogyne* spp. Bionature 6:1586–1592

Rosén S, Sjollema K, Veenhuis M, Tunlid A (1997) A cytoplasmic lectin produced by the fungus *Arthrobotrys oligospora* functions as a storage protein during saprophytic and parasitic growth. Microbiology 143:2593–2604

Rosenzweig WD (1984) Role of amino acids, peptides, and medium composition in trap formation by nematode-trapping fungi. Can J Microbiol 30:265–267

Rosenzweig WD, Ackroyd D (1983) Binding characteristics of lectins involved in the trapping of nematodes by fungi. Appl Environ Microbiol 46:1093–1096

Sanyal PK (2000) Screening for Indian isolates of predacious fungi for use in biological control against nematode parasites of ruminants. Vet Res Commun 24:55

Singh UB (2007) Occurrence, characterization and performance of some predacious fungi. M.Sc. Thesis, Banaras Hindu University, Varanasi, India

Singh UB, Asha S, Nisha S, Singh RK, Renu SD, Manna MC, Sarma BK, Singh HB, Singh KP (2012a) *Arthrobotrys oligospora* mediated induced systemic resistance and biochemical responses in tomato (*Lycopersicon esculentum* Mill.) affects disease dynamics and enhance bio-protection against *Meloidogyne incognita* and *Rhizoctonia solani*. J Appl Microbiol 114:12009

Singh UB, Sahu A, Sahu N, Singh BP, Singh RK, Singh DP, Jaiswal RK, Sarma BK, Singh HB, Manna MC, Rao AS (2013) Can endophytic *Arthrobotrys oligospora* modulate accumulation of defence related biomolecules and induced systemic resistance in tomato (*Lycopersicon esculentum* Mill.) against root knot disease caused by *Meloidogyne incognita*. Appl Soil Ecol 63:45–56

Singh UB, Sahu A, Singh RK, Singh DP, Meena KK, Srivastava JS, Manna MC (2012b) Evaluation of biocontrol potential of *Arthrobotrys oligospora* against *Meloidogyne graminicola* and *Rhizoctonia solani* in Rice (*Oryza sativa* L.). Biol Control 60:262–270

Stirling GR (2018) Biological control of plant-parasitic nematodes. In: Diseases of nematodes. CRC Press, Boca raton, USA, pp 103–150

Swe A, Li J, Zhang KQ, Pointing SB, Jeewon R, Hyde KD (2011) Nematode-trapping fungi. Curr Res Environ Appl Mycol 1:1–26

Tkalenko H, Gadzalo Y, Borzykh O, Horal S (2021) *In vitro* screening of new strains of predacious nematophagous fungi for biocontrol suitability when produced in liquid culture. Agric Sci Pract 8:3–15

Tunlid A, Rosén S, Ek BO, Rask L (1994) Purification and characterization of an extracellular serine protease from the nematode-trapping fungus *Arthrobotrys oligospora*. Microbiology 140:1687–1695

Veenhuis M, Nordbring-Hertz B, Harder W (1985) An electron-microscopical analysis of capture and initial stages of penetration of nematodes by *Arthrobotrys oligospora*. Antonie Van Leeuwenhoek 51:385–398

Wang BL, Chen YH, He JN, Xue HX, Yan N, Zeng ZJ, Bennett JW, Zhang KQ, Niu XM (2018) Integrated metabolomics and morphogenesis reveal volatile signaling of the nematode-trapping fungus *Arthrobotrys oligospora*. Appl Environ Microbiol 84:e02749–e02717

Wang X, Li GH, Zou CG, Ji XL, Liu T, Zhao PJ, Liang LM, Xu JP, An ZQ, Zheng X, Qin YK, Tian MQ, Xu YY, Ma YC, Yu ZF, Huang XW, Liu SQ, Niu XM, Yang JK, Huang Y, Zhang KQ (2014) Bacteria can mobilize nematode-trapping fungi to kill nematodes. Nat Commun 5:5776

Wes PD, Bargmann CI (2001) *C. elegans* odour discrimination requires asymmetric diversity in olfactory neurons. Nature 410:698–701

Worthy SE, Rojas GL, Taylor CJ, Glater EE (2018) Identification of odor blend used by *Caenorhabditis elegans* for pathogen recognition. Chem Senses 43:169–180

Xie H, Aminuzzaman FM, Xu L, Lai Y, Li F, Liu X (2010) Trap induction and trapping in eight nematode-trapping fungi (Orbiliaceae) as affected by juvenile stage of *Caenorhabditis elegans*. Mycopathologia 169:467–473

Xie M, Bai N, Yang J, Jiang K, Zhou D, Zhao Y, Li D, Niu X, Zhang KQ, Yang J (2020) Protein kinase Ime2 is required for mycelial growth, conidiation, osmoregulation, and pathogenicity in nematode-trapping fungus *Arthrobotrys oligospora*. Front Microbiol 10:3065

Yang J, Tian B, Liang L, Zhang KQ (2007) Extracellular enzymes and the pathogenesis of nematophagous fungi. Appl Microbiol Biotechnol 75:21–31

Yang J, Wang L, Ji X, Feng Y, Li X, Zou C, Xu J, Ren Y, Mi Q, Wu J, Liu S (2011) Genomic and proteomic analyses of the fungus *Arthrobotrys oligospora* provide insights into nematode-trap formation. PLoS Pathog 7:e1002179

Yang L, Li X, Xie M, Bai N, Yang J, Jiang K, Zhang KQ, Yang J (2021) Pleiotropic roles of Ras GTPases in the nematode-trapping fungus *Arthrobotrys oligospora* identified through multi-omics analyses. Science 24:102820

Zhu MC, Li XM, Zhao N, Yang L, Zhang KQ, Yang JK (2022) Regulatory mechanism of trap formation in the nematode trapping fungi. J Fungi 8:406

The manufacturer's authorised representative in the EU is Springer Nature Customer Service Centre GmbH, Europaplatz 3, 69115 Heidelberg, Germany. If you have any concerns regarding our products, please contact ProductSafety@springernature.com

Printed and bound by CPI Group (UK) Ltd, Croydon, CR0 4YY
01/10/2026
02225850-0005